THE INDIAN FRONTIER

Horse and Warband in the Making of Empires

JOS GOMMANS

Routledge
Taylor & Francis Group

LONDON AND NEW YORK

First published 2018
by Routledge
4 Park Square, Milton Park, Abingdon, Oxon OX14 4RN
605 Third Avenue, New York, NY 10017

First issued in paperback 2023

Routledge is an imprint of the Taylor & Francis Group, an informa business

© 2018 Jos Gommans and Manohar Publishers & Distributors

Publisher's Note
The publisher has gone to great lengths to ensure the quality of this
reprint but points out that some imperfections in the original copies may
be apparent.

Print edition not for sale in South Asia (India, Sri Lanka, Nepal,
Bangladesh, Afghanistan, Pakistan or Bhutan)

British Library Cataloguing in Publication Data
A catalogue record for this book is available from the British Library

Library of Congress Cataloging in Publication Data
A catalog record for this book has been requested

ISBN: 978-1-03-265259-7 (pbk)
ISBN: 978-1-138-09537-3 (hbk)
ISBN: 978-0-203-71282-5 (ebk)

DOI: 10.4324/9780203712825

Typeset in Adobe Garamond Pro 11/13
by Manohar, New Delhi 110 002

MANOHAR

THE INDIAN FRONTIER

The work of Jos Gommans uniquely knits together the deep structures of human and animal ecology and environmental studies with the history of tribes, states and armies across the Eurasian world through the centuries. Many of his important articles were hitherto hard to obtain, as they were scattered across many books and journals through the decades. This volume makes these conveniently accessible. It is a treasure trove of large ideas and important interdisciplinary insights that should interest historians and social scientists alike.

Sumit Guha
Frances Higginbotham Nalle Centennial
Professor in History, University of Texas at Austin

"From biological to mercantile, military and finally political power, Jos Gommans explores the intersection between humans, horses, war-bands and weapons. These fascinating essays chart the fertile ground between commerce and conquest on the moving frontiers of pre-colonial South Asia."

Nile Green
Professor of History
University of California, Los Angeles

This omnibus brings together some old and some recent works by Jos Gommans on the warhorse and its impact on medieval and early modern state-formation in South Asia. These studies are based on Gommans' observation that Indian empires always had to deal with a highly dynamic inner frontier between semi-arid wilderness and settled agriculture. Such inner frontiers could only be bridged by the ongoing movements of Turkish, Afghan, Rajput and other warbands. Like the most spectacular examples of the Delhi Sultanate and the Mughal Empires, they all based their power on the exploitation of the most lethal weapon of that time: the warhorse. In discussing the breeding and trading of horses and their role in medieval and early modern South Asian warfare, Gommans also makes some thought-provoking comparisons with Europe and the Middle East. Since the Indian frontier is part of the much larger Eurasian Arid Zone that links the Indian subcontinent to West, Central and East Asia, the final essay explores the connected and entangled history of the Turko-Mongolian warband in the Ottoman and Timurid Empires, Russia and China.

Jos J.L. Gommans holds the chair of Colonial and Global History at the University of Leiden. His major publications include *The Rise of the Indo-Afghan Empire, c.1710-1780* (Leiden, 1995; Delhi, 1999; 2017) and *Mughal Warfare: Indian Frontiers and High Roads to Empire, 1500-1700* (London, 2002).

In memory of
J. C. HEESTERMAN

Contents

List of Illustrations   9

List of Original Publications   11

Note on Transliteration   13

Introduction   15

1. The Horse Trade in Eighteenth-century South Asia   25

2. The Silent Frontier of South Asia, *c.* AD 1100-1800   51

3. The Eurasian Frontier after the First Millennium AD:
   Reflections Along the Fringe of Time and Space   78

4. Warhorse and Post-Nomadic Empire in Asia,
   c. 1000-1800   99

5. The Embarrassment of Political Violence in
   Europe and South Asia, *c.* 1100-1800   130

6. Indian Warfare and Afghan Innovation
   during the Eighteenth Century   157

7. Warhorse and Gunpowder in India, *c.* AD 1000-1850   183

8. Slavery and *Naukari* among the Bangash
   Nawabs of Farrukhabad   209

9. The Warband in the Making of Eurasian Empires   249

Index   331

Illustrations

PLATES

1. Muhammad Khan Bangash, *c.* 1730.
2. Nawab Amir al-Umara Zabita Khan by Son of Ganga Ram, Mihr Chand, Faizabad, *c.* 1770.
3. Battle of Panipat 1761. Faizabad, *c.* 1770.
4. Four Afghan Steeds: The Afghan envoy to Beijing presented the Qing emperor with four horses.
5. Tomb of Ali Muhammad Khan Rohilla in Aonla (UP).
6. Tomb of Muhammad Khan Bangash in Farrukhabad (UP).
7. Tomb of Najib al-Daula in Najibabad (UP).
8. Two Yusufzai infantrymen.*
9. Two Rohilla men.*
10. Kabuli horseman in chain mail holding a lance.*
11. Yusufzai horseman.*
12. Rohilla horseman, watercolour by Sita Ram.
13. Watercolour of Patthargarh Fort in Najibabad by Sita Ram (1814-15).
14. Watercolour of the tomb of Hafiz Rahmat Khan in Bareilly by Sita Ram (1814-15).

* British Library London: Add.Or.1347-96
50 drawings bound into a volume; 31 depicting people of the Punjab and neighbouring areas, 11 depicting infantry and cavalry of the Punjab and its environs and 8 depicting rulers of the Punjab and their ministers as well as princes of the Punjab Hill states. By an Indian artist, probably at Amritsar, 1838-9. The majority inscribed with titles in Persian characters and in English. Water-marks of 1837 (e.g. Add.Or.1348).

MAPS

1. South Asia in the Eighteenth Century 30
2. Arid Zone 56
3. Military Zones in Eurasia 253

Original Publications

I HAVE MADE every reasonable effort to locate, contact and acknowledge rights holders and to correctly apply terms and conditions to content. All chapters are slightly revised versions of contributions which were originally published in the journals and books listed below. All are reproduced with the permission of the copyright holders and the publishers. In the event that any content infringes the rights of any third parties, or content is not properly identified or acknowledged I would like to hear from you. I am grateful to Martin Hoekstra who, despite some personal mishaps, helped me out with adapting these publications for the present volume.

Chapter 1: 'Horse Trade in Eighteenth-century South Asia', *Journal of the Economic and Social History of the Orient* 37, 3 (1994): 228-50. Leiden: E.J. Brill Publishers.

Chapter 2: 'The Silent Frontier of South Asia, *c.* 1100-1800 AD', *Journal of World History* 9, 1 (1998): 1-25. University of Hawaii Press.

Chapter 3: 'The Eurasian Frontier after the First Millennium AD: Reflections Along the Fringe of Time and Space', *Medieval History Journal* 1, 1 (1998): 125-45. New Delhi: Sage Publications India.

Chapter 4: 'Warhorse and Post-Nomadic Empire in Asia, *c.* 1000-1800', *Journal of Global History* 2 (2007): 1-21. Cambridge: Cambridge University Press.

Chapter 5: 'The Embarrassment of Political Violence in Europe and South Asia, *c.* 1100-1800', in J.E.M. Houben and K.R. van Kooij (eds.), *Violence Denied: Violence, Non-Violence and the Rationalization of Violence in South Asian Cultural History* (Leiden: E.J. Brill Publishers, 1999): 287-317.

Chapter 6: 'Indian Warfare and Afghan Innovation during the Eighteenth Century', *Studies in History*, 11, 2 (1995): 261-80. New Delhi: Sage Publications India.

Chapter 7: 'Warhorse and Gunpowder in India *c.* 1000-1850', in J. Black (ed.), *War in the Early-modern World 1450-1815* (London: UCL Press, 1999): 105-29.

Chapter 8: 'Slavery and *Naukari* among the Bangash Nawabs of Farrukhabad', in Jos Gommans and Om Prakash (eds.), *Circumambulations in South Asian History: Essays in Honour of Dirk H.A. Kolff* (Leiden: E.J. Brill Publishers, 2003): 179-217.

Chapter 9: 'The Warband in the Making of Eurasian Empires', in Maaike van Berkel and Jeroen Duindam (eds.), *Prince, Pen and Sword: Eurasian Perspectives* (Leiden: E.J. Brill Publishers, 2017).

Transliteration

FOREIGN TERMS NOT in common English use have been italicised.
This book deleted most diacritical marks of earlier versions based on
different systems. Plurals have been indicated by adding the letter *s*.
For the combination of two related words I have used the Persian
izafa -i, and the Arabic *al-,* with the exception of the more simple
personal names like Abul.

Introduction

THIS BOOK BRINGS together some of my research on the warhorse and its impact on medieval and early modern state-formation in South Asia and beyond. Most of these contributions were published following the defence of my PhD thesis at the University of Leiden in 1993, which was itself published a year later as *The Rise of the Indo-Afghan Empire, circa 1710-1780*.[1] Under the guidance of my Leiden supervisors, Jan Heesterman and André Wink, the dissertation was meant to provide one more regional contribution to the hotly-contested debate surrounding the history of India in the eighteenth century; from the very start of my PhD research in 1989, I focused on the neglected history of two Afghan cases of so-called Mughal successor-states in northern India: Rohilkhand and Farrukhabad.[2]

The most important issue seemed fairly obvious: would my research confirm the still dominant view of the established historians of the Aligarh Muslim University, which stressed both political and economic decline, or would it support the revisionist view of upcoming scholars such as Frank Perlin, Christopher Bayly, Muzaffar Alam, and, indeed, my two Leiden supervisors, which denied overall decline and instead stressed political and economic relocations? For them, the Mughal Empire may have declined, but this decline also masked the rise of highly dynamic regional states supported by an increasingly powerful

[1] (Leiden: Brill, 1994). A paperback reprint came out in 1999 at Delhi: Oxford University Press and again in 2016 at Delhi at Manohar. The original title of the 1993 thesis was 'Horse-Traders, Mercenaries and Princes: The Formation of the Indo-Afghan Empire in the Eighteenth Century'.

[2] See also my contributions on 'Afghans in India' (2007), 'Farrukhabad' (2015) and 'Najibabad' (forthcoming) in *Encyclopaedia of Islam*, 3rd ed. (Leiden: Brill, 2007-).

gentry and a new class of so-called portfolio capitalists.[3] The debate was not only about eighteenth-century decline, but also about the administrative reach of Indian empires in general. Had the Mughal Empire, even in its heyday, really been such an exploitative imperial colossus as imagined by Irfan Habib and his Aligarh disciples? My Leiden teachers certainly thought otherwise. In the heat of the moment, the debate even touched the Maurya Empire. Whereas Habib emphasised imperial control, Wink was again extremely sceptical, making comparisons between Ashoka pillars and Heineken beer-cans.

Habib seems to think, contrary to what I think, that there is evidence for imperial-political unification of the Indian subcontinent in the 'fantastic identity' in the texts of Asoka's edicts. Why that should be so he does not explain. But, if we find, for instance, exactly identical Heineken beer-cans all over the world with exactly the same text on them, would that mean that the world is politically unified under the aegis of Heineken?[4]

Such was the bellicose atmosphere in which I started my research on the Indian Rohillas. Hardly surprisingly, my thesis supported the revisionists but, unfortunately, without being able to provide a great deal of fresh evidence on the performance of the local economies of the two states.[5]

Almost simultaneously with the publication of my own work, Aligarh produced its own study on the Rohillas, written by Iqbal Husain. Both works elaborated on the fact that the Afghans were able to carve out their own states by very effectively exploiting the Indian military labour market. However, it was interesting that, at times, it seemed as if the two works were dealing with two different topics. While Husain zoomed in on the local political context of the Rohilla state in particular, my thesis zoomed out, examining interregional aspects of the process of Afghan state-formation in general. What

[3] For an overview of that debate, see S. Alavi (ed.), *The Eighteenth Century in India* (Delhi: Oxford University Press, 2002).

[4] A. Wink, 'A Rejoinder to Irfan Habib', *The Indian Economic and Social History Review* 26, 3 (1989), 363-4.

[5] I. Husain, *The Ruhela Chieftaincies: The Rise and Fall of Ruhela Power in the Eighteenth Century* (Delhi: Oxford University Press, 1994).

had happened? In my case it was sheer serendipity! What I could not have anticipated at the start of my research was the fact that the Rohillas found themselves at the heart of a huge commercial network under the aegis of the new imperial dynasty of the Durranis; across the old border between the Safavid and Mughal empires, this new *shahanshahi* encompassed the worlds of Iran, Turan, and Hind. In Rohilkhand this inspired a true Indo-Afghan renaissance, in which *homines novi* like Hafiz Rahmat Khan were able to study and reshape both their past and their ethnic identity.[6]

What struck me most, though, was Rohilla involvement in the breeding and trading of horses (see Chapter 1). To understand this forgotten aspect of their Indian success story, I had to follow the Rohilla horse trail deep into their original homeland of Roh (Afghanistan) and even further, into what Rudyard Kipling considered 'the back of beyond' of Central Asia.[7] Equally importantly, I had to come to grips with the complicated technicalities of horse-breeding. Soon I discovered that it was more than a hobbyhorse for equine fanatics and military historians.[8] On the contrary, it had been a key ingredient in the making and unmaking of Indian empires, not in the least that of the Mughals.[9] But how to explain the prominent role

---

[6] On the relationship between migration, settlement, and identity among Indo-Afghans, see also N. Green, *Making Space: Sufis and Settlers in Early Modern India* (New York: Oxford University Press, 2012). In nineteenth-century Rampur, this Afghan renaissance triggered new 'vernacular' sentiments of local belonging, as recently analysed by Razak Khan, 'Local Pasts: Space, Emotions and Identities in Vernacular Histories of Princely Rampur', *Journal of the Economic and Social History of the Orient* 58, 5 (2015), 693-732.

[7] This trail was much more thoroughly taken up by S. Levi; see his *The Indian Diaspora in Central Asia and its Trade 1550-1900* (Leiden: Brill, 2002) and his historiographical survey in his edited volume *India and Central Asia* (Delhi: Oxford University Press, 2007).

[8] See also the global perspective taken recently by E. Lambourne, 'Towards a Connected History of Equine Cultures in South Asia: *Bahri* (Sea) Horses and "Horsemania" in Thirteenth-Century South India', *The Medieval Globe* 2, 1 (2016), 57-101.

[9] Here I profited tremendously from the lively discussions with the late Simon Digby, and of course from his pioneering *Warhorse and Elephant in the Delhi Sultanate* (Karachi: Oxford University Press, 1971).

of the north-western parts of the subcontinent in delivering the majority and best of warhorses? Why could India itself not produce good warhorses? In finding an answer to these questions, I was first of all guided by early colonial investigations into the issue.[10] Although being informed by a clear predilection for so-called pure Arabic breeds, colonial vets highlighted what they considered the key ingredient for breeding strong warhorses: the availability of sufficient pastures with certain specific grass varieties. According to the same authorities, the most nutritious grasses for horses are generally to be found in relatively dry regions with lots of space for free grazing. In other words, it was a combination of movement and natural feeding in dry regions that really made the difference.

These nineteenth-century observations are more or less confirmed by modern-day biologists' views on the subject. Although our historical understanding is considerably hampered by the tremendous development in modern breeding and feeding methods, evolutionists have pointed out that the horse has adapted to eat the poorest-quality forage, containing the lowest concentration of protein of any large herbivore: it thrives on grasses that a cow would starve to death on. Unlike other large grazing animals, thanks to its unique digestive organ called the cecum, the horse is able to run and eat at the same time and, more significantly for its poor biological niche, to break down the otherwise indigestible cellulose (and its nutrients) of stems and leaves.[11]

---

[10] Recently, this colonial discourse has been critically analysed by S. Mishra, 'Beasts, Murrains, and the British Raj: Reassessing Colonial Medicine in India from the Veterinary Perspective, 1860-1900', *Bulletin of the History of Medicine* 85 (2011), 587-619, and 'The Economics of Reproduction: Horse-breeding in Early Colonial India, 1790-1840', *Modern Asian Studies* 46, 5 (2012), 1116-44.

[11] S. Budiansky, *The Nature of Horses: Their Evolution, Intelligence and Behaviour* (London: Phoenix, 1998), 7-33. I am grateful to Hans van den Broek for his help in finding a modern scientific answer to this problem. The neglected cultural aspects of the horse in Mughal India was recently discussed by J. Lally, 'Empires and Equines: The Horse in Art and Exchange in South Asia, ca. 1600 – ca. 1850', *Comparative Studies of South Asia, Africa and the Middle East* 35, 1 (2015), 96-116. See also the unpublished PhD thesis (University of Washington, 2013) by Monica Meadows: 'The Horse: Conspicuous Consumption of Embodied Masculinity in South Asia, 1600-1850'.

Engaging myself more seriously by employing everything that I could find on horse-breeding, I became increasingly aware that this was a problem of truly global relevance: in addition to Indian rulers, their counterparts in China and Eastern Europe were simply not able to breed enough good warhorses of their own and had struggled for decades to ensure the supply of warhorses from the closest dry-zones. In fact, these dry-zones together formed one unbroken Arid Zone in which a highly dynamic nomadic way of life intermixed with rich sedentary economies, based on well-irrigated river valleys and oases.[12] Although the sultanates of North Africa and West Asia were able to produce strong enough horses of their own, they lacked the extensive grass pastures to produce the enormous quantities that had supported the more powerful Turkish, Mongol, and Jurchen empires along the Central Asian steppes.

Meanwhile, in the mid 1990s the eighteenth century debate gradually faded out, as most revisionists moved in different, more global directions. Hence, my rediscovery of the Arid Zone as a category of global history conveniently facilitated the opening up of South Asian history towards Central and West Asia. In fact, in Leiden this had already been started in the late 1980s as a result of André Wink's massive *Al-Hind* project, later to be followed by René Barendse's equally voluminous *Arabian Seas* project.[13] At the beginning, my

---

[12] Much later I found out that 'my' biological discovery of the Arid Zone as covering those areas that were suitable for nomadic horse-breeding, came very close to Hodgon's Arid Zone as a key in understanding the expansion of Islam (see M.G.S. Hodgson, *The Venture of Islam: Conscience and History in a World Civilization*, vol. 2: *The Expansion of Islam in the Middle Period* (Chicago and London: The University of Chicago Press, 1974), 71 fol.). The idea of the Arid Zone also informs Victor Lieberman's important dichotomy between exposed and protected zones in his *Strange Parallels: Southeast Asia in Global Context, c. 800-1830*, 2 vols (Cambridge: Cambridge University Press, 2003-9).

[13] A. Wink, *Al-Hind: The Making of the Indo-Islamic World*, 3 vols (Leiden: Brill, 1990-2004); R. Barendse, *The Arabian Seas: The Indian Ocean World of the Seventeenth Century* (Armonk NY: M.E. Sharpe, 2002) and *The Arabian Seas 1700-1763*, 4 vols (Leiden: Brill, 2009). Both the steppes of the Arid Zone and the paddy fields of the Indian Ocean converge brilliantly in R. Palat, *The Making of an Indian Ocean Economy, 1250-1650: Princes, Paddy Fields and Bazaars* (London: Palgrave MacMillan, 2015). See also my own brief survey 'Continuity

thinking of the Afghans as an open, 'conscriptive' group of warriors had profited tremendously from Dirk Kolff's ground-breaking work on military service (*naukari*) in the making of Rajput identity.[14] All these studies sprang from the now almost dissolved Kern Institute, and all were inspired by the work on the so-called *Inner Conflict of Tradition* by the late Leiden Indologist J.C. Heesterman, to whom this work is dedicated.[15] For me, the Arid Zone's extensions into the Indian subcontinent not only provided the basic infrastructure of the Indo-Afghan horse-trade, but also pinpointed a foundational geopolitical inner frontier between (semi-)nomadic wilderness and sedentary centres. As such, it was the *locus operandi* of both the old Indic sacrificial contest and, more generally speaking, premodern state-formation. In the words of J.C. Heesterman:

As regards the early state, what our consideration of the sacrificial contest seems to tell us is that kingship and statehood did not originate with either the itinerant warrior and herdsman or the settled agrarian magnate. Rather it seems to have originated on the interface of their incompatible, yet interdependent world, precariously balanced on the explosive line of conflict that divides as well as joins them.[16]

As I tried to demonstrate in a second book, even the mighty Mughal Empire should be considered a post-nomadic empire which, thanks to its ongoing dependence on warhorses, built its power on a network of imperial highways bridging the empire's various inner frontiers between *raiyati* and *mawas*, settled fields and wilderness.[17] Indeed,

---

and Change in the Indian Ocean Basin, 1400-1800', in J.H. Bentley, M.E. Wiesner-Hanks and S. Subrahmanyam (eds), *The Cambridge History of the World*, vol. 6, part 1: *The Construction of a Global World, 1400-1800* CE (Cambridge: Cambridge University Press, 2015), 182-210.

[14] D.H.A. Kolff, *Naukar, Rajput and Sepoy: The Ethnohistory of the Military Labour Market in Hindustan, 1450-1850* (Cambridge: Cambridge University Press, 1990).

[15] J.C. Heesterman, *The Inner Conflict of Tradition* (Chicago and London: The University of Chicago Press, 1985).

[16] J.C. Heesterman, 'Warrior, Peasant and Brahmin', *Modern Asian Studies* 29, 3 (1995), 650.

[17] *Mughal Warfare: Indian Frontiers and Highroads to Empire 1500-1700* (London: Routledge, 2002). For the idea of post-nomadism, see Wink's comments

the first five chapters of the current volume represent my geographical explorations across the Arid Zone, both in India itself (Chapter 1 to 3) and in comparison with the Chinese and Russian cases (Chapter 4). More or less in line with this approach, Chapter 5 provides a wildly speculative and somewhat immature comparison with Europe. From the first millennium onward, the latter lacked an inner nomadic frontier that paved the way for an entirely different, obviously more rooted process of state-formation.

The military aspects of the warhorse are taken up in Chapters 6 and 7, which should be read in the context of my *Mughal Warfare*. At the time, when I worked on that book the military history of the subcontinent was not at all considered a serious scholarly preoccupation, as it was still part of the national guts-and-glory stories cherished by retired army officers and other hobbyists. Since then a great deal of fresh work has been published that has made it a sophisticated field, and one that is increasingly sensitive to fascinating global comparisons, not only the unavoidable one with Europe, but also with China.[18] Looking back, I feel that in my eagerness to stress the ongoing importance of both light and heavy cavalry against Marshall Hodgson's idea of 'gunpowder empires' I may at times have underestimated the role of gunpowder weaponry.[19] Recent archaeological work on the

---

in A.M. Khazanov and A. Wink (eds), *Nomads in the Sedentary World* (Richmond, UK: Curzon, 2001), 295.

[18] *Pace* the so-called new military history of South Asia represented by authors like Seema Alavi and William Pinch, which is still a fairly recent development and has not focused on the military hard-core of weaponry and tactics. In recent years the field has been dominated by the insightful publications of K. Roy, *War, Culture and Society in Early Modern South Asia, 1740-1849* (London and New York: Routledge, 2011); *Military Manpower, Armies and Warfare in South Asia* (London and New York: Routledge, 2013); Warfare in Pre-British India – 1500 BCE to 1740 CE (London and New York: Routledge, 2015); *Military Transition in Early Modern Asia, 1400-1750: Cavalry, Guns, Governments and Ships* (London and New York: Bloomsbury, 2015). For the China comparison, see K. Roy and P. Lorge (eds), *Chinese and Indian Warfare: From the Classical Age to 1870* (London and New York: Routledge, 2014).

[19] Although mounted archers became much less important fairly soon after the Mughal conquest, cavalry retained its primacy in the seventeenth and much of the eighteenth century. The Mughal case of retarding gunpowder technology

Deccan by Eaton and Wagoner has actually tried to demonstrate that, as early as the fifteenth century, cannon had become an effective instrument in supporting, as well as defending against, sieges. This had considerable consequences on sixteenth-century fortification design, which was adjusted to accommodate a new style for mounting cannons.[20] Although their study provides fascinating new evidence, Eaton and Wagoner specifically highlight this adaptation in the western parts of the Deccan during the sixteenth-century; earlier and later developments, and those in other parts of the subcontinent, remain less clear. For me, despite some indications of a much earlier introduction under Mahmud Gawan, this new evidence seems to confirm that the widespread and effective use of new gunpowder weaponry, both in sieges and on the battlefield, was primarily a phenomenon of the first half of the sixteenth century. As far as fortifications in the Deccan are concerned, adaptations could literally build on some much older adaptations, in line with similar Middle Eastern and North Indian developments of the pre-gunpowder age. Following the sixteenth century, and up to the mid-eighteenth century, it appears that no gunpowder revolutions affected the kind of warfare that continued to depend on the mobility and logistics provided by horses, dromedaries, and bullocks.[21]

Coming to the rapid military developments of the eighteenth century, in another bout of eagerness, this time to stress indigenous sources of military adaptation, I may have exaggerated the innovative

---

seems to be very much in line with the global developments as discussed in K. Chase, *Firearms: A Global History to 1700* (Cambridge: Cambridge University Press), 207-10.

[20] R.M. Eaton and P.B. Wagoner, *Power, Memory, Architecture: Contested Sites on India's Deccan Plateau, 1300-1600* (Delhi: Oxford University Press, 2014).

[21] The various studies by Jean Deloche, although very interesting and detailed, are not conclusive as far as the dating of fortified constructions is concerned: J. Deloche, *Seni (Gingee): A Fortified City in the Tamil Country; Studies on Fortification in India* (Pondicherry: EFEO, 2007); *Four Forts of the Deccan* (Pondicherry: EFEO, 2009). For later developments, see the interesting observations by Pushkar Sohoni in his 'From Defended Settlements to Fortified Strongholds: Responses to Gunpowder in the Early Modern Deccan', *South Asian Studies* 31, 1 (2015), 111-26.

and decisive nature of the camel artillery deployed by the Afghan army during the battle of Panipat in 1761.[22] At the same time, the technological adaptability of eighteenth-century South Asian armies is now widely confirmed, in particular thanks to the innovative work of Randolph Cooper.[23] Of course, making the South Asian armies more compliant to the technological innovations of the eighteenth century raises the important question of what actually gave the European armies the edge over their Indian adversaries. Here, despite some recent criticism by Geoffrey Parker and Sanjay Subrahmanyam, I remain of the opinion that it was not gunpowder technology per se, but its organizational deployment based on standardization, discipline, and drill that really made the difference.[24] This was even more so since these organizational ingredients were by far the most difficult to adapt by rulers who were, up to that point, primarily dealing with permanent threats of sedition, and as such with a kind of warfare that was aimed at avoiding violence through thoroughly negotiated and highly ritualized settlements.

The final two chapters should bear out the fact that European armies had no monopoly whatsoever on effective military organization. Before the mid-eighteenth century, it was Islamicate military slavery and the Turko-Mongolian warband which, from the start of the second millennium CE, if not earlier, offered the two most obvious organizational models for any ambitious ruler in Eurasia. Chapter 8

[22] Despite some earlier examples, the disciplined use of light camel artillery seems to be an Afghan specialty, employed very effectively by them against the Persians in the 1722 Battle of Golnabad (M. Axworthy, *The Sword of Persia: Nader Shah from Tribal Warrior to Conquering Tyrant* (London and New York: Tauris, 2006), 48). See also the comments by Iqtidar Alam Khan on this issue in *Gunpowder and Firearms: Warfare in Medieval India* (Delhi: Oxford University Press, 2004), 122.

[23] R.G.S. Cooper, *The Anglo-Maratha Campaigns and the Contest for India: The Struggle for Control of the South Asian Military Economy* (Cambridge: Cambridge University Press, 2003).

[24] G. Parker and S. Subrahmanyam, 'Arms and the Asian Revisiting European Firearms and their Place in Early Modern Asia', *Revista de Cultura* 26 (2008) 34-5. Repeated in Subrahmanyam's *Courtly Encounters: Translating Courtliness and Violence in Early Modern Eurasia* (Cambridge Mass.: Harvard University Press, 2012), 22.

examines the military slaves of the Bangash Nawabs of Farrukhabad and is partly reproduced from my dissertation. It continues the story, however, into the late eighteenth century when, under British dispensation, slavery had entirely lost its military potential. The final chapter is part of a much larger collaborative work that, from a comparative point of view, discusses the cohesion of medieval and early modern Eurasian empires.[25] This project has enabled me to elaborate further on the connective history of the nomadic warband across the arid frontiers of Central Asia into China, Europe and the Islamicate West, and South Asia. It aims to demonstrate that, for centuries, global history was not so much about the rise and re-production of autonomous cities, capitalism, or nation states – to mention perhaps the three most important institutions of the West – but rather about the real or imagined reproduction of one of the main institutions of Central Asia: the horse-based warband as introduced by Chinggis Khan. As will be shown, though, as with most other nomadic institutions, and despite its undeniable success story, the nomadic warband could not be maintained in the sedentary world which, in the long run, always won out over the nomads.[26]

[25] M. van Berkel, J. Duindam, J. Gommans and P. Rietbergen, *Prince, Pen and Sword: Eurasian Perspectives* (Leiden: E.J. Brill, 2017).

[26] See the conclusions in Khazanov and Wink (eds), *Nomads in Sedentary World*, 292–5.

CHAPTER 1

The Horse Trade in Eighteenth-Century South Asia

DURING THE LAST decade there have been several successful attempts to widen the geographical perspective of pre-modern Indian history. As a result, India now emerges as a core area in a dynamic trading system which stretched from the Atlantic to the Pacific and from Siberia to Sri Lanka. Although this was an area of social and cultural diversity, and rooted in many different civilizations, seaborne trade and continental caravan traffic had created a strong sense of unity. So far, however, scholarly interest has predominantly focused on the maritime relations of trading diaspora groups like the Parsis, the Bohras, the Banias, the Portuguese, the Dutch, the English, and the French.[1] We know far less about the continental trade and its networks of Jewish, Armenian, Kashmiri, Afghan and other merchants.[2] This article will concentrate on one particularly important branch of the latter: the long-distance horse trade in the century preceding the advent of colonialism.[3]

---

[1] See e.g. K.N. Chaudhuri, *Trade and Civilisation in the Indian Ocea: An Economic History from the Rise of Islam to 1750* (Cambridge: Cambridge University Press, 1985).

[2] Very informative, however, is J. Deloche, *La circulation en Inde avant la révolution des transports*, 2 vols (Paris: École française d'Extrême Orient, 1980). For a more recent re-evaluation of these communities during an earlier period, see A. Wink, *Al-Hind: The Making of the Indo-Islamic World*, vol. 1: *Early Medieval India and the Expansion of Islam, 7th-11th Centuries* (Leiden: E.J. Brill, 1990).

[3] Given the importance of the horse in general it is surprising that Simon Digby's *War-Horse and Elephant in the Delhi Sultanate* (Karachi: Oxford Orient

THE PATTERN OF THE HORSE TRADE IN SOUTH ASIA

The Overland Trade

During the eighteenth century, India was still part of a thriving interregional livestock trading system which originated in the major breeding areas of Central Asia, and included eastern Europe and the Middle East. The bulk of the supply was produced by pastoral nomads in the Kalmyk and Kazakh steppes of southern Russia, the Turkoman steppes east of the Caspian Sea, and further to the south-east, in Afghan Turkistan. During the eighteenth century, in the wake of Russian and Afghan expansion into the producing steppe areas, by far the most important markets for sale were those of Russia, mainly for cattle, goats, sheep and horses, and South Asia – the latter mainly for warhorses.[4] At the Indian fairs or *melas* the horses imported from the north-west were generally known as *Kabuli*, *Qandahari*, or *Wilayati*. Actually, these were Turkoman or *Turki* breeds, from the area north of the Hindu Kush around Balkh and the area lower down the Amu Darya and near Andkhui and Meymaneh. They were initially sold at the local markets of Balkh, Bukhara and Herat, which latter place was also an outlet for the minor Iranian market. During the summer, the horses were bought by Afghan merchants, either indirectly through middlemen at the fairs or directly from the breeding nomads themselves. In general, the horses were bought in a rather bad condition for only about one quarter of the ultimate Indian sale price. In order to prepare them for sale they were for 1 or 2 months fattened in the more southern Afghan pastures or *maidans* around Kabul and Kandahar.[5] During October and November, these merchants, joining

---

Monographs, 1971) still stands out as the only larger scale work on the topic. For the Mughal period nothing of the kind is yet available.

[4] For the Russian livestock economy, see I. Blanchard, *Russia's 'Age of Silver': Precious-Metal Production and Economic Growth in the Eighteenth Century* (London: Routledge, 1989), 215-87; see also P. Longworth, *The Cossacks* (London: Constable, 1969), 25-172.

[5] M. Elphinstone, *An Account of the Kingdom of Caubul*, vol. 1 (Oxford: Oxford University Press, 1972), 386-8; *National Archives of India (NAI)*, New Dehli, Military Department Proceedings (MDP), 15-10-1811, no. 80, 'Report W. Moorcroft', fol. 287; MDP, 9-5-1808, 'Letter R. Frith to Secr. Brd. of

the caravans of the *Powindah* trading nomads, moved *en masse* with their horses across the Sulaiman Mountains, either taking the southern routes through the Bolan and other passes to Multan and the Derajat, continuing via Bahawalpur to Bikaner, mainly supplying Jaipur, the Deccan and southern India; or, travelling northwards through the Khyber Pass into Hindustan. After crossing the mountains but before being distributed to the local markets of the Punjab, Rohilkhand, Awadh, Benares and Bihar, the bulk of the horses was kept grazing at the extensive wastes of the Jullundur Doab and the Lakhi Jungle which were created by the recurrent floods of the Indus and its wild tributaries the Beas and the Sutlej. On these inundated alluvial fields the merchants could rest and nourish their horses without much expenditure and free from too much state interference.[6]

At the turn of the eighteenth century the total amount of duties which had to be paid along both routes was around Rs. 40 per horse, which would mean somewhat more than 10 per cent of the sales price.[7] On their way, the horse merchants sold part of their stock at the local fairs and at the same time bought horses from the indigenous breeding centres in Rajasthan, the Punjab and Rohilkhand. Buying and selling *en route*, the merchants proceeded as far as Sonepur-Hajipur in the east and Tirupati and Arcot in the south. In this way, the regional breeding economies were integrated into the long-distance trade with Central Asia.

Many of the local *melas* (in Rajasthan also called *hats*) became important outlets for indigenous horse-breeds. Some important examples in this respect were the fairs at Bhatinda which served as the entrepôt of the Lakhi Jungle, Majalgaon, the main Maratha market for the horses from the Bhima Valley, and Balotra and Pushkar for the indigenous breeds of Sind and Gujarat. All the fairs were held during a few weeks, either during autumn, at the arrival of the foreign

---

Superintendence for the Improvement of the Breed of Cattle', fols. 540-1; MDP, 13-2-1813, 'Letter W. Moorcroft to Secr. Brd.', fol. 150.

[6] For a similar pattern in the fifteenth and sixteenth century, see J. Arlinghaus, 'The Transformation of Afghan Tribal Society: Tribal Expansion, Mughal Imperialism and the Roshaniyya Insurrection, 1450-1600' (PhD thesis, Duke University, 1988), 61.

[7] *NAI*, MDP, 15-10-1811, no. 80, 'Report W. Moorcroft', fol. 287.

horses, or spring, at the end of the grazing season. The timetable of the local fairs was adjusted to the convenience of the traders, enabling them to travel from one fair to the other without losing too much time. For example, the Ummedganj fair in Malwa, which serviced the Kota court with horses, followed neatly 18 days after that of the great Pushkar *mela*, some 200 km to the north-west. Besides, regular postal services kept merchant and customers up to date about the latest developments at the other fairs.

During autumn, Pushkar was the major fair in Rajasthan. It was held in early November and some 5,000 horses were brought to it each year. This was only a minor part of the total quantity on offer because the traders held back the bulk of the horses. Business on the spot was mainly a kind of window-dressing. Customers, mostly army officers or court agents called *chabuk-sawars*, who wanted to buy horses on a large scale, had to purchase strings of horses on being shown only a few specimins.[8] After collecting the horses, the officers resold most of the very best and the very worst animals, whereas the medium quality was reserved for the military.[9] Obviously, the quality within a string of horses varied considerably and only the very best, and best fed and trained, were sent to the actual fair. In general, merchants preferred to dispose of their horses directly but in case there was no ready sale, they retained them, meanwhile fattening, breaking or training them, and at the right time fetching a higher price for them.[10] The prices of the horses were not only related to their quality and to the general market conditions but were also greatly influenced by the local price level of fodders like grain and hay, which affected overall costs of breeding and transport.[11]

Other autumn fairs in the area included Mundwa near Nagaur and Balotra near Jodhpur. The latter, however, had a more important spring fair during March-April, at which season regional breeds

---

[8] *NAI*, MDP, 15-10-1811, no. 80, 'Report W. Moorcroft', fols. 302-3; J.P. Pigott, *Treatise on the Horses of India* (London: James White, 1794), 56-7.

[9] *NAI*, MDP, 15-10-1811, nr. 80, 'Report W. Moorcroft', fol. 287.

[10] Ibid., fol. 303.

[11] *NAI*, MDP, 27-9-1804, 'Report Capt. Nuthull', fols. 74r-84v.

dominated the scene. More to the south, in Malwa, which was the gate to the Deccan market, there was a similar rhythm of alternating autumn and spring fairs. Here Ummedganj, and particularly Bilsa, near Bhopal, serviced the autumnal imports of foreign horses, whereas Manohar Thana and particularly Chand-Kheri (near Khanpur-Kota) held their most important *melas* during spring.[12]

In the north, by far the most important centre was Haridwar, which had a fair both during the autumn and spring, during which time it also served the Himalayan traders from the hill states to its north. The biggest, however, was the spring fair which coincided with the famous religious festival which drew thousands of pilgrims to the banks of the Ganges each year.[13] The combination of trade and pilgrimage was a fairly common phenomenon; it also occurred at the Mundwa, Hajipur, Tirupati and numerous other fairs. Obviously, at the fairs, religious and political interests were two sides of the same coin. For example, the Maratha and Sikh generals and their troops were in regular communication with the Haridwar and other fairs, not only to pay their devotion at the holy places but also to safeguard a secure supply of warhorses.[14] The control of the *melas* was always a cause of intense rivalry. Although at the spring fair there were not as many horses from Afghanistan or Turkistan as during the autumn, customers could buy foreign horses procured during the previous season by speculators who prepared and fattened the horses for the following spring sale or they could order them in advance for the

---

[12] *NAI*, MDP, 10-4-1795, 'Cavalry agent R. Murray to Lieut. A. Green', fol. 169-70.

[13] For Haridwar, see the accounts of Thomas Hardwicke in *NAI*, Field Books, vol. 1, 'Plot of Route from Najibabad to Srinagar and Upper Ganjes (1796)', fols. 36-46 and 'Narrative of a Journey to Sirinagur', *Asiatic Researches* 6 (1799), 309 fol.; also F.V. Raper, 'Narrative of a Survey for the Purpose of Discovering the Sources of the Ganges', *Asiatic Researches* 11 (1812), 450 fol.; *India Office Library and Records* (*IOL&R*), London, Home Miscelaneous, 582, 'Report H. Wellesley (1802)', fols. 2233-2306.

[14] For Marathas at Tirupati: *IOL&R*, Madras Military and Secret Proceedings (MMSP), P/D/45, 26-1-1761, fol. 87. For rivalries at Haridwar fair see Hardwicke, previous footnote.

November fair, since on their way back home many long-distance traders from Afghanistan called or recalled at the swarming Haridwar spring fair. Of course, one could buy indigenous breeds from the Punjab or Rohilkhand or *Gunth* ponies from the Kumaun and Garhwal hills at any time.

More to the east, the autumnal fairs were held at Duddri, answering the big demand from Benares, and Hajipur for the Bihar market. The spring fairs were held at Balrampur and Butwal along the Himalayan

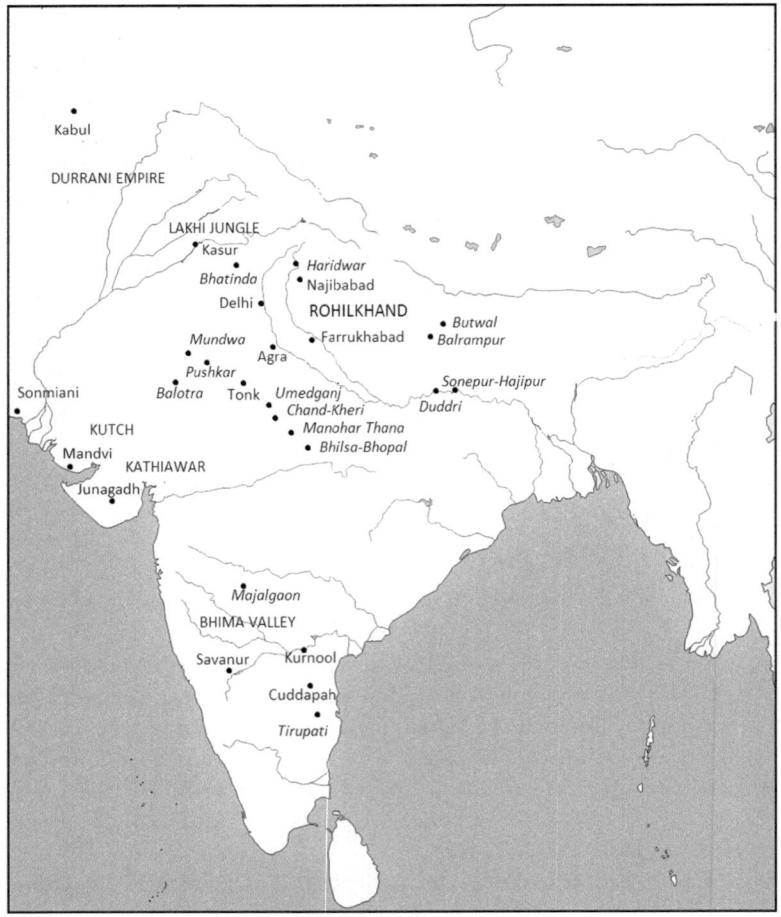

Map 1: South Asia in the eighteenth century

Note: Horse-markets in italics.

fringe, which were also the entrepôts of local indigenous breeds like *Tattus* and hill *Tangans* from Nepal.[15]

Thus, the long-distance overland horse trade connected and integrated several overlapping market areas. In each of these areas were held one or more major spring fairs, mostly an outlet for local produce, and autumnal fairs which specialized in the direct sale of *Wilayati*, i.e. *Turki*, horses which were brought in by foreign merchants. In addition, these fairs served as a market for each of the local states which tried to ensure a stable horse supply for their cavalry. The main fairs of the Rajasthan area were Pushkar in autumn and Bilotra in spring; of Malwa, Bilsa in autumn and Chand-Kheri during spring; of Rohilkhand and the Punjab, Hardiwar was the predominant market, especially in spring.

As far as the eighteenth-century overland horse trade is concerned, a clearly perceptible rhythm and pattern comes into sight. After the horses were bought, they were prepared for the market by letting them graze on the natural *maidans* of Afghanistan and, after crossing the Sulaiman Mountains, on the wastes of the Jullundur Doab and the Lakhi Jungle. In autumn the adjacent Indian markets tuned up for the arrival of this long-distance trade from Central Asia. At the regional level new horses were held for direct use, for fattening, for re-export or for reintroduction into the breeding industry.

THE OVERSEAS TRADE

The overland traffic stood in a complementary relation to the transportation of horses by sea. Whenever the landroute was interrupted by political unrest importation by sea could provide a viable alternative and vice versa. Prior to the eighteenth century, the so called *bahri* or 'sea' horses had usually come from Fars, Iraq or Arabia in large numbers. In the eighteenth century, however, the horse trade with the Persian Gulf was limited and certainly secondary to the overland

---

[15] For information on the fairs: *NAI*, MDP, 15-10-1811, no. 80, 'Report W. Moorcroft', fols. 288-93; B.L. Gupta, *Trade and Commerce in Rajasthan during the 18th Century* (Jaipur: Jaipur Publishing House, 1987), 80-3.

trade with Central Asia.[16] Besides, the bulk of the overseas horses did not originate from the Middle East but from ports in Kathiawar like Porbandar, Gogha, Mandvi or Sonmiani. It was not before the beginning of the nineteenth century that, in order to meet British army demands, sustained horse exports from Iran again reached India by sea. Some decades earlier the government at Madras had already decided to avoid the native overland network and to buy horses in Kathiawar directly through its own channels in Bombay. In order to assure a safe arrival of the horses they were transported by sea to the west coast at Cochin or Mangalore from where they were brought overland to Madras. This was in direct response to the falling off of the overland supply.[17] The Madras territory, like its client the *Walajah* Nawab of Arcot, was located at the other end of the supply lines from the north. At the end of the century the number of northern horses arriving at their regional fair of Tirupati in the south dropped dramatically. The northern horse traders found a growing market in the hands of the agents from the several competing native states in the Deccan. Moreover, Mysorean and Maratha interlopers tried to siphon off the traditional supply lines controlled by the Nizam of Hyderabad and the Afghans of Bhopal, Kurnool and Cuddapah.[18] In fact, the same was true in northern India, where Sikh agents of Ranjit Singh intercepted the supply lines to Haridwar.[19] The situation was aggravated by the fact that whatever numbers were brought to Tirupati – around 500 horses each year – the best of them were first claimed by the local *chabuk-sawars* of the Nawab of Arcot and, only through their mediation, the remainder was left to the Company.[20]

---

[16] S.R. Grumman, 'The Rise and Fall of the Arab Shaykhdom of Bushire 1750-1850' (PhD thesis, John Hopkins University, 1962), 196.

[17] *NAI*, MDP, 27-11-1813, no. 108, 'Report E. Wyatt'; *IOL&R*, MMSP, P/252/18, 3-8-1787, fols. 407-9; MMSP, 14-8-1787, fol. 484; MMSP, 17-8-1787, fol. 536; Madras Military and Political Proceedings (MMPP), P/253/10, January 1793, fol. 565; MMPP, P/253/75, 10-1-1797, fol. 156.

[18] *IOL&R*, MMPP, P/253/12, April 1993, fol. 608; MMPP, P/253/33, 4-10-1794, fol. 4109.

[19] *NAI*, MDP, 13-2-1813, 'Moorcroft to Secr. Brd.', fols. 137-8.

[20] *IOL&R*, MMSP, P/251/72, 2-11-1772, fol. 880; MMPP, P/253/33, 4-10-1794, fol. 4110.

What can be said about the costs of transport of overseas horses in comparison with the overland horses? For the latter we only know the sales prices. At the time the overland horses arrived at Tirupati these had increased considerably. During the 1770s a good cavalry horse at Tirupati would cost from 150 to 200 Pagodas – which was equal to Rs. 500 to 700.[21] The same horse at the northern fairs would be around Rs. 100 to 200 less.[22] At the end of the century the prices had fallen to about Rs. 300 to 400 in the north.[23] At the same time carrying horses from from Kutch to Calcutta by sea would cost around Rs. 760, from Basra to Calcutta Rs. 838. The latter sum consisted of the following subcharges:

purchase at Basra	250
duties	30
fodder	100
freight Basra-Calcutta	300
insurance	80
landing charges	2
casualties on the voyages	76
cost price at Calcutta	Rs. 838[24]

From this we may conclude that the costs of overseas transport of horses were considerably above the level of the overland passage, even at a stage when the latter had become very difficult indeed.

In another reaction to this supply crisis, the Madras government, increasingly aware of the importance of cavalry troops, had to resort to the hiring of native cavalry forces, which was extremely expensive but could not be dispensed with in the protection of British transports

---

[21] *IOL&R*, MMSP, P/251/71, 17-2-1772, fols. 189, 207: horses for officers would cost around 250 Pagodas; MMSP, P/251/72, 14-7-1772, fol. 623: frequently horses were also valued in rupees; *NAI*, Foreign and Political Department (FPD), S, 8-10-1784, no. 12; FPD, S, 19-2-1785, no. 44.

[22] *NAI*, FPD, S, 6-1-1774, no. 1; FPD, S, 3-2-1777, no. 13; Public and Home Department (PHD), 9-7-1782, fol. 1457.

[23] *NAI*, MDP, 9-10-1795, no. 37-39; MDP, 10-4-1795, fols. 169-70; MDP, 15-10-1811, fol. 305.

[24] *NAI*, MDP, 15-10-1811, 'Report W. Moorcroft', fol. 307; see MDP, 27-9-1804 for Kutch-Calcutta.

and communications.[25] Still another attempt to solve this problem was the launching of a breeding operation of their own, as was tried at Pusa in 1793 and at Ganjam in 1795.

VOLUME OF THE TRADE

In order to arrive at a general estimate of the total volume of the horse trade it is necessary first to give an estimate of the total amount of cavalry horses employed in India. Let us start with the relatively small cavalry contingents of the EIC. At the end of the eighteenth century, the Company's officials and army officers became increasingly aware that, in the long run, they could not hold or expand their newly acquired territories without a substantial enlargement of their cavalry establishment. The Maratha and Mysore wars had proven to them the persisting importance of the horse. In 1793 the peace-time Bengal cavalry consisted of only two regiments of native cavalry with an establishment of no more than 500 horses. Only six years later the total Bengal military horse establishment had risen sevenfold to 3,500 animals. In 1809 it had grown twelvefold to 6,000 horses. Similarly, in Madras the cavalry contingent was increased, mainly by incorporating the native regiments of the Nawab of Arcot and the Nizam of Hyderabad. Until 1803, only the Bombay government had no cavalry of its own and had to rely completely on the mounted contingents of its native allies.[26]

Despite these increases, the total British cavalry force was still inferior to the massive cavalry contingents of the native states. In 1778, Haider Ali's army in the Deccan was reported to count 28,000 horse. At the same time, the cavalry of the Marathas, in the Deccan only, numbered 67,000.[27] To Robert Orme we owe the following

[25] *IOL&R*, MMSP, P/251/59, 19-10-1767, fols. 1071-3; MMSP, P/251/61, 29-2-1768, fols. 252-3.

[26] For Bengal: G.J. Alder, 'The Origins of the "Pusa Experiment": The East India Company and Horse-Breeding in Bengal, 1793-1808', *Bengal Past and Present* 98, 1 (1979), 10-2; for Madras, see W.J. Wilson, *A History of the Madras Army from 1746 to 1826 2* (Madras: Government Press, 1888), 149; for Bombay, see B. Mollo, *The Indian Army* (Poole: Blandford, 1981), 16.

[27] *NAI*, FPD, S, 2-2-1778, no. 21, these figures only refer to the Deccan.

evaluation of the strength of the military powers in Hindustan around 1760:[28]

Rohillas	
Hafiz Rahmat Khan	4,000 horse
	20,000 foot
Dunde Khan	10,000 horse & foot
Maula Sardar:	3,000 horse & foot
Najib al-Daula	10,000 horse
	20,000 foot
Ahmad Khan Bangash	10,000 horse & foot
Others	
Rana of Udaipur	20,000 horse
	10,000 foot
Jats	10,000 horse
	30,000 foot
Sikhs under Charhat Singh	20,000 horse
	30,000 foot
Rajputs under Radhu Singh	20,000 horse
	? foot
Chief of Malwar	25-30,000 chiefly horse

These figures, together with those of the Deccan give a total of something over 200,000 horses. These figures do not present a complete picture, however, because many native states are not included in the above list. Obviously, it is not possible to calculate the exact number of the total Indian cavalry but when we only take into account the Maratha army in Hindustan, the cavalry of the rulers of Awadh, Benares and Bengal, and the remaining Rajput chiefs, the figure would already be nearly doubled. Even the Mughal stables still counted some 200,000 horses around 1740. Although this figure is only half of the total of Akbar's cavalry units, given by Abul Fazl, together with the armies of the Mughal successor states, the total sum for the mid-eighteenth century clearly exceeds the sixteenth-century numbers.[29]

Looking at the number of horses of the many local *zamindari* armies in India, even more doubts are raised. When we adopt Habib's

---

[28] *IOL&R*, Orme Mss, ov, 108, fol. 89.

[29] For these figures, see W. Irvine, *The Army of the Indian Moghuls: Its Organization and Administration* (London: Luzac, 1903), 61.

figures for the sixteenth to the eighteenth century, this would result in another 400,000 horses.[30] Considering the increased impact and buying power of the *zamindars* in eighteenth-century India, even this staggering figure would seem to be too low for the later period. All these figures, including those of Habib are, however, very difficult to check and often the contemporary accounts were influenced by the hyperboles of the over-enthusiastic observer. Besides, there was most probably a great deal of overlap in the numbers. For example, Rohilla mercenaries switched sides rather frequently and thus their contribution to the potential strength of native armies could easily be overrated. On the other hand, Kolff rightly observes that even the Mughal inventories of the military labour force did not exhaust the total reservoir of armed men, with or without horses.[31]

Keeping these restrictions in mind, we may still come to a general assessment of the importance of the horse in quantitative terms. As a result of the decentralization of the Mughal Empire and the rise of the regional courts, the total horse population in the first three quarters of the eigthteenth century was brought to a peak. Thus, the total sum of warhorses in India, excluding the Iranian and later Afghan provinces, can be ranged somewhere between 400,000 and 800,000.

What do these figures tell us about the total turnover of the horse trade? Obviously, we have to add another myriad of uncertainties which will leave the end result even more wide-ranging. For example, the price level of horses varied considerably during the eighteenth century. During the first three quarters of the century there was an upward trend in prices because demand continuously exceeded supply and because of an overall trend of inflation. Of course, prices were higher when horses had to be transported to far-off places like Mysore or Bengal. As a convenient alternative, horses could also be imported by sea but this markedly raised the costs of transport, fodder and casualties. As we have seen, the northern horses which arrived on the

---

[30] I. Habib, *The Agrarian System of Mughal India (1556-1707)* (London: Asia Publishing House, 1963), 166-8.

[31] D.H.A. Kolff, *Naukar, Rajput and Sepoy: The Ethnohistory of the Military Labour Market in Hindustan, 1450-1850* (Cambridge: Cambridge University Press, 1990), 3.

southern markets by land fetched a price which was only around 150 per cent of the level on the Hindustani markets. For overseas horses the cost price would be around 200 per cent or more.

The data of the mid-century trade hausse will serve as the basis for the following estimate of the total turnover. At that time, indigenous *Tattu* breeds, the horses that dominated in the local *zamindari* armies, could be procured for only Rs. 15 to 100 each.[32] Larger, stronger and superior foreign and indigenous *Turki* or *Tazi*, i.e. Arabian-like, horses – mainly from Kathiawar, Kutch, Rajasthan, Baluchistan, Afghanistan and Turkistan – fetched a medium price of around Rs. 500 but never under Rs. 400. Rare thoroughbred Arabian horses were even three to four times this price. The enormous price difference between indigenous and foreign horses was due to a persisting high demand for large and strong cavalry horses of the latter category and also indicates the difference in status of both races. The Mughal cavalry during the sixteenth and seventeenth century consisted almost entirely of relatively strong *Tazi* and *Turki* horses. The *Turki* was particularly predominant among the Irani, Turani, Afghan and Rajput contingents, the *Tazi* among the Maratha troops. During the eighteenth century these horses remained the mainstay of the now more decentralized Mughal cavalry.[33] However, in the Deccan at the end of the century, the Marathas started to breed a *Tazi* horse of their own along the Bhima Valley which was a mixture of *Tazi* and local indigenous blood.[34]

Returning to our estimate of the total turnover, it seems reasonable to qualify half of the total amount of cavalry horses as inferior *Tattu* horses and the remainder as varieties of the more expensive *Turki* or *Tazi* breeds. Considering the fact that the annual wastage in peacetime was around 10 per cent (which means that every ten years the existing

[32] Comte de Modave, *Voyage en Inde du Comte de Modave 1773-1776*, ed. J. Deloche (Paris: École Française d'Extrême-Orient, 1971), 327; Pigott, *Treatise*, 43.

[33] Abul Fazl 'Allami, *The A'in-i Akbari*, vol. 1, trans. H. Blochmann and H.S. Jarrett, 245; R.A. Alavi, 'New Light on Mughal Cavalry', *Medieval India: A Miscellany* 2 (New Delhi: Asia Publishing House, 1972), 70-99.

[34] H. Shakespear, *The Wild Sports of India* (London: Smith, Elder and Co., 1862), 298-9.

stock of cavalry horses had to be renewed[35]), and assuming the total amount of cavalry horses in India to be 600,000 around the middle of the century, this produces the following results:

Annual renewal	Annual turnover
Rs. 30,000 × 500 (*Tazis, Turkis*)	Rs. 15,000,000
Rs. 30,000 × 50 (*Tattus*)	Rs. 1,500,000
Total annual turnover	Rs. 16,500,000

Obviously, this is an extremely rough estimate because many of the quantitative data are uncertain. On the other hand the calculation appears to be a conservative one when compared with the contemporary assessment of the French traveller Comte de Modave, who during the 1770s reckoned the annual importation of horses from Turkistan and Iran to be around 45,000 to 50,000.[36] This would result in a total import trade of around Rs. 20 million, which would be more than three times the total of Bengal exports to Europe by the English, Dutch and French EIC's together. Although in terms of regular trade the Frenchman's account seems to be an overestimation, it is certainly possible that occasionally these figures could be realized as a result of a sudden increase of demand. During times of large-scale warfare or epidemics the losses of horses increased from one out of ten to one out of seven or more. Taking the incidental character of Modave's figures into account, they are not so far off from the calculations above.

All in all, we cannot trust these figures to be more than lumpsums which can offer us only an impression of the volume of trade. But even if only 5,000 horses were annually imported into India, this would result in a trade volume which compares still very favourable with the total export trade of, for example, the eighteenth-century

---

[35] This figure from Alder, 'Origins', 12; my own calculations are less optimistic: service of cavalry horse mostly started around 3,5 years, after which it could be active for less than nine years. Other references speak of only seven years, cf.: *NAI*, FPD, S, 6-1-1774, no. 1; C.F. Traver, *Hints on Irregular Cavalry, its Conformation, Management and Use in both a Military and Political Point of View* (Calcutta: W. Thacker, 1845), 60, 85, 88.

[36] Modave, *Voyage*, 327.

VOC in Bengal. Thus, the figures above do not claim any accuracy at all, but they should shed some new light on the overland connections of India and Central Asia and they should have implications for the still current views of an isolated Central Asia. Until far into the eighteenth century, and not least thanks to the horse trade in general, overland commercial relations between India and its northern and western neighbours were still very close and were certainly in a flourishing condition.[37] In northern Afghanistan this resulted in further pastoralist penetration from the late seventeenth century onwards and a general shift from crop-cultivation to pasturage.[38] As we shall see in the following section, in India too the rising demand for war-horses stimulated regional breeding efforts.

THE BREEDING OF HORSES: TWO EXAMPLES

The balance between horse breeding and arable farming could be an extremely delicate one. In areas where conditions of soil and climate were not ideal for crop cultivation and methods of horse breeding required extensive grazing lands, horse-breeding was often at the expense of arable farming. However, in areas of secure and rich harvests the relationship between breeding and farming could be more symbiotic. Therefore, in this section I will pay attention to two examples of local breeding areas. One is Kathiawar, where very good horses had always been bred, but at the cost of a low intensity of settled crop cultivation. In contrast to this predominantly pastoral area, Rohilkhand was traditionally not associated with the breeding of strong horses since it possessed extremely rich conditions for settled cultivation. Nevertheless, during the eighteenth century, the more sedentary area of Rohilkhand not only witnessed an expansion of settled cultivation but also an increasingly flourishing horse-breeding industry.

[37] For the growing horse trade after the decline of the imperial Mauryas and Guptas, see C. Gupta, 'Horse Trade in North India: Some Reflections of Socio-Economic Life', *Journal of Ancient Indian History* 14 (1983-4), 195.

[38] R.D. McChesney, *Waqf in Central Asia: Four Hundred Years in the History of a Muslim Shrine, 1480-1889* (Princeton: Princeton University Press, 1991), 234.

KATHIAWAR

In 1814, the British agent Wyatt was commissioned to the Kathiawar and Kutch area in order to buy horses for the EIC's Bengal army.[39] He wrote to his superiors that this task had become very difficult since some years past there had been a marked decline in the local breed of horses. But during the eighteenth century the local chiefs had greatly encouraged the breeding of horses. Particularly the Kathi tribe had paid great attention to breeding. According to Wyatt, the natural conditions in Kathiawar were ideal for breeding purposes. The face of the country was almost everywhere hilly and mountainous while the soil was generally rich although mixed with stone and sand. According to Wyatt, 'this made the area scantly cultivated and so bare of trees that excepting near the towns and villages there is scarcely a tree to be seen throughout the whole country'. The country abounded, however, in nutritious 'jinjirah' (*ganja*) and 'durru' (*durva*, a variety of *dub* or *Cynodon dactylon*) grasses. Although the climate was relatively dry, there were innumerable small streams and rivulets which took their rise in the hills and ran into the creaks of the sea or lost themselves in subterraneous caverns. In contrast, to the north and east, in Gujarat, where the climate was very moist and the landscape flat, the breed of horses degenerated and became ill-formed. In the early nineteenth century, however, even in Kathiawar not much was left of this once famous breed.

Following an earlier report of Colonel Walker, Wyatt ascribed this sudden decline to the rude system of native government and the several incursions of the Marathas in the area. At the same time, however, he reluctantly had to admit that the earlier political unrest and the predatory politics of the local chiefs were in fact the *raisons d'être* of a blooming local breeding industry. Quoting his own words:

The decline of the breed however amongst the Kattee tribe who were the most famous for Their horses and who to this day possess the best remains of the breed is owing in some degree to a cause which cannot be regretted,

---

[39] This section on Kathiawar is based on *NAI*, MDP, 14-6-1814, no. 76, 'Report E. Wyatt', fols. 75r-83r.

that is to a check having been given to their plundering excursions by which until very lately they almost entirely subsisted.

During the eighteenth century the Kathis were still semi-settled pastoralists who, 200 years before, had migrated from the north-west to Kathiawar. They claimed descent from the mythical Kath who was a robber of cattle, and this, to the indignation of Wyatt, caused them to feel not the least remorse for their earlier predatory way of life. Only after the British had settled the country in 1807, did the Kathis leave their former occupations and did they concentrate exclusively on settled agriculture, which offered more secure profits, given the rise of Bombay and the EIC trade along the coast. While the British had successfully pacified the area, the breeding industry had almost completely disappeared. The area changed from being by far the most important source for supplying the Deccani cavalry troops, to a relatively unimportant agricultural area deprived of its main item of export. Thus, since the breeding industry had been directly linked to a vigorous predatory economy, the imposition of the Pax Britannica took away the last incentive for breeding horses on a large scale.[40]

ROHILKHAND

In Rohilkhand too the breeding industry declined following the end of native encouragement under the Indo-Afghan Rohilla government (1738-74). In the words of the Company veterinary surgeon William Moorcroft:

The spirit of horse-breeding supported by the Rohillas during the period they possessed Rohilkhand has become almost extinguished since their expulsion, and the animals now raised are seldom fit for other service than that of irregular troops.[41]

The Rohilla ruler Hafiz Rahmat Khan (1749-74) stimulated breeding systematically by supervising the regular distribution of stallions to the local *zamindars*. They put them to their own mares and also to those of their circle of relatives and friends. The Rohillas, however,

---

[40] Cf. Shakespear, *Wild Sports*, 313; P. Nightingale, *Trade and Empire in Western India, 1784-1806* (Cambridge: Cambridge University Press, 1970).
[41] *NAI*, MDP, 13-2-1813, no. 156, fols. 137-8.

did not breed horses themselves. Most of the stallions were held by a Hindu caste called Bhat, of mixed origin, which normally served as panegyrists and bards who attended and added lustre to family parties and ceremonies. They are also frequently related to the Banjara caste of highly mobile grain carriers.[42] This caste of Bhats employed the best breeds of horses, which were, however, rejected for cavalry service because they were lame or for some other reason blemished and not fit for warfare. The Bhats regularly took these stallions to the stables of the local *zamindar* who had to pay one Rupee in order to have his mare three times covered by the stallion, with the privilege of a fourth time whenever there would appear to be a necessity for it. Most of the breeding *zamindaris* were situated in the delta of Mehrabad in southern Rohilkhand, between the Ramganga and the Ganges rivers.[43]

Although to a lesser extent than in Kathiawar, the ecological conditions of Rohilkhand were suitable for breeding. The alluvial pastures along the many streams coming from the northern hills had a reputation for their fattening and nutritious qualities for horses. The area in general was extremely fertile and, in Chris Bayly's terms, might be called an area of natural surplus.[44] Although the Rohilla territory became increasingly cultivated, the *zamindars* retained a certain quantity of land for growing the very nutritious *dub* grass, which was cut at the end of the rainy season and made into hay and stacked, and given to the horses during the dry season. Most of the grazing fields were situated along the river shores in order to facilitate easy inundation. Besides, the waterlevel was so near the surface that *dub* grass could grow for over eight months during the year, from the start of the monsoon until the time of the spring fairs. However, the

[42] J. Deloche, *La circulation en Inde*, vol. 1 (Paris: Ecole Française d'Extrême-Orient, 1980), 251, 253.

[43] For breeding in Rohilkhand, see reports of W. Moorcroft, J. Fortescue and Capt. Nuthull in: *NAI*, MDP, 15-10-1811, no. 80, fols. 275-81; MDP, 13-2-1813, no. 156, fols. 137-8; MDP, 2-6-1803, fols. 2r-8v; MDP, 27-9-1804, no. 56, fols. 74r-84v.

[44] C.A. Bayly, *Rulers, Townsmen and Bazaars: North Indian Society in the Age of British Expansion, 1770-1870* (Cambridge: Cambridge University Press, 1984), 80-3.

cultivation of *dub* was extremely labour intensive as it required the close scrutiny by grasscutters who had to keep the ground clear of more savage weeds that naturally suppressed and overpowered the more delicate *dub*.[45] Next to grass and hay, the chaff of other winter crops available in Rohilkhand like gram (*chana*), lentils (*masuri*) and particularly grains like wheat and barley, could serve as a suitable additional source of fodder.

In contrast to the more extensive pastoral production in western India and Afghanistan, the Rohilkhand breeding process represented a more intensive type of mixed farming which confined the animals to stables and small fields of pasture. Besides, Rohilkhand regularly imported horses from abroad, either through the long-distance trade or from the unsettled areas surrounding it to the north and the west. This was the territory of the semi-nomadic and predatory Banjaras, Gujars, Bhats and Bhattis. Although these groups were frequently seen as enemies of settled agriculture and settled government, they played a crucial part in the regional economy of Rohilkhand, especially by providing fresh livestock for transport and breeding.[46]

Although the cultivation area and the population in the region rapidly increased during Rohilla rule (*c.* 1720-70), horse breeding activities did not suffer at all. On the contrary, agricultural expansion facilitated breeding activities because it made fodder and labour readily available and relatively cheap. The high demand and prices for horses further stimulated the breeding economy. In Rohilkhand the demand was so high that advance reservations could be made even before foaling. The horses were sold directly to Rohilla mercenaries or to Afghan horse-traders who mostly purchased them when one and a half or two years old from the local *zamindars*, after which they fattened them well until they were three years of age and fit for military service and resold at the autumn or spring fairs. The quality of the Rohilkhand breed was not only a reflection of local breeding conditions, but also resulted from a regular injection of foreign stallions and mares from the Punjab, Afghanistan and Turkistan, brought by

---

[45] There also exists a tension between making hay and gathering in the *kharif* crops, both of which take place around September.

[46] Bayly, *Rulers*, 29.

Afghan and Sikh traders or by roaming Gujar and Bhatti herdsmen-marauders who exchanged these horses mainly for the Rohilkhand *kharif* cash crops such as indigo and sugar, made ready for the autumn fairs, or the *rabi*ᶜ crops like wheat and barley, entering the market during spring. The invasions of Nadir Shah and Ahmad Shah Durrani gave another impulse to the Indian breeding industry since many horses were bought or stolen from the invading armies by attentive Sikh, Rohilla, Gujar and other *kazaki*. As a consequence, it was claimed that the breeds of the Lakhi Jungle and the Bikaner desert had markedly improved thanks to a sudden influx of *Wilayati* horses.[47]

After the annexation of Rohilkhand by Awadh in 1774, trade, agricultural production and breeding activity declined simultaneously. Most of the Rohillas and their clients concentrated on their remaining territory around Rampur or migrated to other more attractive areas. In order to control his new territory the ruler of Awadh discouraged external trade relations in Rohilkhand, and commercial traffic now began to avoid this more and more depopulated area. The collapse of the Rohilla state and its Indo-Afghan trading network in northern India caused an overall dwindling of the demand for warhorses, exacerbated by increasing exchange problems caused by falling agricultural production and depopulation in the area. In addition, throughout the newly created fallow lands, parasitic grasses could spread, and reports of roaming cattle destroying the crops increased. For a decade the hub of the Hindustan horse trade shifted eastwards to Awadh and Benares. The Nawab of Awadh, Asaf al-Daula (1775-97), enticed Afghan and other horse-traders to move towards his territories instead, circumventing Rohilkhand to the south via Mathura, Farrukhabad and Kanpur to Lucknow.[48] Raja Balwant Singh and his son and successor Chait Singh of Benares every year got hold of many colts and fillies from the Lakhi Jungle which they, like their counterpart *zamindars* in Rohilkhand, distributed amongst their numerous relatives and dependants. In the wake of the expulsion of

---

[47] Pigott, *Treatise*, 43; *NAI*, MDP, 2-6-1803, 'Report J. Fortescue', fols. 2r-8v; similarly the breed of Bihar was known to have been improved through the spoils of the battle of Buxar (1764).

[48] *NAI*, MDP, 15-10-1811, no. 80, 'Report W. Moorcroft', fol. 283.

Raja Chait Singh in 1784, the trade with Benares diminished significantly.[49] The death of Asaf al-Daula (1797) and the increasing political influence of the East India Company dealt the final blow to the remaining horse trade of north-eastern India. Simultaneously, Central Asian horses were progressively drained into the Deccan.[50]

At this point, we may draw some general conclusions. Firstly, the success of the breeding economy was based, on the one hand, on a regular exchange of livestock with north-western India and Central Asia and, on the other hand, on a close relationship of settled with nearby unsettled areas. Secondly, the emergence of new political configurations within a flourishing predatory economy created extremely favourable conditions for both horse breeding and trade. Under these circumstances, when high levels of demand and prices prevailed, an area like Kathiawar specialized in pastoral horse breeding, since ecological and political conditions did not allow an efficient combination with crop cultivation. On the other hand, in Rohilkhand, breeding was tied in to intensified cultivation and was also integrated with the more unsettled fringe areas to its north and west. As such, stability in one area was almost conditioned by instability in the neighbouring area.

DECLINE

At the end of the eighteenth century the Indian provinces under the control of the ascending English East India Company experienced a marked decline in the quality and quantity of the available cavalry horses. The quality requirements for a good British dragoon horse were very high indeed. The British cavalry trooper of the day, together with his saddle, weapons, ammunition and equipment, came to around 115 kg, which was much heavier than the common native horseman.[51] Therefore, if he was to make an effective cavalry charge, the horse needed both height and weight, bone and muscle. Disreputable *Tattu* horses were considered absolutely deficient for

---

[49] *NAI*, MDP, 15-10-1811, nr. 80, 'Report W. Moorcroft', fol. 280.
[50] Ibid., fol. 279; *NAI*, MDP, 13-2-1813, nr. 156, fol. 138.
[51] Cf. Alder, 'Origins', 11.

this purpose. Besides, British agents complained about an overall degeneration of the Indian horse following a marked decline in the trade relations with north-western India and Central Asia.[52] It appeared that every westbound step of British expansion was nearly automatically accompanied by a similar westward retreat of the international horse trade. Consequently, the bulk of the Company's horses could not be bought at the local fairs of Bihar and Bengal but had to be procured, through intermediary agents, from the *melas* of Rajasthan and the Punjab. The resultant import trade, however, caused an increasing drain of specie which the Company could not afford.

To secure an indigenous source of supply the EIC set up a stud-breeding farm of its own at Pusa in Bihar in 1793. Two years later, in reaction to increasing supply problems, a similar project was launched in the Ganjam district in the Madras presidency. Apart from producing horses at the stud-farm itself, the policy aimed at stimulating the breeding industry of the surrounding *zamindars* as well, especially in the traditional breeding districts of Ghazipur, Sarin and Shahabad. Until then, the most prominent figure dominating the local breeding business had been the so called *nalband*. This term literally meant 'farrier' or 'blacksmith' but in fact he was the manager who supervised all aspects of local breeding and selling.[53] As a trader he provided stallions to serve the mares of the *zamindars* after which he was entitled to buy the resultant offspring which he took to the local fairs, or he resold them directly to Afghan and Maratha long-distance dealers who carried the horses to the most lucrative markets of India. At the end of the century, these were certainly not located in the eastern territories under the sway of the EIC. As we observed already, the imposition of the Pax Britannica and the collapse of the northern Rohilla network engendered a dwindling regional demand in Hindustan.

British officers could not afford to pay as much cash for horses as traders in the Deccan or Rajasthan were used to receive. Because the

[52] *NAI*, MDP, 15-10-1811, nr.80, 'Report W. Moorcroft', fol. 295; cf. E. Balfour, *The Cyclopaedia of India and Eastern and Southern Asia* (Graz: Akademische Druck- und Verlagsanstalt, 1967), 'horse'.

[53] Cf. Alder, 'Origins', in Rohilkhand the *nalband* signified a farrier only.

demand from the south was on the increase, the price level was still
under pressure. Anticipating the lack of purchasing power from the
east, the traders from the Punjab and Rampur separated their stock
in two classes. One consisted of the best and most expensive horses
which were earmarked for the Deccani market where they could be
readily sold at high prices. The other group was made up of inferior
horses and sold to local adventurers and mercenaries. Only a small
part of the worst horses reached the eastern *melas* under British control.
As a result, most of the cavalry of the native forces employed horses
far superior to those of the British. The situation was further aggravated
by a European cultural predilection for Arabian horses, which were
directly imported from Iraq to Bengal. This 'Arabomania', as Moorcroft
called it grudgingly, stemmed from Europe and the experience of
breeding English thoroughbreds and race horses. Under British rule
Indian horse breeding became, in effect, an entirely different operation
which required stronger, higher and better trained horses than the
indigenous *Tattus*. On the Bihar scene, however, Arabian crossbreeding
with native stock was not very encouraging. In order to counter the
general lack of qualified indigenous stock, the Pusa stud-farm tried
to control the quality of the breeding stock by supplying stud stallions
and mares to the local *zamindars* and to the *nalbands*. Although these
new breeding investments caused some increase in production and
in commercial interest in the eastern fairs, through the mediation of
the *nalbands*, the majority of these horses were driven to the western
and southern markets.[54]

In general, however, ecological circumstances in Bihar remained
far from ideal for breeding large and strong horses. Because of the
lack of a suitable climate and breeding stock the interest of the stud
officials was redirected to the traditional breeding areas of the north-
west: the Punjab, Rajasthan, Kutch and Kathiawar, and even beyond
to Afghanistan and Turkistan. Here they hoped to procure a bigger

[54] We witness a similar process in the nineteenth-century Dutch East Indies
where the indigenous breed of Batakkers deteriorated as a result of European
interference (W. Groeneveld, 'Het paard in Nederlandsch-Indië; hoe het is
ontstaan, hoe het is en hoe het kan worden', *Veeartsenijkundige Bladen voor
Nederlandsch-Indië* 28 (1916), 195-240).

and bonier parent stock both for crossbreeding and military service. The most enthusiastic example in this respect was William Moorcroft, veterinary surgeon and superintendant of the Pusa stud. From 1810 until his death in 1825, spurred by an inner drive to find the source of the famous Turkistan blood stock, he made long journeys across the Himalayan and Sulaiman Mountains which eventually would take him as far as Bukhara. However, because of the ever progressing pacification of the subcontinent, the Central Asian horse trade with India had shrunk to an utter minimum and Central Asia itself had changed into the imaginary nineteenth-century 'Back of Beyond'.[55]

HORSE TRADE AND STATE FORMATION

Not only for the EIC, but also for the other regional states it was difficult to check the erratic movements of the horse-trade. The highly mobile and experienced horse-trader was in a powerful position vis-à-vis the local consumers. As a result, horse-traders had a particularly bad reputation, to which their wandering life contributed as well. Their mere presence could constitute a serious threat to law and order, as is illustrated in the case of the Rohilla freebooter Daud Khan. Daud Khan was later considered to have been one of the founding fathers of the Indo-Afghan state in Rohilkhand. At the start of the eighteenth century and as an agent of his master Shah Alam Khan, he was sent to the *mela* of Haridwar to buy some horses. After he had bought these he declined to send them to his master and instead distributed them among some fellow Rohillas whom he had gathered around him and thus he began a career as highway robber. A few years later he had collected a following of 80 horsemen and 300 footsoldiers, was able to build his own mud fortress and to defy Mughal rule in the area.[56]

Apart from being a disruptive element, the itinerant horse trader represented a more or less cosmopolitan culture. They were accustomed

[55] For William Moorcroft, see G.J. Alder, *Beyond Bukhara: The Life of William Moorcoft, Asian Explorer and Pioneer Veterinary Surgeon 1767-1825* (London: Century, 1985).

[56] Ghulam ʿAli Khan Naqawi, *ʿImad al-Saʿadat* (Lucknow: Nasiriyyah Libary, 1864), 40-1.

to their own esoteric language, which was a mixture of various local dialects combined with a special jargon and an extensive code of manual signs, exchanged during the actual bargaining at the fair, mostly concealed beneath a handkerchief.[57] Sometimes the local rulers tried to regulate and control the irregular movements of the horse-traders. At the Pushkar fair, for instance, the traders and their horses had to take up their camping ground in the direction of the countries they came from. Through such regulations the authorities hoped to make the trader and his horses more identifiable and to assure that there would always be somebody answerable for frauds or other malpractices.[58]

Partly as a consequence of the extreme mobility of the horse trade, it is very difficult to come very close to the horse-trader himself, who may appear alternately as a merchant, a mercenary, a highway-robber or a Sufi. This multi-faceted role may also explain why the sources remain rather vague about him. Most of them were people from the north-west – Afghans, Sikhs and Punjabis – but at the Hindustani fairs Maratha traders were also active. Even in Kabul the Maratha chiefs had their agents who bought horses for them through bills of exchange.[59] Among the Afghans there were many Lohani *Powindah* traders and Rohillas. Many of them travelled long distances, as, for example, the Rampur merchant Ahmad Ali Khan who, even at the start of the nineteenth century, had a radius of action which extended from Bhatinda to Bundelkhand and far into the Deccan.[60] Some of these people answered the traditional picture of the itinerant pedlar but we should not forget that supplying merely a small number of horses required already huge capital investments for purchase, duties, fodder, and other expenses. Many of the smaller traders at the fairs were agents of the greater men behind the Sulaiman Mountains to the west. An example of such an entrepreneur is one 'Sunderji' who at the end of the eighteenth century had gained a complete monopoly

---

[57] Sa'adat Yar Rangin, *Faras Nama e Rangin or the Book of the Horse by Rangin*, trans. D.C. Phillott (London: Quaritch, 1911), 42-4.

[58] *NAI*, MDP, 15-10-1811, no. 80, 'Report W. Moorcroft', fol. 289.

[59] H.R. Gupta, *Studies in Later Mughal History of the Punjab 1707-1793* (Lahore: Minerva Book Shop, 1944), 259.

[60] *NAI*, MDP, 15-10-1811, no. 80, 'Report W. Moorcroft', fols. 302-3.

of the horse supply from Kathiawar. His agency network extended far to the north from Kalat in Baluchistan to Kabul and Kandahar. During the years 1810-12 he supplied the Company agents at Bombay with 1,800 horses although the quantity and quality of his deliveries had already greatly decreased during the last one or two decades.[61]

The eighteenth century witnessed the rise and expansion of a contiguous series of Afghan states.[62] Like, for instance, Daud Khan Rohilla, many of the new Indo-Afghan rulers started their political careers as horse-traders-cum-mercenaries. As such they helped to promote and widen the Afghan trading network. Afghan mercenaries served in almost every army of the subcontinent and thus acted as valuable contacts, being both agents and customers for the native states of India. Therefore, it is no coincidence that nearly all Afghan states were carved out along the traditional horse-trade routes: the newly emerging Durrani Empire controlled the main breeding areas in Afghanistan proper, while Rohilkhand and Kasur were situated along the northern supply lines, whereas Tonk, Bhopal, Kurnool and Cuddapah supervised the communications to the south, to the Deccan, Hyderabad and Mysore. Even in Kathiawar the Afghan Babi merchants, coming from the Kandahar area, established their *Nawabi* in Junagadh, where they also acquired a stake in the maritime connections at the port of Gogha. Most of the other native states were strongly indebted to Afghan mercenary chiefs and intermediaries for both man- and horsepower. In fact, Afghans rivalled the British in terms of military service. Instead of a disciplined infantry, they could offer horses and cavalry, and during the eighteenth century it was still not clear which of the two was the most important.[63]

---

[61] Letters of E. Wyatt, *NAI*, MDP, 27-11-1813, no. 108; MDP 9-4-1814, no. 89; MDP, 4-6-1814, no. 76.

[62] See my 'Mughal India and Central Asia in the Eighteenth Century: An Introduction to a Wider Perspective', *Itinerario* 15, 1 (1991), 51-71.

[63] E.g. P.J. Marshall, *The New Cambridge History of India*, vol. 2.2: *Bengal: The British Bridgehead* (Cambridge: Cambridge University Press, 1987), 51, 71; S. Chander, 'From a Pre-Colonial Order to a Princely State: Hyderabad in Transition c. 1748-1865' (PhD thesis, Cambridge University, 1987), 136-7.

CHAPTER 2

The Silent Frontier of
South Asia, *c.* AD 1100-1800

THE FRONTIER IN EURASIA

IN THE HISTORY of Eurasia, closed and fixed frontiers or boundaries are a fairly recent phenomenon.[1] Hence, the concept of the premodern frontier, both in its dividing and uniting qualities, stands in need of further elaboration. The historical debate on the frontier still echoes the classical vision of the frontier set out by Frederick Jackson Turner. He imagined the frontier – the American West – as an almost uninhabited wilderness mastered by the gallantry of pioneers. Although Turner's idea often has been depicted as only relevant for the young man going West, there are still many similarities that remain instructive to its eastern counterpart that remain constructive. For example, Turner's frontier rightly contains the notion of frequently violent mission and expansion which so conspicuously determine the identities of all frontier people, be they American cowboys, European crusaders or Muslim *ghazis*. Turner's idea of the frontier also highlights the common purifying and ennobling capacities of the frontier.[2]

Nevertheless, the differences stand out more clearly. The idea of

---

[1] See e.g. the illuminating article by E.R. Leach, 'The Frontiers of "Burma",' *Comparative Studies of Society and History* 3, 1 (1960), 49-73. See also, A.T. Embree, 'Frontiers into Boundaries: From the Traditional to the Modern State', in R.G. Fox (ed.), *Realm and Region in Traditional India* (Durham: Duke University, 1977), 255-81.

[2] For a general debate on frontiers, see e.g. the special frontier issue of the *Journal of World History* 4, 2 (1993).

the frontier as ever shifting in only one direction, located in almost empty, wide-open space, within common ecological parameters, contradicts the general Eurasian experience. For Eurasia, the frontier that has moulded its history consisted of a broad overlapping area in which two different ways of life encountered each other, the one predominantly pastoral-nomadic, the other predominantly sedentary-agrarian. Both were conditioned by distinct ecological spheres and circumscribed by laws of diminishing returns. But this ecological frontier never served as a fixed or closed borderline. Rather, both sides of the frontier witnessed flourishing mixed economies of wandering pastoralists coexisting with settled peasants. Apart from the seasonal variation, the frontier could be shifted more permanently by the common efforts of peasants to bring wasteland under the plough, often with the help of artificial irrigation. But the roaming pastoralists of the steppes, deserts and jungles of Eurasia were far from being exclusively at the receiving end of the encounter. On the contrary, there was an almost constant eagerness to enjoy – either by plunder or investment – the riches of the settled world. This led to frequent inroads of pastoral nomads, often holding the peasantry to ransom or even replacing agrarian fields with pastures. Looking at the way Eurasia has dealt with this always disrupting capacity of the frontier, one may reflect that the settled authorities in Europe succeeded in keeping the nomads out of their societies, whereas Central Asia and the Middle East remained very much dual economies undergoing regular tribal invasions produced by the ongoing interaction between the settled centres of society and its more mobile fringes – a process so aptly described by the historian Ibn Khaldun (1332-1406) and, more recently, by one of his latest epigones, Ernest Gellner.[3] On the other hand, it appears that China, although frequently harassed and

[3] The idea of ecological conditions dividing pastoral nomadism and sedentary agriculture derives from the Chinese experience as analyzed by O. Lattimore. See e.g. his *Inner Asian Frontiers of China* (Oxford: Oxford University Press, 1988). For Gellner's thought-provoking Ibn Khaldunian views on the historical significance of the frontier in the Middle Eastern context, see e.g. his 'Tribalism and State in the Middle East', in Ph.S. Khoury and J. Kostiner (eds), *Tribes and State Formation in the Middle East* (London and New York: I.B. Tauris, 1991), 109-27.

even ruled by nomads, managed to keep its outer frontier with the pastoral-nomadic world in place. But how should South Asia be envisaged in this broad characterization?

Until very recently, only a few historians were really interested in the role played by pastoral nomads in South Asian history. Still, some classical Indologists, mostly inspired by the Vedic era, are perfectly aware of the ongoing relevance of the opposition between the settled agricultural community (*grama*) and the alien outside sphere of the jungle (*aranya*) and its mobile people. To paraphrase a recent argument by one of them, the interaction between the two different realms was always rather ambivalent and did not necessarily imply that the settled community should at all times be on its guard against the dangers that threatened it from the alien jungle world. In fact, the two spheres complemented each other in a number of ways. The jungle and its tribal inhabitants looked to the settled areas as a source of agricultural produce, young cattle and other riches as well as a source of employment, mainly as soldiers and transporters. On the other hand, the *grama* needed the *aranya* for grazing ground; as a source of new land, manpower and forest products; as a link between settled areas; and finally as a refuge. In other words, *grama* and *aranya* were always in complementary opposition to each other. Whenever for one reason or another the frontier became permanently sealed off, the relationship lost its most violent aspect but could continue in a ritualized form. Obviously, the whole idea of close interaction and symbiosis between the settled and unsettled realm all over South Asia, required the frontier to be very near at hand; to state it differently, there should exist an *inner* frontier. Hence, although to a different degree, the dual economy of South Asia appears to belong to the Middle Eastern and Central Asian pattern.[4]

Coming to the so-called classical and medieval stage in history,

[4] See J.C. Heesterman, 'Warrior, Peasant and Brahmin', *Modern Asian Studies* 29, 3 (1995), 637-54. Historians on 'ancient' South Asia like D.D. Kosambi, R. Thapar, H. Kulke and S. Ratnagar also stress the significance of the sedentarization process in South Asia. For a brief introduction, see e.g. D.D. Kosambi, 'Living Prehistory in India', *Scientific American* 216, 2 (1967), 105-14. See also the special issue on pastoralism in South Asia in *Studies in History* 7, 1, n.s. (1991).

however, scholars tend to highlight the sedentary characteristics of South Asian society. Most focus on the traditional agrarian nuclei of the subcontinent, mainly along the fertile alluvial plains of the Indus and the eastern rivers Ganges, Mahanadi, Godaveri and Kaveri, stressing the expansion of agriculture through states and temples. As a result, South Asia, like Europe and China, is perceived as an almost exclusively sedentary domain, only occasionally disturbed by ferocious nomads from the outside world of Central Asia and the Middle East. Besides, encouraged by the modern claims for separate national or communal legacies, scholars too often imagine the frontier as a fixed dividing line between insiders and outsiders of different religions, cultures and languages. In this light, empires like that of the Mauryas, Guptas and Mughals in the north or of the Cholas and Vijayanagara in the south become falsely depicted as precursors of modernity and hence as durable and centralizing entities capable of pruning away most of the agrarian surplus through a highly efficient land revenue system. Although others have emphasized that these states were at best patrimonial, segmentary or ritualistic configurations, their focus still remains on the settled peasant, while the pastoral nomad vanishes from the medieval scene. Even though most historical sources, especially those in Sanskrit and Persian, were written at the capitals of the sedentary society and tended to criminalize or even deny the ongoing role played by pastoral nomads, the nomadic existence can still be pieced together, both by reading between the lines of these sources and also by studying the few, mostly oral traditions produced by the so-called marginal groups themselves.[5]

Working along these lines, we see medieval South Asia emerge as still very much a mixed economy, provided with extensive jungles,

[5] Perhaps the two most outstanding synthetical works on agrarian society in South Asia are still: I. Habib, *The Agrarian System of Mughal India* (London: Asia Publishing House, 1963) for northern India and B. Stein, *Peasant State and Society in Medieval South Asia* (Oxford: Oxford University Press, 1980) for southern India. Scholars who have stressed the ongoing importance of mobile groups of pastoralists and warriors include Chetan Singh, A.R. Khan and D.H.A Kolff. Showing the potential of 'tribal' studies is G.-D. Sontheimer, *Biroba, Mhaskoba and Khandoba: Ursprung, Geschichte und Umwelt von Pastoralen Gottheiten in Maharashtra* (Wiesbaden: Steiner, 1976).

deserts, savannas and forests, and full of pastoral nomads and animals, either wild or domesticated. As hinted already, pure nomads, like pure peasants, hardly existed anywhere. Pastoralists used to be more settled, while peasants were far more mobile than their ideal types would allow for. South Asian pastoralists, in particular, were mostly involved in a kind of herdsman husbandry and often had semipermanent dwelling places in villages. Compared to their Central Asian and Middle Eastern counterparts, they were more closely integrated into their settled surroundings and had a more limited radius of action.[6] But despite this difference in degree, it is my contention that, even in the case of South Asia, the frontier between pastoral nomadism and sedentary agriculture remained a crucial one until the nineteenth century. Moreover, from the start of the second millennium, South Asia became more closely linked to that much wider, frequently broken, ecological continuum sometimes called Saharasia, which included all the dryer zones of Eurasia, stretching from the Atlantic coast of northern Africa to the eastern and southern extremes of the Indian subcontinent. This so-called Arid Zone receives relatively low annual rainfall (less than 1,000 mm) and constitutes the territory where pastoral nomadism had a natural advantage over sedentary agriculture.[7] From this perspective, South Asia as a whole serves as a wide transitional zone between this Arid Zone, where extensive pastoral nomadism thrives, and the more humid area of monsoon Asia, where intensive settled agriculture flourishes. From about the twelfth century onward – a time when all the dryer areas of Eurasia began to converge as a result of nomadic expansion – the Arid Zone in South Asia emerged as a vibrant frontier region that widened the

[6] For a good typology of pastoral nomads, see A.M. Khazanov's classic, *Nomads and the Outside World* (Cambridge: Cambridge University Press, 1983).

[7] Often a distinction is made between arid and semi-arid zones. For the present purpose I have brought them together under the heading Arid Zone since this is the most relevant category in terms of stockbreeding. Obviously, even under the more humid conditions of monsoon Asia many areas were not or only temporarily cultivated. These lacked, however, the vast mobile resources of the Arid Zone. For the ecological perspective of the Arid Zone, see the various publications of UNESCO's research project on the Arid Zone during the 1950's and 1960's, in serials such as *Arid Programme*, *Arid Research* and *Arid Zone*.

horizon and opened new channels for highly mobile pastoralists, warriors, merchants, pilgrims, and others.

This sudden upsurge of the Arid Zone found its cultural counterpart in the spread of what recently has been labelled the Turko-Persian ecumene: a cultural mix of Arabic, Persian and Turkic elements that melded in ninth- and tenth-centuries Khorasan and Transoxania, from where it was carried by conquering horse warriors throughout the Arid Zone, towards northern and central India. Even in southern India, warriors from the north, both Muslim and Hindu, carved out their own principalities on the basis of their increased stakes in the growing resources of the Arid Zone. As a result, it appears that South Asia became both more mobile and more open to the outside world. Taking these considerations as a point of departure, I present here a first attempt to put the Arid Zone on the historical map by articulating its spatial and temporal dimensions as the premodern inner frontier of South Asia (see Map 2).[8]

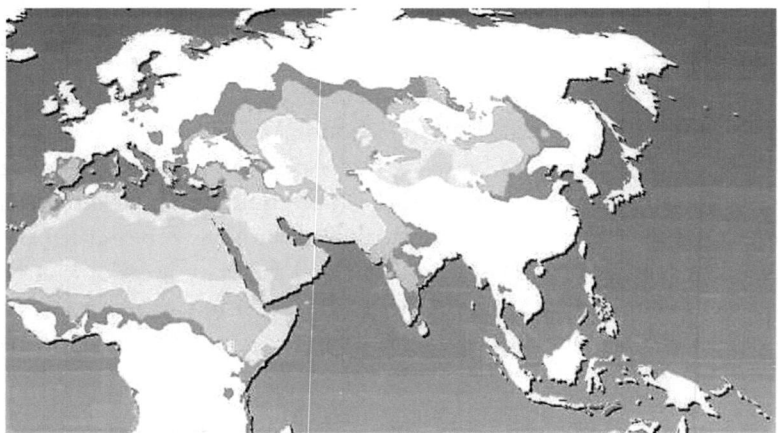

Map 2: Arid Zone

Note: For the purpose of the present study the Arid Zone includes (if zooming out from the central deserts) hyper arid, arid, semiarid and dry subhumid areas.

Source: Millennium Ecosystem Assessment, 2005 – Ecosystems and Human Well-being: Desertification Synthesis (Washington DC. World Resources Institute, 2005).

[8] The concept of Turko-Persia was introduced by R.L. Canfield (ed.), *Turko-Persia in Historical Perspective* (Cambridge: Cambridge University Press, 1991).

THE ARID ZONE IN SOUTH ASIA

From the Middle East and Central Asia the Arid Zone stretches deeply into the Indian subcontinent. In South Asia, the area remains relatively untouched by either the south-west monsoon, occurring from June to August, or the northeast monsoon, taking place from December to February. From the very dry tracts of Sind and Rajasthan the Arid Zone branches off in eastern and southern direction. Eastward, it embraces the large rivers and subsequently narrows down along the southern banks of the Ganges until, near Benares, it reaches the fertile and more humid heartland of ancient Magadha. Southward, it stretches from the extremely arid Thar Desert across the Aravalli Range into Malwa. To the lee side of the Western Ghats, it continues towards the dry western Deccan plateau. The eastern slopes of the mountains receive substantial amounts of rainfall and serve as a transitional zone between the humid coasts of the Konkan and Kerala – where annual rainfall of 4,500 mm permits intensive rice cultivation – and the dry interior of the *desh* – whose rainfall of less than 500 mm annually makes this the natural habitat of pastoral groups. The high aridity of this area continues in a southeastern direction into the so-called Rayalaseema (literally 'the frontier of the kingdom'), between the Tungabhadra and Krishna Rivers but extending towards Kurnool and Cuddapah in the south. This became a hotly contested border region between the prevalent powers in the Deccan to the north and Karnataka to the south. South and east of the Rayalaseema, rainfall increases again, partly profiting from the northeast monsoon, but usually keeps well under 1,000 mm. In the south-west, the Mysore Plateau, with rainfall of 600-900 mm, still belongs to the Arid Zone, and from there several dry outliers reach

---

Eaton has masterfully analyzed the coinciding frontiers of settled cultivation and Islam in Bengal. As it appears, sedentarization in this humid area was far less hindered by superior military resistance beyond the frontier (see R.M. Eaton, *The Rise of Islam and the Bengal Frontier, 1204-1760* (Berkeley and Los Angeles: University of California Press, 1993). As such, the Bengal experience more resembles sedentarization in Java than in other parts of South Asia (cf. D. Lombard, *Le carrefour Javanais: Essai d'histoire globale*, vol. 2: *Les réseaux asiatiques* (Paris: Éditions de l'École des Hautes Études en Sciences Sociales, 1990).

the extreme southern end of the subcontinent. In the southeast, the Arid Zone descends the relatively gentle slopes of the Eastern Ghats towards the more humid Coromandel Coast. Although rainfall is far from excessive, the numerous east-west rivers and their deltas in the Coromandel region made this another core area of intensive rice cultivation.

Throughout the Arid Zone the natural deserts and savannas intersected with well-watered river valleys where sedentary agriculture could prevail. In the north, of course, the Indus, Ganges and Yamuna Rivers provided its fertile alluvial soil with additional sources of water which made the settlement pattern different from that of the surrounding arid countryside. To the south, the Chambal, Narmada and Tapti Rivers furnished Malwa with an extensive agricultural base. For this reason Malwa, situated at the heart of the subcontinent, always served as an important staging post and storehouse for passing armies and caravans. To a lesser extent, south of Malwa, the Godaveri, Bhima and Krishna Rivers had a similar effect on their surrounding valleys. But apart from creating favourable conditions for cultivation, these rivers could equally provide rich pasturage for pastoralists, especially during the dry seasons. The east-west trade that connected the coasts with the interior also followed the south-eastern course of the main rivers, although the latter were unsuited for commercial river traffic. Hence, the rivers provided meeting places where the peasant, the pastoralist and the trader all encountered each other. Not surprisingly, the main trading emporia of the region grew out of former ox camps, a process to which numerous old *viragal* (hero-stones) that commemorate the successful struggle for pasturage and cattle bear witness. Obviously, apart from these rivers and their irrigation systems there were other sources of alternative water supply. For instance, large parts of the land in the Arid Zone could be cultivated with the help of tank and well irrigation.

The prevalent moisture regime of the Arid Zone largely determined the natural vegetation and the way human beings and animals adapted to it. For example, grain and millet, such as *jowar* and *bajra*, where the main field crops, while scrub prevailed in the open forests. In general, the small quantity and high variability of rainfall rendered peasant life rather precarious. For this reason, the Arid Zone was often viewed as a famine tract in which the inhabitants had to resort to wide-ranging survival strategies in case of failing monsoons. This partly explains the prominence of stockbreeding in the Arid Zone, since pastoral nomads could move with all their belongings to more favourable grazing areas, even at great distances. Furthermore, pastoralists, especially those with herds of camels and goats, could survive from the produce of their herds in the driest of circumstances. But it was mainly the supply of nutritious natural grasses and fodder crops that made the Arid Zone suitable for stockbreeding, especially the raising of oxen and horses. Apart from grasses and forest scrub, the stalk and leaf of dry millet, one of the major crops of the Arid Zone, could serve as excellent supplementary fodder. Hence, stock-breeding in South Asia took root in its dryer regions most liable to scarcity.

The pastoral nomads of South Asia moved their herds according to the rhythm of the monsoons from one pasture to the other. Since arid and humid areas (i.e. summer and winter pasturage) were within relatively short distance from each other, the range of these grazing movements was less extended than in the Middle East and Central Asia. Most animal herds of South Asia consisted of cattle, sheep, and goats. The herds moved either vertically up and down the hills – for example, along the Western Ghats and the Himalayas – or horizontally back and forth the various riverbeds, many of which turned dry during the arid season. Riverbeds such as that of the Sutlej and the Bhima were particularly good breeding and rearing grounds, especially for horses. And although long-distance grazing of full-time pastoral nomads was rare, recurrent draughts could, of course, considerably widen the pastoralist's range. In general, though, wandering tribes preferred to conduct their herds along the fringes of the Arid Zone, within close range of the sedentary economy. At the same time, in spite of its well-watered river valleys full of settled life, the rich grazing lands of the Arid Zone stimulated the north-south mobility of both humans and animals, if not from grazing necessity, then for military campaigns and trade. As we will see, the larger armies and caravans that criss-crossed the subcontinent always had to stay close to the nutritious marches and supply lines of the Arid Zone.[9]

[9] For the geographical data, see O.H.K. Spate and A.T.A. Learmonth, *India and Pakistan: A General and Regional Geography* (London: Methuen, 1984) and

Only after the twelfth century did the Arid Zone reach full maturity as a major conductor of people, animals, goods, and ideas. Like the Indian Ocean and the navigable rivers of the north, the Arid Zone became a major artery linking the still narrow territorial arenas of South Asia to each other and to the wider world of Iran and Central Asia. From this time onward, South Asia witnessed breathtaking changes that involved large-scale migration of mostly Muslim and Telugu warrior groups and the emergence of previously marginal tribal groups, such as the Yadavas in the Deccan and the Kallars and Maravars in south-east India. Migrating warriors, pastoralists and cultivators carried their various power divinities with them into new territories, thereby helping to build complex interregional networks of devotion and pilgrimage.[10]

At the same time, the pace of land reclamation all over the Arid Zone increased. Temples and other religious institutions remained important instruments for extending cultivation but gradually changed from small sets of shrines focused on the local community to wide-ranging socio-religious redistribution centres, such as Haridwar and Tirupati, attracting rich endowments from far-off rulers and long-distance merchants. Access to these temples was gained by extremely

---

the numerous regional gazetteers covering South Asia. Apart from insights through the oral traditions of pastoralists themselves, information on stockbreeding derives mainly from early-nineteenth-century travel accounts of animal lovers such as William Moorcroft and Francis Buchanan, and, also, the lengthy British accounts concerning the attempts to improve of the breed of cattle in South Asia. For this, see the separate sections in the Bengal Military Consultations at the Oriental and India Office Collections in London. Informative is also the work done by R.O. Whyte, especially *Land, Livestock and Human Nutrition* (London: F.A. Praeger, 1968).

[10] For an account of this process in the Deccan, see Sontheimer, *Biroba*. For southern India, see the excellent survey of D. Shulman, 'Die Dynamik der Sektenbildung im mittelalterlichen Südindien', in S.N. Eisenstadt (ed.), *Kulturen der Achsenzeit II: Ihre institutionelle und kulturelle Dynamik*, vol. 2: *Indien* (Frankfurt am Main: Suhrkamp, 1992), 102-29. In general, it appears that from the twelfth century, all over South Asia, religious expression becomes more political and agressive. See e.g. the spread of the royal cult of Rama as analysed by S. Pollock in his 'Ramayana and Political Imagination in India', *The Journal of Asian Studies* 52, 2 (1993), 261-97.

mobile sectarian groupings, often consisting of ascetic warriors or long-distance traders, such as Lingayats and Gosains, groups that were crucial intermediaries in the making and legitimation of newly emerging conquest states. In general terms, the new warrior elite, consisting of both Muslims and Hindus, redirected the state's investments from Brahman colonies (*brahmadeyas*) to these new religious centres and to caravanserais and feeding houses along the trade and pilgrimage routes.[11] Extensive military, commercial, and religious networks across the Arid Zone began to constitute a level of power and authority not before seen in South Asia. It was only during the nineteenth century that the dry regions were permanently pushed back into their former subservience to the settled world.

Historiography has often attributed these changes to the Islamic onslaught from the north, which forced Hindu people to either defend themselves or to seek safety by moving toward the south. Hence, it is often suggested that Islam made the supposedly innate peaceful and sedentary Indian society more violent and more mobile. To refute this, I would suggest that the spread of Islam was a part of a more general process in which people from the Arid Zone, all over Eurasia, increased their hold on settled society. Not Islam in itself, but a general enlivening of the Arid Zone appears to have increased its people's mobility and widened their horizons. These people, mostly well equipped with arms and cash, undermined the existing political order, which as a whole became more dynamic and volatile. Military recruitment involved an ever wider area in which highly mobile Afghans, Rajputs, Marathas, Nayakas and other martial groups could deliver military superiority and, in due course, political dominion. At the same time, horizontally oriented sectarian movements – Muslim Sufis, Hindu sectarians and others – violated and reversed the existing,

---

[11] For the increased pace of settlement in southern India, see e.g. N.B. Dirks, *The Hollow Crown: Ethnohistory of an Indian Kingdom* (Cambridge: Cambridge University Press, 1987), 32-5 and D. Ludden, *Peasant History in South India* (Delhi: Oxford University Press, 1989), 42-59. For the impact of sectarian movements in southern India, see A. Appadurai, 'Kings, Sects and Temples in South India, 1350-1700 AD', *The Indian Economic and Social History Review* 14, 1 (1977), 47-73 and Stein, *Peasant State*, 413.

now rather parochial, Brahmanical order.[12] Moreover, extensive commercial networks of Banjara ox owners, Gosain ascetics, and Afghan horse dealers linked fairs and pilgrimage centres of the southern interior to the coast and to the north. Everywhere the scale and pace of human activity was clearly on the increase.[13]

Without claiming to offer a full alternative explanation, I will point out that these well-known, far-reaching developments would have been inconceivable without the necessary animal power to sustain them. Actually, the capacity of animals – primarily bullocks, horses, and dromedaries – as the major premodern source of locomotive energy, must have been one of the driving forces in these developments.

THE SPREAD OF DROMEDARIES, HORSES AND OXEN

As mentioned already, the Arid Zone of South Asia possessed a distinct zoogeographical identity. Its animal husbandry was based on oxen, horses, sheep, goats, and dromedaries, while the more humid territories of eastern India, especially at places where wet paddy monoculture or dense forests preclude grazing animal, husbandry was restricted to pigs, buffaloes, ducks, chickens, and geese. Although domesticated animals, both for traction and dairying, have always played a significant role in South Asia's society, it appears that during the first half of the second millennium, the economic role of horses, dromedaries, and oxen, greatly increased. The natural habitus of all three species overlapped to varying extents in the Arid Zone of the Indian sub-continent.[14]

[12] It is very well possible that the south Indian polarity between right hand and left hand castes, most probably stemming from the eleventh century, is partly related to the ascendancy of mobile groups from the Arid Zone.

[13] Thanks to the work done by Stein and Wink there appears to be a growing consensus on the main developments of the period. See, e.g. B. Stein, *Peasant State* and the first two volumes of A. Wink, *Al-Hind: The Making of the Indo-Islamic World* (Leiden: E.J. Brill, 1990/6). So far, though, explanations of these developments widely differ. Interesting in this respect is the recent contribution by D. Ludden, 'History outside Civilisation and the Mobility of South Asia', *South Asia* 17, 1 (1994), 1-23.

[14] Very helpful as a first introduction to the economic role of animals in South

In South Asia the conditions for horsebreeding were relatively poor as a result of the general want of nutritious fodder grasses. But although the best warhorses originated from Central Asia and Iran, some good Indian horses were bred along the fringes of the Arid Zone – for example, in Kathiawar, in the Lakhi Jungle in the Punjab, and along the Bhima River in the Deccan.[15] Sometime during the second half of the first millennium a whole range of technological, biological and climatological changes culminated in what has been properly called a horse-warrior revolution. In South Asia this revolution arrived relatively late – in the north from about the eleventh century, in the south from about the thirteenth century – but the successful campaigns of Muslim armies of Ghaznavids, Ghurids, and Khaljis proved the might of mobile warfare, especially of well-trained archers on horseback.[16]

Initially, most horses and warriors were drawn from Iran and Central Asia. For the Indian population the most prominent feature of these warriors was not their distinctive religion or ethnic identity but their military and equine expertise, which was always for hire. Apart from being loosely labelled Muslims or Turks (*Turushka*), these mounted men were known as *ashvapatis* (lords of horses) or *irauttars* (horsemen). In due course, Indian armies incorporated these martial groups and adopted their cavalry techniques. Indeed, the military success of warrior groups such as Rajputs, Marathas and Nayakas was a direct consequence of their experience with and imitation of the early Muslim armies. From the twelfth century onward military superiority of both Muslims and Hindus depended on easy access to, and employment of, warhorses. At about this time, horses emerged in religious cults of warrior heroes, such as Khandoba or Lord Aiyanar,

---

Asia is J. Deloche, *La circulation en Inde avant la révolution des transports*, vol. 1: *La voie de terre* (Paris: Ecole Française d'Extrême-Orient, 1980).

[15] For the South Asian conditions of horse-breeding, see my *The Rise of the Indo-Afghan Empire c.1710-1780* (Leiden: E.J. Brill, 1995), 69-79.

[16] For an analysis of the horse-warrior revolution in South Asia, see A. Wink, *Al-Hind: The Making of the Indo-Islamic World*, vol. 2: *The Slave Kings and the Islamic Conquest 11th-13th Centuries* (Leiden: E.J. Brill, 1997). For southern India, see J. Deloche, *Horse and Riding Equipment in Indian Art* (Pondicherry: Indian Hertiage Trust, 1990).

and other martial traditions of mobile pastoralists in which, at times, even Shiva could appear as a Muslim horse-dealer. All this suggests the enormous impact of the mounted warrior on both politics and ritual in South Asia.[17]

The dromedary fared best in the driest parts of the Arid Zone. Although it was domesticated before 1000 BC, only around the beginning of the present era did it become the most efficient means of transport in the Middle East. According to the pioneering work done by Richard Bulliet, this was partly the result of the invention of the new, more convenient, north Arabian saddle, and partly due to the increased integration of camel-breeding nomads in settled society. The spread of the dromedary entailed the gradual disappearance of the transportation on wheels and on two-humped or Bactrian camels. The latter retreated to the equally arid but colder climate of Central Asia. Later, in Khorasan, where the habitats of dromedaries and camels overlapped, strong and cold-resistant hybrids (*bukhtis*) could be bred that were able to link Central Asia with the Middle East.[18] Apart from these hybrids, a new breed of one-humped camels was developed that was perfectly adapted to the colder conditions of Iran and southern Central Asia. According to Bulliet, these developments in Khorasan took place well into the Islamic period. He also believes it likely that in South Asia the one-humped camel became known in significant number only after AD 1000, which

[17] For the reception of Muslim warriors in South Asia, see e.g. S. Bayly, *Saints, Goddesses and Kings: Muslims and Christians in South Indian Society 1700-1900* (Cambridge: Cambridge University Press, 1989), 1-241; C. Talbot, 'Inscribing the Other, Inscribing the Self: Hindu-Muslim Identities in Pre-Colonial India', *Comparative Studies in Society and History* 37 (1995), 692-723; G.-D. Sontheimer, 'The Mallari/Khandoba Myth as reflected in Folk Art and Ritual', *Anthropos* 79 (1984), 155-70. For Shiva's disguise as a horse-dealers, see G.-D. Sontheimer, 'Dasara at Devaragudda: Ritual and Play in the Cult of Mailar/Khandoba', in L. Lutze (ed.), *Drama in Contemporary South Asia: Varieties and Settings* (Heidelberg: South Asia Institute, 1984), 27.

[18] Early Arabic geographers like Yaqut, Ibn Hauqal, Al-Idrisi and others refer to an area called Nudiyah in eastern Baluchistan where pastoralists exported young camels for breeding purposes to Khorasan (see H.G. Raverty, 'The Mihran of Sind and its Tributaries: A Geographical and Historical Study', *Journal of the Asiatic Society of Bengal* 61 (1892), 216-17, 224, 230).

suggests that the introduction of the dromedary occurred simultan-eously with the horse-warrior revolution. Thus, it appears that Indian dromedaries, mainly those from Baluchistan and Rajasthan, got a late start as the most efficient beasts of burden in the dryer parts of the subcontinent.[19] Indeed, as the eighteenth-century Afghan and Gosain long-distance traders knew, caravans of dromedaries could travel as far south as the famous fair of Tirupati. Thus their radius of action more or less paralleled that of the warhorse, both coinciding with the climatological parameters of the Arid Zone.[20]

Although the best warhorses and dromedaries were always produced outside the subcontinent, especially along the steppes and deserts of Central Asia and the Middle East, the ox was and still is the South Asian animal *par excellence*. Actually, the breeding of oxen does best in the dryer areas of the subcontinent, while the buffalo thrives in the more humid eastern parts. Although some ox breeds – for example those of Gujarat and Mysore – possessed a distinctive reputation, all breeds were in fact mixed varieties. There is much evidence to suggest that cattle from different regions were continuously exchanged, both for breeding and for rearing purposes. From nineteenth-century observations it appears that we can distinguish between two interacting circuits of cattle-husbandry. First there was the breeding of so-called village cattle, often unattended and freely grazing and mating at the local commons of waste and fallow land. Although this inbred variety constituted the bulk of the agricultural stock and was the main source of dairy supply, it was unfit for heavy ploughing and transport services over long distance. Second, a stronger

[19] This fits in well with the historical background of the famous epic of Pabuji. Some time during the twelfth or thirteenth century, this Rajput hero is supposed to have introduced 'reddish-brown she-camels' to Marwar from Lanka (J.D. Smith, *The Epic of Pabuji: A Study, Transcription and Translation* (Cambridge: Cambridge University Press, 1991), 324-80). The Sindhi Muslims of Barmer District believe that Pabuji took the camels from a hamlet called Sairo Bagani in village Lankaye, 12 miles from Umarkot (Vinay Kumar Srivastava, 'The Rathore Hero of Rajasthan: Some Reflections on John Smith's *The Epic of Pabuji*', *Modern Asian Studies* 28, 3 (1994), 610-1).

[20] See R.W. Bulliet, *The Camel and the Wheel* (New York: Columbia University Press, 1990).

variety of cattle was produced and raised by professional pastoralists, both nomadic and sedentary, on the basis of selective breeding and extensive grazing in the open and broken jungles of the Arid Zone. The essential elements for successful breeding were the keeping of large herds of cows, often supplied by the village breeders; the selection of superior bulls, often dedicated to a regional temple; and the castration of the lesser bulls, often sold at the regional market. Clearly, the grazing of selected cattle across extensive pastures considerably improved the stamina and strength of the breed. This second, more extensive breeding circuit served as the major supplier of the stronger plow-oxen for the cultivator and the hardier beasts of burden for the long-distance trader.[21]

To start with the beasts of burden used in long-distance trade, from at least the twelfth century onward so-called Banjara ox-owners began to dominate overland long-distance conveyance. They possessed thousands of oxen that covered thousands of kilometres across the subcontinent. Both Muslim and British sources acknowledge that the success of their military campaigns in the interior of the subcontinent hinged on the willingness of these Banjaras to supply their armies with grain and other necessities. Not surprisingly, the Banjaras themselves were involved in stockbreeding, as the range of their movements tended to improve the quality of their pack animals. Again, their natural circuit of operations coincided with the rich grazing and fodder facilities of the Arid Zone.[22]

The ox was perhaps even more important for agrarian purposes,

[21] There are numerous, mostly nineteenth-century, British regional studies that deal with cattle-breeding in South Asia. Very instructive, though, proved R.W. Littlewood, *Livestock of Southern India* (Madras: Government Press, 1936).

[22] The significance of the Banjaras for the British campaigns in southern India is well-known. See e.g. R.G. Varady, 'North Indian Banjaras: Their Evolution as Transporters', *South Asia* 2, 1 (1979), 1-19. For their crucial role during the Mughal campaigns in the south, see e.g. Nawab Samsam-ud-Daula Shah Nawaz Khan and Abdul Hayy, *The Maathir-ul-Umara*, vol. 2, part 1, trans. H. Beveridge and B. Prashad (Calcutta: Asiatic Society, 1979), 21. See also I. Habib, 'Merchant Communities in Precolonial India', in J.D. Tracy (ed.), *The Rise of Merchant Empires: Long-Distance Trade in the Early Modern World, 1350-1750* (Cambridge: Cambridge University Press, 1993), 371-400.

to pull the plow. It also provided manure, raised the water from the village wells, trod out the cut crop on the threshing floor, and transported the produce to the local market. Oxen must have played an essential part in the massive expansion of cultivation throughout the Arid Zone from the twelfth century onward. This was even more so in those territories, mostly in the dryer zones of the Deccan and Karnataka, that consisted of so-called black-cotton soil (*regur*). Although this type of soil was relatively fertile, it was also extremely heavy, so that could hardly be worked by the ill-bred cattle raised in the village. Probably only the cattle produced from extensive grazing and tended by professional pastoral nomads could sustain the large-scale reclamations of such land.[23] In this way, stockbreeding linked the growing mobility to the increased pace of land reclamation throughout the Arid Zone. Taking all the circumstantial evidence together, it appears that the spread of warhorses, dromedaries and oxen must have significantly increased South Asia's potential for warfare, transportation, and cultivation.

FRONTIER AND STATE FORMATION

Especially after the twelfth century, new, more powerful warrior elites emerged on the arid peripheries of the old agrarian centres. Their military superiority enabled them to appropriate rights to protect these centres, which in turn justified the collection of various fees and taxes. The emergence of the Muslim sultanate in northern India, the Yadavas in Maharashtra, the Kakatiyas in Andhra and the Hoysalas in Mysore clearly marked the rapid rise of the Arid Zone polities. In the longer term, however, the power of new rulers hinged on their capacity to attract both man and animal power. New capitals, such as Delhi, Devagiri, Warangal, and Dvarasamudram, and at a later stage similar headquarters, such as Bijapur, Golkonda, and Vijayanagara, functioned as major recruitment centres for military men, preferably without local roots but with extensive service networks of their own.

---

[23] On the agrarian conditions of black-cotton soils of South Asia, see Spate and Learmonth, *India and Pakistan* and E. Simkins, *The Agricultural Geography of the Deccan Plateau* (London: G. Philip and Son, 1926).

They were often assigned frontier territories to be brought under control or under the plow. At the same time rulers were also eager to find a safe and regular supply of war-horses and, as needed, bullocks and dromedaries to maintain their growing armies. All these mobile resources were to be found along the fringes of the Arid Zone, where warriors and pastoralists preferred to display their services and where long-distance traders could find wholesale consolidation and transhipment facilities. The modern onlooker might regard these eccentric new capitals as extraordinary misfits in the middle of relatively poor and precarious countryside, but actually they were ideally situated at the interface of unsettled marches and more settled agrarian fields. This frontier delivered its rulers the best of both worlds: mobile power and sedentary investment (reminding us again of that perfect ox camp located in between summer and winter pastures).

Initially, like their trained riders, most warhorses were imported through the Arid Zone from the north-western breeding tracts. Especially the rulers of southern and eastern India always felt the threat of being cut off from the overland supply lines. At the same time, however, they were perfectly aware that the overland trade bore considerable risks. The sheer size of warrior bands, herds of animals and trading caravans could always undermine the precarious order of new conquest states. To counter the unpredictability and destabilizing effects of the northern trade, some of the more ambitious rulers founded their own breeding studs. For example, the rulers of Mysore reserved extensive areas exclusively for grazing and induced peasants to maintain a minimum number of cattle or horses. An example is the Amrit Mahal establishment of the eighteenth-century ruler Tipu Sultan, which built on the much earlier efforts under Wodyar and Vijayanagara rulers.[24] Both Vijayanagara and Mysore attempted to open new channels of alternative supply, mainly through the coast to overseas breeding tracts, such as in Kathiawar, Iran, and Arabia. Even

---

[24] For the breeding-stud in Mysore, see the informative account by Francis Buchanan, *A Journey from Madras through the Countries of Mysore, Canara, and Malabar*, 2 vols (London: T. Cadell and W. Davies, 1807) and *Oriental and India Office Collections (OIOC)*, London, L/Mil/5/465, 'Copy correspondence and proceedings on breeding cattle and horses in India', appendix 3. For horse-breeding in northern India, see my *Rise of the Indo-Afghan Empire*, 92-6.

warriors were directly recruited from these overseas sources. Not surprisingly, the Karnataka states especially attempted to stimulate the east-west trade across the subcontinent, which made them less dependent on the more dangerous northern supply lines. At the same time, these southern states were keen on keeping the rugged Rayalaseema as a frontier zone with the north. Consequently it was frequently assigned and reassigned to all kinds of unruly armed retainers, mostly from the north, who in due course carved out their own power bases in the area. For example, during the seventeenth and eighteenth centuries Afghan traders and warriors succeeded in establishing a string of principalities along the dry east-west belt of the Rayalaseema by playing on the ongoing rivalry between the Vijayanagara successor states and the various Muslim sultanates of the Deccan. From their strategic centres in Savanur, Kurnool, and Cuddapah they not only controlled the supply lines of horses and mercenaries from the north, but also the important pilgimage route from the Deccan to Tirupati.[25]

But despite their efforts, the rulers of the newly risen polities knew that they could never dispense with the overland services of military entrepreneurs, merchants, and pastoralists. Most of them had themselves emerged from a tribal background, some of them full pastoral nomads, but all of them warriors, always trading their goods and services to the highest bidder. Even as late as the seventeenth and eighteenth centuries, new dynasties, such as the Rohillas in northern India, the Marathas in the Deccan, and the Maravars in the deep South, had longstanding experience with the breeding, rearing, and trading of animals.

The rapid political careers of these self-made men brings us back to the central theme of this article. From the twelfth century onward, state formation more and more revolved around the frontier between the Arid Zone and the agrarian cores of the subcontinent.[26] In the

[25] For a political account of these Afghan nawabs, see *OIOC*, Mir Husain Ali Kirmani, 'Tazkirat al-bilad wa'l hukkam', Add.10,582, fols. 51a-88a. For the control of the trade and pilgrimage route to Tirupati, see *OIOC*, Eur. F.149, 'Malet Papers', vol. 5, fol. 398.

[26] The vast potential for state-formation along the Arid Zone is also suggested by the preponderance of western and north-central states in the geopolitical

words of Burton Stein, 'the new regimes set up their authority in interstitial political zones, which brought about a shift of dominance in peninsular politics from the old riverine core kingdoms of the earlier medieval age to the large dry zone of the upland'.[27] This also should put in proper perspective the idea that India's identity stems from longterm external frontiers, such as the Sulaiman Range or the Indus River, which set it apart from the 'other' Iran or Central Asia. It was, in fact, the inner frontier of the Arid Zone that molded South Asian history.[28] But especially after the twelfth century, following the spread of war-horses, dromedaries, and oxen, this frontier came fully into its own as the 'political womb' of South Asia.[29] Mainly through pastoralists, merchants, and warriors, this frontier not only brought its surrounding territories of South Asia together but also made them part of that even wider ecumene of Turko-Persia.

Admittedly, the foregoing remains speculative and can only be substantiated by more detailed, regional studies that will pay due attention to all those elements that could keep the frontier together by bridging the mobile with the sedentary: pastoral nomads, traders, warriors, and ascetics. As a first step in this direction, the following section takes a closer look at one of the longstanding centres of political dominion in India, Delhi.

---

synopsis presented by Schwartzberg, although he might have overrated the effective range of the ancient pan-Indian powers. See J.E. Schwartzberg, *A Historical Atlas of South Asia* (Oxford and New York: Oxford University Press, 1992), 254-62.

[27] B. Stein, *Vijayanagara* (Cambridge: Cambridge University Press, 1989), 13-21.

[28] See also A.R. Khan's interesting presidential address at the Punjab History Conference, Patiala, 1981.

[29] Term derived from E. Gellner in his 'War and Violence', in E. Gellner, *Anthropology and Politics: Revolutions in the Sacred Grove* (Oxford: Blackwell Publisher, 1995), 164. Interestingly, D.D. Shulman, referring to the Upanisads, uses the same wording in describing the position of the Brahman as being the 'womb of kingship' (see his *The King and the Clown in South Indian Myth and Poetry* (Princeton: Princeton University Press, 1985), 110). In these terms, it appears that the new political regime closely resembles the janus-faced position of the Brahman.

THE DELHI FRONTIER

One of the most lasting frontiers of India and the most significant in terms of delivering political dominion was the jungle immediately bordering on the western city walls of Delhi. This was the extensive semi-arid jungle of the Indo-Gangetic divide between the Sutlej and Yamuna Rivers. It corresponds roughly to present-day Haryana but extends into the southern Punjab. For convenience I will use the term *Delhi frontier*.[30]

In ecological terms the Delhi frontier is a broad transitional zone between the arid area of Rajasthan to the west and the more humid area to the east. In the north-west, the lack of sufficient rainfall in the Punjab *doabs* is more than compensated by the annual supply of meltwater from the Himalayas. In the north, the submontane strip of the Siwalik Hills receive sufficient winter rain. In the east begins the medieval heartland of north Indian settled society, the fertile and well-watered Gangetic basin. To the south-west the Delhi frontier grades into the extreme aridity of the Thar Desert, while in the south and south-east the semidry conditions extend towards the forests of Mewat and Malwa into the Deccan plateau. As we have already seen, in these relatively dry areas agriculture could persist only precariously. Largescale agricultural development was possible only by artificial irrigation.

What set the Punjab and Haryana jungles apart from the other dry tracts was their constantly shifting riverbeds. At about 1000 the Sutlej River appears to have flowed far to the south-east of its present course. Around 1250 it was still connected to the Ghaggar River, but about 400 years later it had joined the Beas River to its north. After their first confluence, the Sutlej left the Beas again, only to rejoin it after some 300 km. In about 1700 their separate channels in the north again joined. At the same time, losing most of its tributaries,

---

[30] For the present introductory purpose, the geographical data of the area have been mainly gathered from mostly eighteenth- and nineteenth-century sources. Interesting are the reports following the early British attempts to map the area. See e.g. the detailed reports by Mughal Beg (*OIOC*, Eur. F.22) and White (*OIOC*, F/4/304: 6997, 'Board's Collections'). See also fn. 37. I am grateful to W.R. Pinch for the reference to Mughal Beg.

the Ghaggar River gradually dried up, and at the end of the eighteenth century it ceased to flow, as a result of the dams that were put up by the upstream Patiala Sikhs. As a result, the landscape around Sirsa and Bhatnir became completely desolate. The same phenomenon had happened frequently at the nearby Hisar area following the failing water supply of the Chittung River. All in all, from the existing evidence, it appears that from *c.* 1200 the Sutlej shifted its course to the north-west, which left the southern Punjab even more dependent on the scanty, unreliable, and unequally distributed supplies of the monsoon rains. The only remaining river was the rainfed Ghaggar – often identified with the mythical Sarasvati – which still reached Sirsa but near Bhatnir was lost in the sand. The gradual desiccation of the area was exacerbated by the extremely deep water table. As agriculture was precarious and crucially dependent on scarce water supplies, regional conflicts often erupted over access to water through tanks, wells, and dams.[31]

Nevertheless, some of the abandoned riverbeds could still become active during the monsoon, which created the huge temporary lakes and the extensive jungles of the area. Whenever sufficient water was available – for example, during one of the plentiful monsoons or through artificial irrigation – the light sandy soil (*bagar*) could suddenly produce the most exuberant vegetation. In fact, irrigation could generate some extremely fertile, oasislike settlements such as Hansi, Sirsa, and Bhatnir. At some of these places agriculture was possible all year round and yielded crops such as rice and sugar cane.

These settlements also served as staging posts for long-distance caravans *en route* from Delhi to the Indus valley. Although the main trade route to Lahore and Kabul was skirting the area along its north, at times it could become an attractive alternative route. Its extensive pastures, producing very nutritious grasses, also made the area an important thoroughfare for pastoral-nomadic traders, such as Afghan horse dealers, Banjara cattle tenders, and Baluchi camel drivers. Even the trade routes to its north and south could hardly dispense with

---

[31] On the shifting of the Punjab rivers, see H.G. Raverty, 'The Mihran of Sind', 155-508.

the excellent transport facilities provided by the tribes based in the Punjab jungles.

It follows from this that the area was celebrated for its first-class stockbreeding, which was the foremost occupation of the local population. Since the twelfth century the area has been linked to the Bhattis (hence names such as Bhatnir, Bhatiya, and Bhathinda – the Tabarhind of the early Muslim sources), a widely distributed Rajput tribe also connected to the ruling lineage of Jaisalmer. Large numbers of Bhattis wandering in the Punjab and Haryana accepted Islam and became attached to regional Sufi saints, such as the famous Farid al-Din (d.1265) and the legendary Baba Ratan.[32] Like most mobile groups of the Arid Zone, the Bhattis were an open ethnic category consisting of all kinds of Rajputs, Jats, and various other groups. According to local tradition, one of the Bhatti chiefs, Rana Lakhi, settled in the Lakhi Jungle, the extensive jungle produced by the ever shifting Sutlej, in the reign of Mahmud of Ghazni (998-1030). He subsequently became a Muslim and founded 300 villages. He also procured a great number of Iranian horses and ordered his subjects to begin horse-breeding. Other evidence also confirms the high quality of the local horse breed, a standard that was kept up by the foreign horses that staged the perennial invasions from the north-west. Sultan Balban (1266-87), being cut off from foreign sources, procured most of his cavalry horses in the Bhatti area. The whole of the eastern Punjab, but the Lakhi Jungle in particular, remained a major horse-breeding area up to the nineteenth century.[33]

---

[32] The meeting of saint and bandit in the wilderness, both escaping the constraints of society, is of course a well-known theme in Indian hagiography. See e.g. W.L. Smith, 'The Saint and the Bandit', in A.W. Entwistle and F. Mallison (eds), *Studies in South Asian Devotional Literature* (Paris: Ecole Française d'Extreme Orient, 1994), 363-70.

[33] Cf. S. Digby, *War-Horse and Elephant in the Delhi Sultanate: A Study of Military Supplies* (Karachi: Oxford Orient Monographs, 1971), 22. Although Digby refers to Punjabi horses, he equates the so-called *baladasti* horses with horses from Central Asia (34, 38, 42). I agree with Digby that *mulk-i baladast* means 'the land on the higher side' but this term probably refers to the elevated plateaux of the Punjab doabs and not to Central Asia. See also Ranabir

Apart from horses, the area brought forth excellent oxen and also some dromedaries. Although both were employed as pack-animals throughout the Arid Zone, the Delhi region served as a transhipment area between the Rajasthani transportation network, mainly based on dromedaries, and the eastern and southern transportation system, which relied on oxen and river transport across the Ganges and Yamuna Rivers. For example, the caravan track from Multan or Bikaner to Hisar was convenient only for camels. Beyond Hisar, however, trade was preferably conducted by bullocks and carts. This also emerges from the discussions of British officials regarding their transport system while crossing the area during the Maratha wars. At the time the army train arrived at Delhi, the British authorities decided to replace most of their bullocks for dromedaries, which were considered far more efficient in the arid surroundings north-west of Delhi. Only a few years later, however, following their first inroads into the Maratha territories of the Deccan, the British returned to oxen, having decided that the dromedary was less fit for maintaining the supply lines to the more humid south.[34]

Related to their pastoral occupations, the Bhattis had a long-time reputation as ferocious rebels who plundered or ransomed neighbouring villages and passing caravans. As so often, this was a logical additional source of income to pastoralists. Both these attributes made them fit for mercenary service. Hence, in addition to their supply of excellent livestock, the Bhattis were known for providing cheap military labour. Although the obstreperous Bhattis posed a permanent threat to the authorities in Delhi, they also endowed them with the indispensable mobile resources of the Arid Zone.[35]

Chakravarti, 'Horse Trade and Piracy at Tana (Thana, Maharashtra, India): Gleanings from Marco Polo', *Journal of the Economic and Social History of the Orient* 34, 2 (1991), 172.

[34] This follows from studying the Bengal Military Consultations for the years 1794-1820. For the discussion at the early stage, see e.g. *OIOC*, P/20/34 (1802), no. 39; for the later considerations, see *OIOC*, P/29/13 (1820), no. 192.

[35] A somewhat similar story might be written about the so-called Mewatis in the arid wastes south-west of Delhi. See e.g. the interesting paper by Shail Mayaram, 'Mughal State Formation: The Mewati Counterperspective' (paper presented at the third international seminar on Rajasthan, 14-18 December 1994).

POLITICAL DOMINION IN DELHI

Returning to the imperial capital itself, Delhi was situated very near the rebellious jungles of the Bhattis to the west, which made the position of its rulers rather precarious. In a worst-case scenario, the area could serve not only as a seedbed for rebels but also as a slide for invaders from the northwest. As in the case of Timur (1398-9), its rich pastures funnelled the invading armies automatically in the direction of Delhi. Not surprisingly, the great battles of the north Indian plains – for example, at Tarain (1191, 1192) and Panipat (1526, 1556, 1761) – were all fought in the wastes facing Delhi to the west. Any invader who managed to gain control of the extensive pastures in Bhatinda and Haryana had a very good chance of controlling Delhi as well. Consequently the Delhi frontier provided the key to political dominion in northern India. The area's alleged permanent turbulence was only another indication of its enormous military potential. Although the Bhattis themselves never succeeded in taking over power in the capital, nearly all the early ruling dynasties in Delhi started their career exactly in this area, as *muqtas* and so-called wardens of the marches: the Khaljis in Samana, the Tughluqs in Dipalpur, the Sayyids in Haryana, the Lodis in Sirhind.[36]

Once established, the sultans of Delhi tried everything in their power to curb the ingrained turbulence of the area. The Tughluqs were most ambitious in this respect. Firuz Shah Tughluq (1351-88), whose mother was a local Bhatti, decided to bring the area under tight control. He erected the forts of Muhammadpur, Zafarabad and Radabad; strengthened the former district capital of Hansi; and

---

[36] For the early Muslim period, see M. Shokoohy and N.H. Shokoohy, *Hisar-i Firuza: Sultanate and Early Mughal Architecture in the District of Hisar, India* (London: Monographs on Art Archaeology and Architecture, 1988) and the articles 'Bhattinda' and 'Hissar Firuza' by A.S. Bazmee Ansari and 'Hansi' and 'Hariyana' by J. Burton-Page in the *Encyclopaedia of Islam*, 2nd ed. (Leiden: E.J. Brill, 1960-2005). The history of the area during Mughal rule derives from the works of Chetan Singh [mainly his *Region and Empire: Panjab in the Seventeenth Century* (Delhi: Oxford University Press, 1991)] and Irfan Habib [mainly his *Agrarian System* and *An Atlas of the Mughal Empire* (Delhi: Oxford University Press, 1982)]. Most of the Persian and European material for the area, however, still waits to be fully explored.

founded the new cities of Fathabad and Hisar Firuza. He also attempted to sedentarize the area by laying out two irrigation canals that prolonged the Chittung River. In due course, Hisar grew into a large and prosperous city surrounded by numerous fruit and vegetable gardens. This allowed it to serve as a kind of civilizing pier stretching from Delhi into the wastes of Haryana. Nevertheless, beyond Hisar, the area again and again reverted to rebellion and was hardly ever fully integrated into the Delhi administrative system.

The Mughal rulers were more successful in keeping the area at bay. This was done by the usual means, such as regular hunting expeditions, forced migrations, and large-scale irrigation programmes. Akbar (1556-1605), for example, restored one of the Tughluq canals. Perhaps more important was the Mughal policy of divide and rule. They imposed several overlapping jurisdictions on the area, as a result of which it belonged to several *subas*, *sarkars*, *dastur*-circles and *faujdaris*, all without clearly marked rights or boundaries. Nevertheless, even for the Great Mughals, control from Delhi remained rather superficial. At the beginning of the eighteenth century the whole area came under the almost autonomous control of the Khweshgi Afghans, the local rulers of Qasur and the Lakhi Jungle. After Nadir Shah's invasion (1739) the Bhattis again cropped up, while the Sikhs in the north, the Marathas in the south, the Afghans in the west, and their Rohilla allies in the east contended for control of the area. After the final collapse of Mughal rule during the late eighteenth century, the area again became a hotbed of military adventurers, such as George Thomas, Walter Reinhardt, Louis Bourquin, and James Skinner.[37] The East India Company followed in their footsteps by exploiting the region's capacity for the breeding and trading of horses and oxen. Not surprisingly, during the Mutiny of 1857 the troops at Hansi were the first to rise against their British masters. Only late in the nineteenth

---

[37] A great deal of information on the Bhattis may be gathered from the accounts of these adventurers. See the Thomas papers of the British Museum (Mss Add.13.579-80) published in W. Francklin's, *Military Memoirs of Mr George Thomas* (Calcutta: Hurkaru Press, 1803) and his *Tracts, Political, Geographical and Commercial* (London: T. Cadell and W. Davies, 1811). See also James Skinner's manuscripts: *OIOC*, Add.27.255, 'Tashrih al-Aqwam', fol. 453; and *OIOC*, Add.27.254, 'Tazkirat ul-Umara', fol. 267.

century did the British authorities manage to fully pacify the area. This was achieved partly by reducing the demand for warhorses and free military labour and partly by isolating the area from its settled and unsettled environment.[38] Modern forms of coercion and cultivation have finally changed the marches of Bhattiyana into the highly productive green fields of Haryana.

CONCLUSION

The aim of this explorative survey was to introduce the Arid Zone as a category of South Asian history. The limits of this ecological zone broadly coincided with the always fluid frontier between pastoral nomadism and sedentary agriculture. From about the twelfth century onward, this inner frontier was invigorated by the increased input of stronger warhorses, dromedaries, and cattle, all of which tremendously enhanced the existing capacity for warfare, trade, and land reclamation. By fully exploiting these resources, new warrior states emerged at this interface of jungle and arable land. The Delhi frontier is an example of the historical significance of the Arid Zone. The continued centrality of Delhi as the *dar al-sultanat* of Hindustan hinged on its own peripheral location at the junction of the unsettled areas to its west and the more stable territories to its east. The dilemma for the Delhi rulers was both a classic Islamic and Indic one: although their rule belonged to the settled order of sedentary society, their actual power derived from the arid wastes.

[38] See e.g. *OIOC*, Eur. D.163, 'Bhatti Papers'.

CHAPTER 3

The Eurasian Frontier after the First Millennium AD: Reflections Along the Fringe of Time and Space

Should it happen that the community where they are born is drugged with long years of peace and quiet, many of the high-born youths voluntarily seek those tribes which are at the time engaged in some war; for rest is unwelcome to the race.

TACITUS on Germany (*c.* AD 98)

The kings of India hunt the elephant. They will stay a whole month or more in the wilderness and in the jungle

ABD AL-RAZZAQ on India (*c.* AD 1450)

TOWARDS THE FRINGE

IN HISTORICAL PERIODIZATION temporal and spatial categories are closely interconnected. This is particularly true of the classical three-piece paradigm of ancient-medieval-modern which exclusively relates to western-European developments. Nevertheless, for a very long time, eurocentric periodization has been very influential in moulding the historical imagination of so-called non-western regions such as the Middle East and South Asia. Paradoxically, historians of these areas attempted to thwart this eurocentrism by linking them as closely as possible to the conventional European time-frame. In this perspective, the whole world began to share in once exclusively European experiences such as medieval feudalism and modern enlightenment. But while most of the rest of the world was increasingly dressed in this traditional three-piece suit, Europe itself was gradually stripped from it. By now it appears that most historians of Europe take the view that the most fundamental changes in Europe – the advent of the city, the agrarian village, the university, literacy, the lineage-based aristocracy, the revival of Greco-Roman art and thoughts, Parliament, the Western legal and constitutional traditions, large scale international commerce, and much else – occurred not after the waning of the Middle Ages in 1500 but, much earlier, after AD 1000. In this view, the next truly fundamental turning point occurred in the eighteenth and nineteenth centuries with the coming of the French and Industrial revolutions. In other words, it was after about 1000 that Europe came fully into its own as a spatial category distinctive of other regions.[1]

Part of the rethinking on European periodization was informed by the increased tendency to transgress the conventional spatial boundaries of civilizations by focusing on interstitial regions where various 'civilizations' converge and overlap. Sea-based projects of historians like Fernand Braudel, Archibald Lewis and K.N. Chaudhuri are well-known examples which show what might be achieved.[2] Another related example where shifting spatial attention leads to the adoption of new time-frames is the growing interest in cross-cultural interaction. Interestingly, from this global perspective, the year 1000 emerges again as an important caesura which inaugurates an age of

[1] As most lucidly described by C.W. Hollister, 'The Phases of European History and the Nonexistence of the Middle Ages', *Pacific Historical Review* 61, 1 (1992), 1-22. D. Gerhard, *Old Europe: A Study of Continuity, 1000-1800* (New York: Academic Press, 1991) is a late example of a German tradition to see long-term continuities in European history. See e.g. O. Brunner, *Sozialgeschichte Europas im Mittelalter* (Göttingen: Vandenhoeck und Ruprecht, 1978). The French Annales medievalists also tend to follow this line but most of them keep focusing on the classical Middle Ages. See J. le Goff, *L'imaginaire médiéval* (Paris: Gallimard, 1985), 7-13.

[2] F. Braudel, *La méditerranée et le monde méditerranéen à l'époque de Phillippe II* (Paris: A. Colin, 1949); A.R. Lewis, *Naval Power and Trade in the Mediterranean AD 500 to 1100* (Princeton: Princeton University Press, 1951) and *The Northern Seas: Shipping and Commerce in Northern Europe AD 300-1100* (Princeton: Princeton University Press, 1958); K.N. Chaudhuri, *Trade and Civilisation in the Indian Ocean: An Economic History from the Rise of Islam to 1750* (Cambridge: Cambridge University Press, 1985) and *Asia before Europe: Economy and Civilisation of the Indian Océan from the Rise of Islam to 1750* (Cambridge: Cambridge University Press, 1990).

transregional nomadic empires. After 1000 migrations, conquests and empire-building efforts guaranteed that cross-cultural interaction would take place in more intensive and systematic fashion than in earlier eras. This so-called conquest of distance promoted the diffusion of technological innovations and as such stimulated economic growth throughout Eurasia.[3]

In this short preliminary and speculative essay I will attempt to build on these earlier findings by focusing on the most extensive interstitial area situated at the fringes of the settled cores of China, South Asia, the Middle East and Europe. This is the so-called Eurasian Arid Zone which centres on Central Asia but, in various ways, extends into the major sedentary societies surrounding it. Indeed, it will again highlight the year 1000 after which the Arid Zone began to experience the recurrent rise of mounted warriors carving out new conquest states along the inner frontiers of sedentary society.[4] The following will broadly set out the various ways in which in particular its most western and southern neighbours, i.e. Europe and South Asia, came to terms with this sudden dynamism. Both societies derived much of their spatial identity from this encounter.

THE ARID ZONE

In the early history of Eurasia, frontiers have always figured more prominently as lines of open communication than as lines of rigid separation. As such, they rarely circumscribed compact and bounded blocks of territories – characteristic of modern nation states – but

[3] J.H. Bentley, 'Cross-Cultural Interaction and Periodization in World History', *American Historical Review* 101, 3 (1996), 766-8. See also W.H. McNeill, *The Pursuit of Power: Technology, Armed Force and Society since AD 1000* (Chicago: Chicago University Press, 1982).

[4] Highlighting frontier societies may bring about a renewed acceptance of cyclical time, very much in the spirit of the intelligent, fourteenth-century observations of the Arabic historian Ibn Khaldun. Unfortunately, cyclical time-frames far too long have been relegated to the realm of past myth whereas the linear, tripartite compartment of history is still seen as one of the major achievements of enlightenment and modernity. Actually, the 'modern' concept of time stems from equally mythical views such as the sequence of the four world monarchies and the tripartite apocalyptic tradition of Latin Christendom.

rather – like the Roman *limes*[5] – ran outwards as radial lines connecting various centres. In this case, the spatial image should not be a centre surrounded by widening concentric circles in which the influence of the centre gradually decreases – this is the Von Thunen model as adopted by Braudel, Wallerstein and other world-system historians – but a network of arteries connecting several rural and urban centres. At the same time, across the verges of these arteries, an *inner* frontier divides settled centres from wild peripheries.[6]

The frontiers that molded Eurasian history most were all part of that much wider, frequently broken, ecological continuum – sometimes called Saharasia – that included all the arid and semiarid zones of Eurasia, in the north spanning from the eastern outskirts of Vienna to the Chinese Wall, in the south stretching from the Atlantic coast of the Maghreb to the south-eastern extremes of the Indian sub-continent. This so-called Arid Zone receives relatively low annual rainfall – less than 1,000 mm – and, as such, marks the territory where pastoral nomadism had a natural advantage over sedentary agriculture, the latter primarily in the more humid areas of Europe and monsoon Asia.[7] In other words, the natural frontiers of the Arid Zone often marked broad overlapping areas in which two different ways of life encountered each other, the one predominantly pastoral-nomadic, the other predominantly sedentary-agrarian. Both were

---

[5] See e.g. C.R. Whittaker, *Les frontières de l'empire romain* (Besançon: Université de Besançon, 1989) and B. Isaac, *The Limits of Empire: The Roman Army in the East* (Oxford: Oxford University Press, 1992), 408-18.

[6] See especially J.C. Heesterman, 'Two Types of Spatial Boundaries', in E. Cohen, M. Lissak and U. Almagor (eds), *Comparative Social Dynamics: Essay in Honour of S.N. Eisenstadt* (Boulder: Westview Press, 1985), 59-72. Also L. Febvre, 'Frontière', *Revue de Synthese Historique* 45, 5 (1928), 31-44.

[7] Often a distinction is made between arid and semi-arid zones. For the present purpose I have brought them together under the heading Arid Zone since this is the most relevant category in terms of stockbreeding. Obviously, even under the more humid conditions of monsoon Asia, many areas were not cultivated, or were cultivated only temporarily. These lacked, however, the vast mobile resources of the Arid Zone. For the ecological perspective of the Arid Zone, see the various publications of the UNESCO research project on the Arid Zone during the 1950s and 1960s, in serials such as *Arid Programme*, *Arid Research*, and *Arid Zone*.

conditioned by distinct ecological spheres and, as such, were circumscribed by forces of diminishing returns. Moreover, this ecological divide never served as a fixed or closed borderline. Rather, both sides of the frontier witnessed flourishing mixed economies of wandering pastoralists living in co-existence with settled peasants. With the help of artificial irrigation, peasants often succeeded in crossing arid frontiers by bringing land under the plough. On the other hand, nomads felt an almost permanent eagerness to enjoy the riches of the settled world. Through the input of mobile resources – mainly cash and cattle – this could lead to substantial agrarian investments but, at the same time, it could easily undermine the precious sedentary balance and lead to the renomadization of agrarian fields. Hence the chequered pattern of the sedentarization process along the inner and outer frontiers of the Arid Zone in which sudden outbursts of agrarian investment alternated with tremendous nomadic devastation.[8] To put it differently, the arid frontiers of Eurasia contained both uniting and dividing capacities. As commercial corridors they connected sedentary centres but, at the same time, along its verges, marked the transition from the settled world of the peasant to the mobile world of the nomad. Hence the strategic location of such long-standing imperial capitals like Vienna, Tabriz, Beijing and Delhi which all served as major commercial and pastoral crossroads and transshipment areas, comparable to seaport-capitals such as London, Amsterdam or Calcutta.[9] Their rulers realized that in order

[8] This is most clearly witnessed in Iran. See e.g. A.K.S. Lambton, *Continuity and Change in medieval Persia: Aspects of Administrative, Economic and Social History, 11th-14th Century* (London: Tauris, 1988). For Balkh, see R.D. McChesney, *Waqf in Central Asia: Four Hundred Years in the History of a Muslim Shrine, 1480-1889* (Princeton: Princeton Universith Press, 1991), 234. The (semi-)arid zones of South Asia were also characterized by repeated renomadization. See e.g. H.G. Raverty, 'The Mihran of Sind and its Tributaries: A Geographical and Historical Study', *Journal of the Asiatic Society of Bengal* 61 (1892), 155-508, and, more recently, Chetan Singh, 'Forests, Pastoralists and Agrarian Society in Mughal India', in D. Arnold and R. Guha (eds), *Nature, Culture, Imperialism: Essays on the Environmental History of South Asia* (Delhi: Oxford University Press, 1995), 21-48.

[9] See the crucial importance of respectively the Alföld (see A.N.J. den Hollander, 'The Great Hungarian Plain: A European Frontier Area', *Comparative*

to stay in power they had to keep their stakes on both sides of the frontier.

It is from about the eleventh century onward that the Arid Zone emerged as a huge continental *méditerranée*, a vibrant interstitial region that widened the horizon of all its adjoining societies and opened new channels for pastoralists, warriors, merchants, pilgrims and other restless wanderers. To a large extent, this unprecedented globalization of the Eurasian economy was the result of large-scale conquests of nomadic warriors, primarily Turks and Mongols.[10] Their success story should be related to the more effective use of the war-horse, both by light archers and heavy-armoured cataphracti. Apart from earlier innovations such as the stirrup and the composite bow, there were some major improvements of the horse-harness and the saddle which made the mounted warrior far superior to the undisciplined and undrilled footslogger.[11] In addition, new develop-ments in siege technology during the thirteenth century gave fresh impetus to the offensive capabilities of mounted warfare.[12] Although

---

Studies on Society and History 3 (1960), 74-88, 155-69); the Moghan plains (see R. Tapper, 'Nomads and Commissars on the Frontiers of Eastern Azerbaijan', in K. McLachlan (ed.), *The Boundaries of Modern Iran* (New York: St. Martin's Press, 1994), 21-37); the Mongolian steppe (see O. Lattimore, *Inner Asian Frontiers of China* (Oxford: Oxford University Press, 1988); and Haryana (see my 'The Silent Frontier of South Asia, *c.* 1100-1800 AD', *Journal of World History* 9, 1 (1998).

[10] Creating a cultural ecumene recently labeled Turko-Persia. See R.L. Canfield (ed.), *Turko-Persia in Historical Perspective* (Cambridge: Cambridge University Press, 1991).

[11] See C. Uray-Kühalmi, 'La périodisation de l'histoire des armements des nomades des steppes', *Études Mongoles* 5 (1974), 145-55. For southern India, see the interesting studies of J. Deloche, *Military Technology in Hoysala Sculpture (Twelfth and Thirteenth Century)* (New Delhi: Sitaram Bhartia Institute of Scientific Research, 1989) and *Horses and Riding Equipment in Indian Art* (Madras: Indian Heritage Trust, 1990).

[12] For the improved use of early gunpowder in China, see J. Needham et al., *Science and Civilization in China*, vol. 5: *Chemistry and Chemical Technology, part 7: Military Technology: The Gunpowder Epic* (Cambridge, 1986), 1-18, 161-92; for the Middle East, see A.Y al-Hassan and D.R. Hill, *Islamic Technology: An Illustrative History* (Cambridge: Cambridge University Press, 1986), 106-20;

by about 1100 almost all Eurasia supported a socially and politically dominant horse-riding elite, especially along the extensions of the Arid Zone, the military revolution of the war-horse gave a huge military advantage to the nomadic warrior, not only in the heartlands of Central and West Asia but, with some delay, in the semi-arid tracts of South Asia as well. This is most clearly illustrated in the rise of new conquest states along the rim of the Arid Zone: during the eleventh century, the Khitan and Jurchen in northern China, the Seljuqs in northern Iran and the Ghaznavids in northern India, and during the twelfth and thirteenth centuries, the Yadavas, the Kakatiyas and the Hoysalas in southern India.[13]

Looking at the way the sedentary societies of Eurasia have dealt with this sudden upsurge of the Eurasian frontiers one may reflect that the settled authorities in Latin Europe succeeded in keeping the nomads out of their societies, whereas, the Middle East and, to a lesser extent, South Asia remained very much dual economies undergoing regular tribal break-ins produced by the ongoing in-teraction between the settled centres of society and its more mobile

---

for South Asia, see Iqtidar Alam Khan, 'Coming of Gunpowder to the Islamic World and North India: Spotlight on the Role of the Mongols', *Journal of Asian History* 30, 1 (1996), 27-45. For the introduction of the trebuchet in China, see J. Needham and R.D.S. Yates, *Science and Civilization in China*, vol. 5: *Chemistry and Chemical Technology*, part 4: *Military Technology: Missiles and Sieges* (Cambridge: Cambridge University Press, 1994), 203-40; for the Middle East, see D.R. Hill, 'Trebuchets', *Viator* 4 (1973), 115 and under the heading 'hisar', in *Encyclopaedia of Islam*, new edition (Leiden: E.J. Brill, 1960-2005); for India, see several reference in M. Habib, *The Campaigns of 'Ala'u'd-din Khilji being the Khaza'inul Futuh (Treasures of Victory) of Hazrat Amir Khusrau of Delhi* (Madras: Taraporewala, 1931), 39-42, 48, 53-4, 65-6, 89-90. See also my 'War-horse and Gunpowder in India, *c.*1000-1850', in J. Black (ed.), *War and Warfare, 1450-1815* (London: UCL Press, 1999).

[13] See e.g. W.E. Kaegi, 'The Contribution of Archery to the Turkish Conquest of Anatolia', *Speculum* 39, 1 (1964), 96-108; M. Chamberlain, *Knowledge and Social Practice in Medieval Damascus, 1190-1350* (Cambridge: Cambridge Universit Press, 1994), 29-47 and A. Wink, *Al-Hind: The Making of the Indo-Islamic World*, vol. 2: *The Slave Kings and the Islamic Conquest 11th-13th Centuries* (Leiden: E.J. Brill, 1997), 79-111.

fringes. Although China was also repeatedly conquered and ruled by nomadic invaders, nomads always had to make a crucial choice between either ruling China by becoming sedentary or staying mobile by leaving China.[14] In contrast with the Middle East and South Asia, China's nomads, like its arid territories, were located on its outer fringes, near or beyond its Great Wall. In these terms, China stands somewhere midway between Europe, on the one hand, and South Asia on the other. Let us take a closer look at the way the latter two coped with the increasingly vigorous inhabitants of the arid frontiers in their often overlapping capacities of nomads, warriors and ascetics.

NOMADS

Being a rather eccentric appendix of the huge Eurasian landmass, Europe had always been relatively immune for massive nomadic inroads since its climate favoured broken forests in stead of semi-arid steppes or savannah which were more suitable for pasture.[15] This always posed considerable logistical problems for nomadic invaders, as witnessed in the Mongol invasion of the thirteenth century.[16] At the very time most of Eurasia experienced new nomadic incursions, Europe began to enjoy an even greater immunity for them. At the same time, it managed to join in with the almost global expansion of trade and agriculture. Indeed, through an extensive web of commercial and financial networks it converged with the still far more advanced and monetized economies of the Islamic Middle East, South

---

[14] D. Morgan, *The Mongols* (London: Wiley-Blackwell, 1990), 84.

[15] The English *forest*, the German *Forst* or the French *forêt* derive from the Latin *forestis* which refers not to the modern use of forest but more in general to waste. The same goes for the German *Wald* or the Anglosaxon *weald* (F. Vera, *Metaforen voor de wildernis* (The Hague: Ministerie van Landbouw, Natuurbeheer en Visserij, 1997). Hence, it appears that there was some space for grazing animals but certainly less than in the Arid Zone. Similarly, in India the word *jungle* or *jangal* from Sanskrit *jangala* also refers not to dense forest but to uncultivated land.

[16] D. Sinor, 'Horse and Pasture in Inner Asian History', *Oriens Extremus* 19 (Wiesbaden: O Harrasowitz, 1972), 181-2. Cf. R.P. Lindner, 'Nomadism, Horses and Huns', *Past and Present* 92 (1981), 14-15.

Asia and China.[17] This commercial build-up was closely tied in to rapid agrarian expansion both internally and externally. Everywhere was a massive thrust to bring uncultivated forests – and even seas – under the newly developed heavy plough. Most spectacular was the extension of the agrarian frontier towards the east. This *ostsiedlung* partly replaced earlier Slav settlements but also effectively reclaimed land which formerly had been either waste or pasture. The same was true of the Anglo-Saxon and Norman expansion into the Celtic west and, to a lesser extent, of the Spanish *reconquista* of the Muslim south.[18]

The increased cerealization of Europe was characterized by rulers bestowing 'boundaries and rights' to all those who came to settle new land. This mostly involved well-defined freedoms and immunities from external tolls and taxes. Potential conflicts between feuding parties were preferably solved by systematically defining and demarcating their respective rights. This parcelization of space increasingly produced self-contained, mutually exclusive, territorial units that came to be supervised by more permanent and more uniform political and judicial institutions, propped up by Roman law and the extraordinary legal authority of the European town. Gradually, by being exempt from nomadic invasions, European states could develop as well-defined, contiguous territories with a relatively centralized governments differentiated from other organizations.

European sedentarization was accompanied by major alterations in the organization of cattle breeding. First of all, cattle production was pushed ever more outwards. The existing pastoral base of regional markets in western Europe retreated before an expansive system of arable cultivation. This gave birth to a new primary network of interregional exchanges involving more and more producers on the European periphery, such as of Denmark, Hungary, Poland and the Ukraine. This network interacted with surviving regional market structures, served to provide the great cities of west-central Europe

---

[17] See e.g. J. Abu-Lughod, *Before European Hegemony: The World System AD 1250-1350* (Oxford: Oxford University Press, 1989).

[18] The best general account of the period is R. Bartlett, *The Making of Europe: Conquest, Colonization and Cultural Change 950-1350* (London: Penguin, 1993).

with a regular supply of meat. Meanwhile, in the western and central parts of Europe breeding and fattening gradually gave way to dairying. Generally speaking, pastoralism in western Europe became increasingly incorporated into the expanding sedentary order.[19] Also in the south, cattle production was pushed outward towards Europe's most southern Extremaduras (i.e. furthermost pastures) and, eventually, to the Pampas of South America. However, this was not nomadic pastoralism but rather semi-sedentary ranching that was closely regulated by the settled authorities of the Mesta.[20] Thus apart from bringing its inner marches to the plough, Europe created a new cattle frontier beyond the outer extensions of agricultural settlement. In other words, Europe canted its nomadic frontier towards the arid fringes to its east and south.[21]

Like Latin Europe, after about AD 1000, South Asia went through a phase of rapid economic expansion and even became the hub of world trade.[22] But unlike Europe, agricultural expansion went hand in hand with the emergence of a new elite of tribal warriors from the drier parts of western and central South Asia. This new gentry often had a nomadic or semi-nomadic background and most of them had gained power from the increased military potential of the arid frontier, in particular through the improved exploitation of the warhorse. As intermediaries these tribal warriors were extremely well placed to direct the moveable wealth of the frontier, especially cattle and cash, into productive channels of agrarian investment. Hence, their conquests considerably enhanced the sedentarization of the dry interior of the Indian subcontinent, especially its fertile but stiff black soils which could only be cultivated through tremendous investments of

[19] I. Blanchard, 'The Continental European Cattle Trades, 1400-1600', *Economic History Review* 39, 3 (1986), 427-60.

[20] See the studies of C.J. Bishko, especially his 'The Castilian as Plainsman: The Medieval Ranching Frontier in La Mancha and Extremadura', in A.R. Lewis and Th.F. McGann (eds), *The New World Looks at its History* (Austin: University of Texas Press, 1963), 255-77.

[21] Cf. A.R. Lewis, 'The Closing of the Medieval Frontier, 1250-1350', *Speculum* 33, 4 (1958), 475-83.

[22] Wink, *Al-Hind*, vol. 2, 1-8. Cf. Abu-Lughod, *Before European Hegemony*, 261-91.

capital and cattle.[23] Also different from Europe, South Asian sedentarization proceeded by fits and starts. Until the late-nineteenth century, the subcontinent retained a potent population of mobile warriors and pastoralists who once and again undermined the sedentary order, in particular along the drier fringes which continued to be attractive for both grazing and raiding.[24] What we see are Ibn Khaldunian cycles of tribal conquest followed by agrarian expansion. As such, nomadic warriors always played a dual role, one sustaining and one interrupting the sedentarization process.

Apart from this internal interaction, through its semi-dry extensions of the Arid Zone, South Asia remained closely linked to the wider nomadic world of Central and Western Asia. For example, Indian horse-breeding and trading strongly depended on the continuous input of Central Asian horses which were produced by nomadic tribes and transported by mobile merchant-warriors.[25] Like the breeders and traders of dromedaries and oxen, they declined to be muzzled

[23] B.J. Murton, 'The Evolution of the Settlement Structure in Northern Kongu to 1800 AD', in B.E.F. Beck (ed.), *Perspectives on a Regional Culture: Essays about the Coimbatore Region of South India* (Delhi: Vikas, 1979), 1-33 and B. Stein, *Thomas Munro: The Origins of the Colonial State and his Vision of Empire* (Delhi: Oxford University Press, 1989), 120-1. Also the monograph from Cambridge Univeristy Press of Prasannan Parthasarathi on the textile industry of southern India.

[24] The so-called hero-stones or *viragal* bear witness of the increased tension between nomads and peasants after AD 1000. See Ajay Dandekar, 'Landscapes in Conflict: Flocks, Hero-stones, and Cult in Early Medieval Maharashtra', *Studies in History* 7, 2 (1991), 301-24. For the ongoing impact of nomads and pastoralism in the dry tracts of South Asia, see e.g. G.-D. Sontheimer, *Biroba, Mhaskoba and Khandoba. Ursprung, Geschichte und Umwelt von Pastoralen Gottheiten in Maharashtra* (Wiesbaden: Steiner, 1976); the contributions of Chetan Singh, 'Forests'; N. Bhattacharya, 'Pastoralists in a Colonial World' and Atluri Murali, 'Whose Trees? Forest Practices and Local Communities in Andhra, 1600-1922', in D. Arnold and R. Guha (eds), *Nature, Culture, Imperialism: Essays on the Environmental History of South Asia* (Delhi: Oxford University Press, 1995); B.J. Murton, 'Land and Class: Cultural, Social and Biophysical Integration in Interior Tamilnadu in the Late Eighteenth Century', in R.E. Frykenberg (ed.), *Land Tenure and Peasant in South Asia* (Delhi: Orient Longman, 1977), 81-99.

[25] See my *The Rise of the Indo-Afghan Empire, c.1710-1780* (Leiden: E.J. Brill, 1995), 68-104.

by the settled authorities. By moving along the many inner frontiers of the subcontinent, their extensive radius of action was clearly beyond the more restricted reach of the regional rulers. Although South Asia never witnessed the huge tribal disturbances of Central Asia and the Middle East, its tribal history neatly followed the rhythm of the Arid Zone's pulse.[26]

WARRIORS

Before 1000 both Europe and South Asia consisted of multiple inner frontiers consisting of all kinds of waste and pasture. Although not all of these wild territories were conducive to nomadic life, they certainly served as the ideal home and refuge of the nomad's alter ego: the free warrior periodically raiding the surrounded countryside. Most of this raiding was perfectly legitimate as long as it could be considered a fight for one's rights, for the sake of vengeance or redress. In other words, there was still no fundamental difference between the warlike activity of the state and the pillaging and feuding of the free warrior; both were perfectly normal means of collecting tribute or obtaining supplies. This did not mean that there were no countervailing ideas that aimed to limit uncontrolled eruptions of violence and destruction. For example, the killing of one's opponent was something rather to be avoided. Of course, this could be out of fear of supernatural punishment but, more importantly, killing would spoil the lucrative ransom and evoke the blood wrath of the victim's relations. Frequently the parties of conflict entered into periodic parleys to end recriminations so that life could continue more or less normally.[27] Nevertheless, both the European and Indian authorities

---

[26] Interestingly, indigenous horse-breeding in Ming China was conciously disconnected from the breeding tracts of the Arid Zone and situated in the, for horses, rather unhealthy south (Tani Mitsutaka, 'A Study on Horse Administration in the Ming Period', *Acta Asiatica* 21 (1971), 73-97).

[27] See e.g. M. Bloch, *Feudal Society*, vol. 1 (London-New York: Routledge, 1989), 125-30; O. Brunner, *Land and Lordship: Structures of Governance in Medieval Austria* (Philadelphia: University of Pennsylvania Press, 1984), 1-95; G. Duby, *The Early Growth of the European Economy: Warriors and Peasants from the Seventh to the Twelfth Century* (Ithaca: Cornell University Press, 1992), 48-

devised various additional measures which aimed to contain and to direct the dangerous drive of the free warrior.

In medieval Europe, both kings and Church leaders not so much agitated against the warrior's aggression itself but, more in particular, against its erratic nature. During the eleventh and twelfth centuries several Peace of God movements and other peace associations emerged which aimed to counter the very irregularity of raiding and its dreadful side-effects. These produced no general ban on raiding but only periods of general abstention in which one attempted to suppress the daily violence, exactions and unlawful fines exacted by plunderers and other troublemakers. Although this is still a long way from the state's monopoly of violence, these peace movements clearly represent the first attempts to establish some kind of public order in European society. In attempting to limit erratic raiding, kings and Church leaders had at their disposal a long tradition of Roman and scholastic thinking – Gratian and Thomas Aquinas following in the footsteps of St Augustine – on the legitimacy of political violence. In the end, this led to the criminalization of internal feuding whereas legitimate violence gradually became the prerogative of the sovereign state.[28]

The Peace of God movement and scholastic thinking linked up well with the ideology of the crusades. In all these cases, the state and Church authorities sought to mobilize warriors into the fight for a higher goal, to liberate Latin Christendom from the 'internal' feud and the 'external' infidel. On several occasions 'knights of peace' were instituted, and benefited, in exactly the same way as the crusaders, from the privilege of sacred immunity. Inspired by the example of the crusading orders, this kind of martial activity combined aggression with self-abnegation and, as such, acquired a spiritual and redemptive value. As a consequence, both the peace movements and the crusades favoured the definition and promotion of a new chivalric ideal that gradually looked down upon the now rather pusillanimous violence

9; F. Redlich, 'The Praeda Militari. Looting and Booty 1500-1815', *Vierteljahrschrift für Sozial- und Wirtschaftgeschichte*, Beiheft, 39 (1956), 1-79.

[28] R.W. Kaeuper, *War, Justice and Public Order: England and France in the Later Middle Ages* (Oxford: Oxford University Press, 1988), 134-84; Ph. Contamine, *War in the Middle Ages* (Oxford: Blackwell, 1984), 270-84.

of feuding. The latter became demonized as the excess of the lawless, as a form of 'civil' disobedience.[29]

The crucial point is that by both taming and siphoning off its more radical elements into the colonies, Latin Europe succeeded in taking its first steps towards what later came to be the monopolization of legitimate violence by the state. To use the words of Robert Bartlett, the creation of conquest states on the periphery of Latin Christendom was a consequence of intensified political control in its core.[30] Hence, the locus classicus of the free warrior shifted from the inner frontiers of Europe to its increasingly sacrosanct, external boundaries. Gradually, the juridical concept of international war emerged victorious, to cover external conflicts between sovereign states whereas within these states the self-help of the free warrior was criminalized and suppressed. One other illustration of this phenomenon is the creation of West and East India Companies which often combined seaborne raiding beyond, and peaceful trading within the outer European boundaries.[31] In any case, the domestication and expulsion of Europe's free warriors could never have been achieved without its inner wildernesses being tamed simultaneously.

As we have stressed already, the preconditions of the frontier in South Asia were entirely different. After 1000 it experienced various nomadic inroads of which the coming of Islam under the Ghaznavids and Ghurids is just one well-known example. Similarly along the southern extensions of the Arid Zone, in the Deccan and the Carnatic, the recurrent emergence of new tribal dynasties clearly mark the

[29] The classic analysis on the so-called domestication of the European aristocracy is of course N. Elias, *Über den Prozess der Zivilisation: Soziogenetische Untersuchungen*, 2 vols (Frankfurt am Main: Suhrkamp, 1991). Elias, however, situates the more radical changes much later, during eighteenth-century and follows Burkhardt and Huizinga in their traditional view of the European Middle Ages.

[30] R. Bartlett, 'Colonial Aristocracies of the High Middle Ages', in R. Bartlett and A. MacKay (eds), *Medieval Frontier Societies* (Oxford: Oxford University Press, 1989), 30.

[31] For an enlightening 'continental' view of the European trading companies, see Thomas A. Brady Jr., 'The Rise of Merchant Empires, 1400-1700: A European Counterpoint', in James D. Tracy (ed.), *The Political Economy of Merchant Empires* (Cambridge: Cambridge University Press, 1991), 117-61.

increased impact of the drier tracts of the subcontinent. In due course, these areas developed as huge zones of military entrepreneurship supplying the Indian armies with specialized warriors of Rajputs, Marathas and Nayakas.[32] More than ever before these inner frontiers became invigorated by intensified contacts with the larger nomadic world of Eurasia and its enormous military potential of horse- and man-power. For the Turkish, Iranian and Afghan free warrior South Asia proved to be the ultimate paradise since it combined, within short distance, exorbitant riches with excellent opportunities for refuge. Hence, South Asia counted an almost endless amount of iconoclastic, ghazi-like warriors, continuously shifting their loyalties and recurrently marching out for new lucrative conquests.[33]

Obviously, under these volatile circumstances it was nearly impossible for rulers to establish long-term internal stability and public order, let alone to impose a full monopoly on the use of violence.[34] Taking the political flux for granted, we should keep in mind that for the settled ruler the almost endless availability of free warriors held many blessings in disguise. Apart from raiding and plundering, which in itself mobilized resources and increased their velocity of circulation, these warriors could equally inject enormous amounts of movable wealth into the settled world. For the Indian ruler, the main challenge was not to tame or to expel, but to incorporate these free warriors, albeit temporary, into his realm but, at the same time, without being overwhelmed by them. Therefore, the Indian ruler had to invent a whole range of political mechanisms which aimed at both containing and exploiting the huge military potential of the frontier.

Of course, one Indian way to accommodate the frontier warrior

---

[32] S. Gordon, *Marathas, Marauders, and State Formation in Eighteenth-Century India* (Delhi: Oxford University Press, 1994), 182-208.

[33] Cf. the iconoclasm of the Virasaiva sect described by V. Narayana Rao, *Siva's Warriors: The* Basava Purana *of Palkuriki Somanatha* (Princeton: Princeton University Press, 1990).

[34] Even during the high point of Mughal rule, the imperial roads remained rather insecure. See e.g. the many examples provided by Fransisco Pelsaert in D.H.A. Kolff and H.W. van Santen (eds), *De geschriften van Fransisco Pelsaert over Mughal Indië, 1627. Kroniek en remonstrantie* (The Hague: Nijhoff, 1927).

was through the time-honoured bestowal of temporary and concurrent, that is shared, rights in the proceeds of the agrarian exploitation. In the existing Indian situation of interlaced personal shares and territorial holdings the king could give away much of his realm without definitely loosing his stake in it, the more so since his gifts often involved rebellious, unconquered or uncultivated lands. Although being just one of many shareholders, the king managed to keep his stake in each field and every threshing floor in order to ensure the collection of his share. At the same time, through distributing these overlapping rights the king was able to arbitrate between the ongoing conflicts of rivalling parties. Actually, ruling in South Asia came down to this kind of management through conflict; the crux was not to tame or expel but to incorporate existing conflict.[35]

In this light, the policy recommendations of the *Arthashastra* proved perfectly adequate. These amount to no more than stratagems which the king should employ to keep the delicate political balance intact. According to that logic, the wise application of conciliation, gift-giving and sedition takes precedence over the use of political force. In practical terms ruling in India came down to what André Wink has labelled *fitna*: sowing dissension and forging alliances.[36] Actually, it was the persistence of open, *inner* frontiers which conditioned South Asia's interpretation of political sovereignty in which the king ruled through *fitna* on the basis of shared rights. Conversely, the European king increasingly ruled by solving inner conflicts on the basis of demarcating rights. As consequence, the European sovereign state developed an ever more fanatic zeal to defend its, ever more sacred, *outer* frontiers.

ASCETICS

Both in Europe and South Asia, the wild frontier was not only the home of the mobile nomad and warrior but also of the wandering

[35] Based on the unpublished paper by J. Heesterman, 'Traditional Empire and Modern State', IDPAD symposium on State and Society, New Delhi, 5-9 March 1990.

[36] A. Wink, *Land and Sovereignty in India: Agrarian Society and Politics under Eighteenth-century Maratha Svarajya* (Cambridge: Cambridge University Press, 1986), 9-51.

ascetic. In ancient Europe the latter is most vividly represented by Celtic druids and Irish monks who preferred to wander the desolate wastes of forests and seas. It was this wilderness which gave access to the gods and which later evoked apocalyptic expectations of the end of times and the coming of the messiah. By renouncing worldly pleasures and 'becoming wild' man was believed to find his inner self.[37] The European ideal of the noble savage is not merely a late invention of European romanticism but clearly stems from the long-standing attraction of life in the forest.

Surely, this life was not only considered more authentic and adventurous, its unconventional dynamic went right against the settled order of town and village. At the time Europe lost its inner frontier, the settled world was gradually cut off from the ennobling faculties of the wilderness. Its remaining inhabitants, such as wandering monks and witches, lost their mediating function between the village and the forest and found themselves increasingly demonized by the settled authorities. Meanwhile, the purifying capacity of Europe's internal wilderness were projected towards its new external frontier of an ever more 'outside' world. From about AD 1000, new religious orders – such as the Templars and Hospitallers – translated the earlier ideal of ascetism into a calling for holy war against Islam. Other parts of reformed European monkhood – such as Cistercians and Franciscan – found a way to preserve ascetism within an increasingly sedentary community. For some, these monks – many were still active along the retreating inner frontiers of Europe – continued to represent a new order which was to lead mankind into the glories of the final age of the Spirit. More and more, however, apocalyptic expectations were focused on the outer European world. While Jerusalem became the prime place where the coming of the messiah was to be expected,[38] Europe relinquished its own inner potential for this-worldly ascesis. With the gradual disappearance of Europe's remaining wildernesses the religious establishment was convicted to focus on settled life and

---

[37] J. le Goff, *L'imaginaire*, 59-75 and H.P. Duerr, *Traumzeit: Über die Grenze zwischen Wildnis und Zivilisation* (Frankfurt: Syndikat, 1978).

[38] Cf. N. Cohn, *The Pursuit of the Millennium: Revolutionary Millenarians and Mystical Anarchists of the Middle Ages* (London: Harper Torchbooks, 1961). For the medieval fascination with liminal areas, see H. Pleij, *Dromen van Cocagne: Middeleeuwse fantasieën over het volmaakte leven* (Amsterdam: Prometheus, 1997).

to vie with the political authorities for earthly power and resources. In the end, this historical struggle helped to produce the rigid separation between the earthly domain of man and the heavenly domain of God.

South Asia never really experienced the intense conflict between the 'secular' and the 'sacred' as witnessed in Europe. Hence the transcendental world continued to play an active part in this world which deprived earthly power from possessing an ultimate meaning in itself. The ultimate aim for both Brahmins and kings was not earthly power but eternal salvation; the way thither was not conquering the world but renouncing it. Interestingly, while the Western state and Church stood face to face competing for earthly power, the Indian king and Brahmin had turned their backs against each other competing for being relieved of it. This pursuit of renouncement was easy enough for the ascetic who could always leave the settled world but for the Indian king this proved rather problematic.[39] Nevertheless, since South Asia retained large areas of desolate wilderness, even the worldly king found ample space to partly live up to this ideal. For example, as in the Indian royal consecration, the king could combine settled life and ascetic life as two phases of the ritual year cycle. In more mundane words, the king gained legitimacy by seasonal campaigning. In this light, the latter was not primarily meant to conquer the world but, exactly the opposite, to publicly renounce it.[40]

Another option for the king was, of course, to distribute his realm to Brahmins or 'gods' in the form of *brahmadeyas* and *devadeyas* which often involved lands beyond the agrarian frontier.[41] Here the king's eagerness to give found its counterpart in the Brahman's reluctance to receive. Actually the religious establishment – in Hinduism as well

[39] Primarily based on D.D. Shulman, *The King and the Clown in South Asian Myth and Poetry* (Princeton: Princeton University Press, 1985) and J.C. Heesterman, *The Inner Conflict of Tradition: Essays in Indian Ritual, Kingship, and Society* (Chicago: Chicago University Press, 1985).

[40] For the cyclical migration between sedentary society and wilderness in South Asian tradition, see Th. Parkhill, *The Forest Setting in Hindu Epics: Princes, Sages, Demons* (Lewiston: Mellen University Press, 1995).

[41] Actually during the fourteenth century royal gift-giving shifted from Brahmins to sectarian leaders (B. Stein, *Peasant State and Society in Medieval South India* (Oxford: Oxford University Press, 1994), 413).

as in Indian Islam – came increasingly under attack for their worldly commitments. New sectarian movements – such as the Lingayats or the Chishti Sufis – dissociated from settled society, favouring the immediate experience of the divine through both sensual excess and worldly renunciation.[42] Not surprisingly, these iconoclastic sects often drew their following from the same arid tracts whence Indian rulers recruited warriors for their armies.[43] Their leaders were crucial intermediaries for the introduction, extension and institutionalization of the new warrior regimes of the Indian interior.[44] As in the case of the Sikhs or the Gosains, ideals of devoted service to a spiritual master became transmuted into those of dedicated military service to a warrior chief.[45] For the peasant of the dry tracts of South Asia military service was another way to escape the compromising commitments of settled life and to live up – albeit seasonably – to the ideal of worldly renouncement.[46] All this is not to deny that there were many ways

[42] For a general, albeit strongly biased, account of the religious mood of the eleventh and twelfth centuries, see Buddha Prakash, *Aspects of Indian History and Civilization* (Agra: Shiva Lal Agarwala, 1965), 218-353, especially 310-41. Referring to the new sects of 'Siddhas, Nathas and Yogins', Prakash speaks in terms of 'crusade against caste', 'efflorence of the seeds of revolt and levelling' and 'movements of revolt and heterodoxy' which brought about a 'hectic urge to escape from the world and plunge in the enjoyment of pleasures, all sense of moderation and decency disappears and its place is taken by obscenity and vulgarity'. For a similar but more recent and more balanced analysis of these cultural redirections, see V. Narayana Rao, D. Shulman and S. Subrahmanyam, *Symbols of Substance: Court and State in Nayaka Period Tamilnadu* (Delhi: Oxford University Press, 1992).

[43] D.D. Shulman, 'Die Dynamik der Sektenbildung im mittelalterlichen Südindien', in S.N. Eisenstadt (ed.), *Kulturen der Achsenzeit II: Ihre institutionelle und kulturelle Dynamik, Teil 2: Indien* (Frankfurt am Main: Suhrkamp, 1992), 102-29. Cf. S. Fuchs, *Rebellious Prophets: A Study of Messianic Movements in Indian Religions* (London: Asia Publishing House, 1965).

[44] A. Appadurai, 'Kings, Sects and temples in South India, 1350-1700 AD', *The Indian Economic and Social History Review* 14, 1 (1977), 47-73.

[45] See e.g. D. Gold, 'The Dadu-Panth: A Religious Order in its Rajasthan Context', in K. Schomer et al. (eds), *The Idea of Rajasthan: Explorations in Regional Identity*, vol. 2: *Institutions* (Delhi: Manhohar, 1994), 242-64.

[46] D.H.A. Kolff, *Naukar, Rajput and Sepoy: The Ethnohistory of the Military Labour Market in Hindustan, 1450-1850* (Cambridge: Cambridge University Press, 1990).

to sustain an ascetic ideal in the sedentary world – both Europe and South Asia provide many examples of this – but it certainly appears that the persistence of the inner frontier in South Asia made this ideal both more practicable and more widely acceptable, even for the main representatives of settled life the king and the peasant.

PARADISE LOST

By taking the Arid Zone of Eurasia as a starting point, the year 1000 appears to be a crucial watershed in the history of both Europe and South Asia. It clearly marks the diverging responses of both regions to the new challenges of the invigorated nomadic frontier. As a spatial category, the Latin West came fully into its own by canting its frontier outwards. Its ecological conditions had made it relatively immune for the uprooting effects of the horse-warrior revolution. With the taming of its inner frontier, it also expelled or domesticated the frontier's prime inhabitants of nomads, warriors and ascetics. In a way, one could say that Europe turned its zeal towards its outer frontiers, both at land and sea, which not only brought massive wealth and power but also a sense of *mission civilisatrice*. It was only in the late nineteenth century, when the last external frontier were closed, that one was forced to find salvation in the self-contained, settled world of the here and now. To paraphrase the well-known words of William Blake, the New Jerusalem was to be built in Europe's own green and pleasant land.

South Asia's experience with the frontier was rather different. After about 1000 it became, more than ever before, part of that much larger Arid Zone which witnessed the rapid rise of the nomadic warrior. In the words of Ernest Gellner, the arid frontiers of Eurasia continued to serve as 'political wombs' from which various new warrior elites emerged.[47] Concomitantly, South Asia's inner frontier very much remained a wide, open-ended zone which not only favoured the circulation of people, animals, goods and ideas but also agricultural expansion. The importance of the inner frontier should warn us to see too easy dichotomies, such as Muslims *versus* Hindus or Indians

---

[47] E. Gellner, *Anthropology and Politics: Revolutions in the Sacred Grove* (Oxford: Blackwell, 1995), 164.

versus foreigners.[48] It is my view that a far more relevant dichotomy for South Asia is that between the sedentary world of peasants, soldiers and 'priests' and the mobile world of nomads-cum-warriors-cum-ascetics. This crucial divide was neutralized during the late nineteenth and twentieth centuries when the sedentarization of South Asia came to its final conclusion. Indeed, after one millennium of diverging developments, South Asia joined Europe again by being faced with the challenge of an inner paradise lost.

[48] Similar warnings are expressed in C. Talbot, 'Inscribing the Other, Inscribing the Self: Hindu-Muslim Identities in Pre-Colonial India', *Comparative Studies in Society and History* 37 (1995), 692-723 and D. Ludden, 'History Outside Civilisation and the Mobility of South Asia', *South Asia* 17, 1 (1994), 1-23.

CHAPTER 4

Warhorse and Post-Nomadic
Empire in Asia, *c.* 1000-1800

Earlier generations have said that China is unable to defend itself
against the mounted soldiers of [the Jin]. The enemy's strength derives
from its cavalry; their ability lies in riding and archery. . . . Their lands
in Henan and the northern marches are broad in expense, and there
are numerous pastures. In our country there is not four- or five-tenths
of the number of pasture. The Danchang and Hengshan horse markets
are extremely far from the front, and the convoy-relay depots are
spaced irregularly. Therefore our nation does not even have two- or
three-tenths as many horses as our enemy, the Jin.

HUA YUE in 1206[1]

For when India, from the Sutlege to the Sea shall be entirely under
British protection, an event which whether desirable or not cannot
be far distant, that immense Empire will have no protection against
the predatory hordes of horse from the Panjab, from Afghanistan, or
even from Persia or Tartary, except what it shall derive from cavalry,
the produce of India itself.

J. SALMOND in 1816[2]

INTRODUCTION

THIS ESSAY AIMS to explore the relationship between the production,
trade and maintenance of warhorses and processes of empire building

---

[1] P.J. Smith, *Taxing Heaven's Storehouse: Horses, Bureaucrats, and the Destruction
of the Sichuan Tea Industry, 1074-1224* (Harvard: Council on East Asian Studies,
1991), 303.

[2] G.J. Alder, 'The Origins of "the Pusa Experiment": The East India Company and
Horse-Breeding in Bengal, 1793-1808', *Bengal Past and Present* 98 (1979), 28.

in medieval Asia. The perspective will be that of comparative world-history. At times this includes Europe, Africa and Southeast Asia, but most of the attention will be paid to India and China. These two regions are rarely compared to each other; when they are, both are usually measured against a still-dominant European paradigm which is primarily focused on the rise of 'modern' industry and state-formation. In our investigation, though, Central Eurasia will play the role of Europe, while the warhorse replaces the machine in the making and unmaking of states. But the interactions of India and China are rarely compared either, since at first glance India – like Europe and Southeast Asia – appears to be far too distant from the steppes to have experienced a substantial nomadic impact. As we will see, however, both China *and* India saw the emergence of extensive post-nomadic empires that were built on the exploitation of the frontier between their increasingly productive sedentary heartlands and the semi-nomadic and nomadic marches bordering Central Eurasia. The persistent role of the Central Eurasian warhorse prevented the conquering rulers of India and China from becoming entirely settled – in fact, much of the organisation, the ideology and the ritual in both regions remained thoroughly nomadic in character. In my view, the two absolute highpoints of the development of such post-nomadic frontier empires were the roughly simultaneous Mughal Empire (1555-1858) in India and the Manchu Qing Empire in China (1636-1911). Despite the enormous expansion of agriculture and overseas commerce, their rulers knew all too well that to reap the fruits of this expansion required a large and steady supply of strong Central Eurasian warhorses ridden by nomadic warriors. In my view, this partly explains the often ignored persistence of Central Eurasian political institutions in the at first sight mainly settled societies of India and China.

Obviously the relationship between warhorse and the political organization of empires is an old one. Recently, Bernard Lewis reminded us of the various equine metaphors associated with the horse, the accoutrements, the stirrups, the reins and even the tail to be in use for political authority. Even that magic word of modern-day leadership, *management* derives from the Italian *maneggio*: 'the handling or training of a horse'. The same holds true of Islamic political language in, for example, *siyasa*, meaning both 'horse-handling' and 'statecraft' or simply 'politics'.[3] The idea that the horse belongs to a specific pattern of state-formation, because of the different economic demands that it involves, goes back at least to Aristotle, who believed that cavalry states were likely to be oligarchies since only the rich could afford horses. By contrast, states whose power depended on heavy infantry would be moderate democracies, since service as heavy infantry involved a lesser but still considerable outlay. Naval powers would be radical democracies, since it was the poorest citizens who served as rowers in the galleys.[4]

Apart from these more general associations, thanks to the studies of Lynn White and others, we all know how to connect the highly de-centralized European state system called feudalism to the introduction of the stirrup and the spread of heavy cavalry.[5] In the case of West Africa, drawing inspiration from White, Jack Goody and, in particular, Robin Law have likewise sought to show that the savannah states that depend on the use of warhorses tend to be less centralized than the coastal states, which lack horses and instead trust primarily on gunpowder technology. Both agree that:

Both in cavalry and in firearm states, the core of specialist soldiers who formed the army's principal strength regularly comprised the bands of retainers contributed by the major chiefs. Rulers regularly had private armies, usually recruited from palace slaves, even in cavalry states; and non-royal chiefs supplied contingents of soldiers even in firearm states, such as Dahomey. But in cavalry states, although the royal contingent can be seen as 'one of the pillars of the power' of the ruler, it is clear that the king did not normally provide more than a small proportion of the total strength of the army, and remained dependent upon the chiefs both militarily and politically. In firearm states, on the other hand, it was often (though by no means always) the case that the royal slave regiments formed the principal fighting force, so that

[3] B. Lewis, *The Political Language of Islam* (Chicago-London: University of Chicago Press, 1988), 19.

[4] R. Law, *The Horse in West African History: The Role of the Horse in the Societies of Pre-Colonial West Africa* (Oxford: Oxford University Press, 1980), 184-5.

[5] L. White, *Medieval Technology and Social Change* (Oxford: Oxford University Press, 1962).

the king could to some extent dispense with the need to depend upon their chiefs for the supply of military forces.[6]

Lynn White on Europe, and Goody and Law on Africa are not too far off from Marshall Hodgson's very influential thesis that the use of new gunpowder technology – mainly artillery – actually created the far more centralized, 'early-modern' Asian empires of Ottomans, Safavids, Mughals and, as many would quickly add, Manchus.[7] Although, generally speaking, I would agree with this impressive list of scholars that gunpowder weaponry required or facilitated a different, perhaps even more centralized political organization, it is also my contention that, until the late eighteenth century, warhorses were more, or at least equally important in exercising imperial control, at least along the empires' most dangerous, arid frontiers. Under the prevailing circumstances, to dispense with the horse in favour of gunpowder may have stimulated centralized rule in some parts of the empire but, at the same time, would have undermined that rule itself along the fringes. Here I will not elaborate on the ongoing role of the warhorse since it is the overall drift of the most recent literature on the topic both for Mughal India as for Manchu China.[8] It raises,

---

[6]Law, *The Horse*, 188; see also J. Goody, *Technology, Tradition and the State in Africa* (London: Oxford University Press, 1971).

[7]M.G.S. Hodgson, *The Venture of Islam*, vol. 3: *The Gunpowder Empires of Modern Times* (Chicago-London: University of Chicago Press, 1974).

[8]Although during the seventeenth century the Manchus used more and better quality gunpowder weaponry against the Ming and the southern Feudatories, mainly in connection with sieges, one can still fully agree with Kenneth Chase who, from a comparative perspective, states that 'the Manchus were able to use the resources of the steppe to conquer the steppe, and they never suffered from the same shortage of cavalry that plagued the late Ming' and 'the solution to the problem of the steppe border had little to do with better firearms anyway' [K. Chase, *Firearms: A Global History to 1700* (Cambridge: Cambridge University Press, 2003), 170-1]. For India, see J.J.L. Gommans, *Mughal Warfare: Indian Frontiers and High Roads to Empire, 1500-1700* (London-New York: Routledge, 2002); For China, apart from Chase, see H. van de Ven (ed.), *Warfare in Chinese History* (Leiden: E.J. Brill Publishers, 2000); N. Di Cosmo (ed.), *Warfare in Inner Asian History (500-1800)* (Leiden: E.J. Brill, 2002); N. Di Cosmo, 'Did Guns Matter? Firearms and the Qing Formation', in L.A. Struve (ed.), *The Qing Formation in World Historical Time* (Cambridge Mass.-London: Harvard

however, various questions about the production, the trade and the maintenance of warhorses and, following this, about the way the use of warhorses has continued to influence the political organization of these empires. Let us start, however, by taking a closer look at the specific temporal and spatial coordinates of our present investigation. Why to focus on 1000-1800 and why compare India to China?

THE POLITICAL ECOLOGY OF FRONTIER EMPIRES
(1000-1800)

The most obvious reason for focusing on the timeframe 1000-1800 is the so-called horse-warrior revolution of the eleventh to thirteenth centuries. So far, however, such a revolution has been demonstrated for India only since it is so closely associated with the tremendous success story of Turkish armies creating powerful new sultanates all over the subcontinent. In appears that this period saw the somewhat belated culmination of various earlier, technological developments such as the use of stirrups, deeper saddles and new horse tack, which markedly improved both the stability of the rider and the manoeuvrability of the horse. Together with new, mainly ox- and dromedary-based capacities of supply, this increased the effectiveness of the warhorse, now used in a devastating new tactical combination of heavy cavalry at the centre and wheeling mounted archers on the flanks.[9]

Simultaneous with the rise of horse-based Indo-Turkish sultanates – first at the India's north-west frontier under Ghaznavids (977-1186) and Ghurids (c. 1000-1215) but than extending into northern India and the Deccan (c. 1200-1400) – we see in northern China the emergence of the great steppe empires of the Khitan Liao (907-1125), Tangut Xi Xia (990-1227) and Jurchen Jin (1115-1234), culminating

---

University Asia Center, 2004), 121-67; P.C. Perdue, *China Marches West: The Qing Conquest of Central Eurasia* (Cambridge Mass.-London: The Belknap Press of Harvard University Press, 2005).

[9] J.J.L. Gommans, 'The Silent Frontier of South Asia, c. 1100-1800 AD', *Journal of World History* 9 (1998), 1-2. More in particular, see A. Wink, *Al-Hind: The Making of the Indo-Islamic World*, vol. 2: *The Slave Kings and the Islamic Conquest, 11th-13th centuries* (Leiden: E.J. Brill, 1997).

in the Mongol empire of the Yuan (1260-1368). In the Chinese case, however, technology appears to have been far less important in explaining the sudden success of these steppe empires. Peter Golden stresses that the armament and tactics of the nomads did not change appreciably over time although the (earlier) spread of the stirrup may have facilitated more heavily armed men and armoured cavalry.[10] Even with regard to the Mongols, Thomas Allsen claims that they did not enjoy technological superiority over their sedentary enemies but rather that 'their success was due to the fighting qualities of their soldiers and to the tactical, logistical, and organizational abilities of their leaders'.[11]

Whatever their relative importance, all these technological, tactical, logistical and organizational aspects, both in the case of India and China, do suggest that at the turn of the first millennium Central Eurasian cavalry armies had become more effective than ever before. Of course to be effective both the quantity and quality of the warhorse were important. Even western and central Europe, Southeast Asia and northern Africa, areas which could raise strong warhorses but for ecological reasons could not support great numbers of them, in one way or the other did experience its effects. But, generally speaking, their failure to develop units of effective mounted archers diminished the usefulness of their cavalry, which left considerable room for infantry warfare. Consequently, they were the first to experience the full impact of new gunpowder technology.[12] Somewhat surprising from the ecological point of view, however, is the sheer mass of warhorses in both India and China – in both cases from about 1 to 3 million, mostly held in the northern regions – for reasons that we still have to take account of.

---

[10] P. Golden, 'War and Warfare in the Pre-Cinggisid Western Steppes of Eurasia', in N. Di Cosmo (ed.), *Warfare in Inner Asian History (500-1800)* (Leiden: E.J. Brill, 2002), 150, 157.

[11] Th.T. Allsen, *Mongol Imperialism: The Policies of the Grand Qan Möngke in China, Russia and the Islamic lands, 1251-1259* (Berkeley: University of California Press, 1987), 189.

[12] B.S. Bachrach, 'Animals and Warfare in Early Medieval Europe', *Settimane di studio del centro italiano di studi sull'alto medioevo (L'Uomo di fronte al mondo animale nell'alto medioevo)* 31 (1985), 707-52.

The quality and the quantity of warhorses were matched in importance by logistical support, facilitated by the increasing availability monetization and credit. Actually the crucial condition that made such a thing as a horse-warrior revolution possible was neither technology nor any other military condition but rather the rapidly changing medieval economy, i.e. the tremendous expansion of agrarian and commercial resources in the sedentary societies immediately surrounding the main Central Eurasian production centres of the warhorse. Hence, we cannot understand the horse-warrior revolution, or, connected to this, the rise of various new border states all across Eurasia, if we do not take into account the coinciding *agrarian* revolutions in Europe, the Middle East, and, indeed, most spectacularly, in India and China.[13] Much of the fruits of this expansion were invested in larger and better-equipped cavalry armies, mostly recruited at the arid borders but with a great deal of clout on the fertile interior.

Although this expansion started from about the eighth century and achieved its maximum levels in the wetter, rice-producing areas in coastal India and southern China, it was the more effective use of more numerous warhorses that enabled an unprecedented degree of integration between the sedentary growth economies of Monsoon Asia and the nomadic and semi-nomadic areas of the so-called Eurasian Arid Zone.[14] Here one could very well agree with Nicola di Cosmo who writes that while the basic tactical principles of nomadic armies may not have changed dramatically over time, the way in which

[13] M. Elvin, *The Pattern of the Chinese Past* (Stanford: Stanford University Press, 1983); B. Stein, *Peasant, State and Society in Medieval South India* (Delhi: Oxford University Press, 1980); A.M. Watson, *Agricultural Innovation in the Early Islamic World: The Diffusion of Crops and Farming Techniques 700-1100* (Cambridge: Cambridge University Press, 1983); R.A. Palat, 'Historical Transformations in Agrarian Systems Based on Wet-Rice Cultivation: Toward an Alternative Model of Social Change', in Ph. McMichael (ed.), *Food and Agrarian Orders in the World-Economy* (Westport, Conn-London: Praeger, 1995), 55-80.

[14] Gommans, 'Silent Frontier'; A. Wink, 'India and the Turko-Mongol Frontier', in A.M. Khazanov and A. Wink (eds.), *Nomads in the Sedentary World* (Richmond: Curzon, 2001), 211-34.

resources were obtained and turned into military supplies did change considerably. Hence a nomadic army with access to greater resources (such as those that could be provided by neighbouring sedentary states or provinces) was able to sustain longer and more distant campaigns.[15] This was exactly what happened all across the Middle East, India, and China in the period of *c.* 900-1400 when rapid nomadic conquests could feed on the economic growth of the neighbouring sedentary economies. Later, in the period 1500-1800, this was repeated at a still higher level with the establishment of larger, far more stable and extremely powerful frontier empires of the Ottomans, Mughals and Manchus. Assisted by the increasing influx of cash and bullion, they became post-nomadic, trans-frontier rulers who proved more successful than ever in bridging the old but now also shifting frontier between (semi-)nomadic horsepower and agrarian expansion.

Some of the major political developments of Asian societies for our period can simply be traced to their common ecological frontier between the arid and semi-arid nomadic horse-breeding areas and the wetter grain- and rice-producing centres. What really strikes the careful observer when focusing on Central Eurasia is the contiguous belt of relatively dry desserts and steppes that extends from northern Africa to northern China. Although desert and steppe are different ecological zones, supporting different nomadic economies, this so-called Arid Zone roughly indicates the natural habitat of nomadic-pastoralism in general, and nomadic horse breeding in particular, and, as such, also denotes the natural range of operation of nomadic armies.[16] It shows, for example, that Central Eurasia and Iran are the most liable to repeated horse-based nomadic incursions. What it does not show, however, is that the Middle Eastern deserts cannot support

---

[15] N. Di Cosmo, 'Introduction: Inner Asian Ways of Warfare in Historical Perspective', in N. Di Cosmo (ed.), *Warfare in Inner Asian History (500-1800)* (Leiden: E.J. Brill, 2002), 10.

[16] Gommans, 'Silent Frontier.' See also Chase's excellent analysis that connects ecological zones with the spread and effectiveness of firearms (Chase, *Firearms*, 197-211). For an earlier interpretation, see M.G.S. Hodgson, *The Venture of Islam*, vol. 2: *The Expansion of Islam in the Middle Periods* (Chicago-London: University of Chicago Press, 1974), 69-85.

as many horses as Central Eurasia or northern Iran, an ecological fact that determined the natural, thirteenth-century boundary between Mamluk and Mongolian power.[17] For similar reasons, the Carpathians marked the far western European frontier of nomadic armies.[18]

From the ecological point of view the sharpest frontier between the predominantly nomadic Arid Zone and surrounding sedentary economies occurs in China where the Great Wall neatly demarcates the transition from steppes to sown.[19] On the Indian subcontinent the two semi-arid extensions flow into the open jungle and scrub of the far east and deep south and make the transition far more gradual, but at the same time, far more intrusive. These eastern and southern extensions of the Arid Zone never occasioned the building of a defensive system like the Chinese one, but instead, facilitated the creation of India's *longue-durée* road axis of northern (*uttarapatha*) and southern highways (*dakshinapatha*). As a result, through these inner frontiers-*cum-limites*, the humid but very productive South and East in India are more closely linked to (semi-)nomadic Central Eurasia than is the humid and equally productive South in China.[20] Finally taking a look at the other end of the Arid Zone, the transition between Europe and Central Eurasia was in ecological and historical terms the least rigid, the more so since the deciduous forests of Eastern Europe did not yet support the rich economic and demographic centres so characteristic of the Indian subcontinent. Thus in India the encounter between agrarian prosperity and nomadic dynamism is comparable to China but it is also much less restricted to some external border as it is almost omnipresent (with the exception of the coastal regions of the Southwest and in Orissa).[21]

The ecological circumstances of the Arid Zone can explain much

---

[17] J. Masson Smith Jr., 'Nomads on Ponies vs. Slaves on Horses', *Journal of the American Oriental Society* 118 (1998), 54-63.

[18] D. Sinor, 'Horse and Pasture in Inner Asian History', *Oriens Extremus* 19 (1972), 171-84; R.P. Lindner, 'Nomadism, Horses and Huns', *Past and Present* 92 (1981), 3-20.

[19] O. Lattimore, *Inner Asian Frontiers of China* (Boston: Beacon Press, 1962).

[20] Gommans, *Mughal Warfare*, 8-23.

[21] Cf. J.C. Heesterman, 'Warrior, Peasant and Brahmin', *Modern Asian Studies* 29 (1995), 637-54.

of the degree of havoc the nomads of Central Eurasia produced in its surrounding sedentary societies: at its greatest in the arid Middle East, Iran and Russia, at its least in the more distant parts of Europe and Southeast Asia.[22] Perhaps, the most interesting middle position is taken up by India and China. Beyond a very dynamic nomadic frontier, both cover the world's two richest medieval sedentary economies. Even more than in the case of the Middle East, the post-nomadic Mughal and Manchu conquerors of these regions were probably the most sensitive to ongoing forces of assimilation – the same for indianization as for sinification – and in response were perhaps the most keen in maintaining as well as reinventing their nomadic outlook and organization. They knew perfectly well that only such a post-nomadic stance would enable them to get both cultural and material access to the Central Eurasian supply-lines of nomadic warriors and warhorses.

EURASIAN HORSE-ECONOMIES

The warhorse was the one essential element in warfare that both the Indian and Chinese states could not breed in sufficient numbers for their own need. What they lacked most were extensive grazing facilities, especially in India's east, south and south-west and in China's south-east; those areas that had experienced a medieval agricultural breakthrough on the basis of more intensive paddy cultivation. In addition, like most of the hot and humid parts of Monsoon Asia, these areas possessed a hostile disease and reproduction environment for the horse. Insufficient grazing was not compensated by sufficient quantities of alternative and equally nutritious fodder crops such as oats in Europe or barley in the Middle East, both of which integrated horse-breeding more tightly with the agrarian economy and stimulated the breeding of relatively high quality warhorses such as the European

---

[22] This may very well be the crux behind Victor Lieberman's strange parallels between the political integration of Southeast Asia and Europe in the period 800-1830 (V. Lieberman, *Strange Parallels: Southeast Asia in Global Context, c. 800-1830*, vol. 1: *Integration on the Mainland* (Cambridge: Cambridge University Press, 2003).

destrier, a mixture of indigenous with Spanish and Arabian stock.[23] From the Mamluk experience, Masson Smith Jr. concludes that although nomads can produce more horses, sedentary people can produce better ones.[24] In the Indian and Chinese cases, indigenous horses of adequate quality were bred in the drier areas of northern and central India, and northern and western China – but the quality of the Indian and Chinese breeds remained critically dependent on regular crossbreeding with Central Eurasian horses.[25] In terms of quantities, Turkish and Mongolian warhorses tended to dominate the market but in southern India there was, especially during the Bahmani sultanate and Vijayanagara (1300-1500), also an important influx of more expensive Arabian and Iranian horses from overseas sources.[26] In this period, Ming China also imported horses by sea – mainly from Southeast Asia but also from Bengal and further west – but the quantities were far from sufficient to make a real impact on the total demand for good warhorses.[27] Moreover, longer-distance

---

[23] R.H.C. Davis, *The Medieval Warhorse: Origin, Development and Redevelopment* (London: Thames and Hudson, 1989); A. Coates, *China Races* (Hong Kong: Oxford University Press, 1994); J.J.L. Gommans, *The Rise of the Indo-Afghan Empire, c. 1710-1780* (Leiden: E.J. Brill, 1995), 68-104; Gommans, *Mughal Warfare*, 111-22; M. Mitterauer, 'Roggen, Reis under Zuckerrohr: Drei Agrarrevolutionen des Mittelalters im Vergleich', *Saeculum* 52 (2001), 245-66; W.G. Clarence-Smith, 'Horse Breeding in Mainland Southeast Asia and its Borderlands', in P. Boomgaard and D. Henley (eds.), *Smallholders and Stockbreeders: Histories of Foodcrop and Livestock Farming in Southeast Asia* (Leiden: KITLV-Press, 2004), 189-211.

[24] For this, see in particular the following contributions by J. Masson Smith: 'Ayn Jalut: Mamluk Success of Mongol Failure?', *Harvard Journal of Asiatic Studies* 44 (1984), 307-46; 'Mongol Society and Military in the Middle East: Antecedents and Adaptations', in Y. Lev (ed.), *War and Society in the Eastern Mediterranean, 7th-15th Centuries* (Leiden: E.J. Brill, 1997), 249-67; 'Nomads on Ponies.'

[25] Mitsutaka Tani, 'A Study on Horse Administration in the Ming Period', *Acta Asiatica* 21 (1971), 73-98; Smith, *Taxing Heaven's Storehouse*, 17-31; Gommans, *The Rise of the Indo-Afghan Empire*, 68-79.

[26] S. Digby, 'The Arab and Gulf Horse in Medieval India', unpublished ms.

[27] R. Ptak, 'Pferde auf See: Ein vergessener Aspekt des maritimen Chinesischen Handels im frühen 15 Jahrhundert', *The Journal of the Social and Economic History of the Orient* 34 (1991), 199-233.

overseas transport meant higher death tolls and prices for horses. Anyway, during both Mughal and Manchu times the overseas importation of horses declined significantly.[28]

The most important difficulty facing sedentary horse-breeders in India and China was the competition with other agrarian activities that supported large populations. In India, for example, the busy agrarian seasons allowed little time for haymaking. In northern Song China, a region of low economic productivity and high population density, peasants tended to chip away at the fringes of the government's grasslands.[29] In the mid-Ming period many of the pastures earmarked for horses were converted into manors and other private estates involving a shift from pasturage to stable-breeding which was accompanied by an increased burden of expenses for fodder – rice- or millet-straw, black- or yellow-beans and other low quality substitute forage – which caused the quality of horses to deteriorate.[30]

In general, the state authorities proved reluctant to stimulate private production as they, for obvious reasons of security, preferred to keep a close eye on both the production and the imports of warhorses. For this reason, the Song and Ming, for example, tended to prefer a policy of self-sufficiency by attempting to produce as many indigenous horses as possible. This policy usually failed, mainly because limited space and bad climate prevented the production of a sufficient quality and quantity of warhorses. During the Ming period, despite territorial control that encompassed the most northern parts of China, the policy of private stock-farming that at first provided the foundation of the dynasty's horse supply was transformed in about a century into a monetary tax used to buy horses from the Mongols. Thus, following the conclusions of Paul J. Smith, ultimately any dynasty that did not possess substantial tracts of steppe land was forced to buy horses from the pastoralists who did.[31]

[28] For the Manchus this was the crux of a presentation by Angela Schottenhammer during the conference 'The International Horse Economy in Iran, India and China' held on 19-20 October 2006, organized by the Institute of Iranian Studies of the Austrian Academy of Sciences at Vienna.

[29] Smith, *Taxing Heaven's Storehouse*, 20-2.

[30] Tani, 'A Study on Horse Administration', 80-2.

[31] Smith, *Taxing Heaven's Storehouse*, 23.

In and along the semi-arid extensions of northern and central India, private, nomadic and semi-nomadic horse-breeders often had more favourable breeding conditions; these included better grazing facilities and more contact with the breeding centres of Central Eurasia, Iran and the Middle East. These mostly Afghan or west-Indian breeders supplied the studs of the political courts, sometimes as revenue or tribute paid in kind but mostly through trade at market prices. Although the Indian governments shared the horse anxieties of their Chinese counterparts, horse-breeding remained closely associated with nomadic and semi-nomadic free grazing and, nonetheless, remained a more durable and far more integrated part of the Indian agrarian economy than in the case of China.

It should be noted, though, that compared to any other part of the world, India and China not only imported but also required far more warhorses – about 25,000-50,000 a year – as both regions encountered a far more immediate nomadic threat. In both cases, there is no doubt whatsoever that the most, and the best, warhorses came from abroad. Even more than breeding, however, the interregional trade in warhorses involved enormous security risks for the settled political authorities. For example, the Mughal emperor Aurangzeb warned his purchasing officers in Kabul to take care that the horse-traders imported their horses without riders. He knew perfectly well that India had a tradition of large and small Afghan horse-traders leading armed caravans eastwards and southwards across India, carving out principalities of their own, or as in the case of the Lodi Afghans, per-haps even creating a true sultanate.[32] In India, horse-traders could easily turn into warlords and warlords easily turn into sultans. This is also shown by the fact that many of the Delhi sultans started their careers as so-called wardens of the marches (*marzban*), i.e. as governors of the north-western border districts, which not only had easy access to the horse-markets of the north-west but also experienced a marked improvement of the horse-stock thanks to the recurrent Mongol incursions of the thirteenth and fourteenth centuries. For the same

[32] S. Digby, *War-Horse and Elephant in the Delhi Sultanate* (Karachi: Oxford Orient Monographs, 1971); Gommans, *The Rise of the Indo-Afghan Empire*; Digby, 'The Arab and Gulf Horse.'

reason, the Indian capital of Delhi itself, in this case not unlike the Chinese capital of Beijing, developed as a kind of frontier town that remained strategically close to these marches.[33]

For the sultans in Delhi, as for the later Mughal emperors, the outside borders of the empire were relatively porous. What they really controlled was not a well-defined external border but, at best, the main urban centres, the agrarian heartlands surrounding and the main routes connecting these centres. All this accounts for the specific Indian pattern of the horse trade: only at times of relatively tight imperial control, horses were bought at border towns by imperial officers but, in general, there always remained a vigorous private market, or actually a string of markets which, following India's two semi-arid extensions, stretched from the far north-west deep into the east and south of the subcontinent, where the seasonable supplies of mostly Afghan and, in the south also, Portuguese horse-traders could meet the combined imperial, regional and local demand.

With respect to the horse trade the Indian case appears to be somewhat similar to the Russian one. For Muscovy the Nogai – a purely nomadic confederacy extending east from the Volga to the Irtush River in Siberia – were an important source of warhorses; Muscovy being the main source of income for the Nogai. In the sixteenth and seventeenth centuries the Nogai horse trade appears to have been strictly controlled by the Russian authorities and took place at a designated place near Moscow or in several Russian towns along the Volga. At this time, the Nogai traders brought as many as 30,000-40,000 horses to the capital annually.[34] Hence, compared to the Indian situation, the Nogai trade appears to be more centrally supervised, based on a more direct, tribute-like, exchange between nomadic breeders and the government. By contrast, in India we see well-functioning market-forces dominated by specialised transfrontiersmen acting as intermediaries between nomadic supply and

[33] Gommans, 'Silent Frontier'; Cf. P. Jackson, *The Delhi Sultanate: A Political and Military History* (Cambridge: Cambridge University Press).

[34] M. Khodarkovsky, *Russia's Steppe Frontier. The Making of Colonial Empire, 1500-1800* (Bloomington-Indianapolis: Indiana University Press, 2002), 26-7; Perdue, *China Marches West*, 79.

sedentary demand. It should be no surprise that these wealthy intermediaries turned out to be far more threatening to the political establishment than the Nogais, giving rise to that enduring Indian rivalry between Afghans and Mughals.

Returning to the Chinese situation, the contrasts are indeed striking. As indicated already with regard to breeding, Chinese governments always attempted to confine the trade in horses to the place where they were most needed, the western and northern frontiers. This trade mainly involved Chinese tea for Mongolian horses. But the Chinese transported the tea from the interior to the borders instead of having the barbarians bring their horses to the interior (mainly Szechwan and Shensi). In this way, the government not only anticipated tremendous security risks but also avoided the expenses of lodging and feeding the barbarians on their trip through the interior. After purchase, the horses were sent directly to the frontier garrisons. Under the Song, horses from as far as Tibet were transported along a belt of relay posts that ran parallel to the border. At the northern frontier, imported mares were transported to the royal pastures or, in Ming times, to the non-governmental studs near Beijing and, sometimes, as far south as Nanking. For reasons of security, the Manchus, who almost entirely depended on imports, declined to procure their warhorses from the Zunghars, their main Mongol rivals, and instead preferred to purchase their horses from the smaller Mongolian tribes immediately beyond the Great Wall. The biggest security problem of the trade was only solved at the second half of the eighteenth century when, following the conquest of modern-day Sinkiang and Mongolia, these tribes were incorporated into the empire.[35] What is really striking in the Chinese case, however, is not only the degree of supervision and command mobilization through endless government agencies and offices but also the rigid demarcation between nomadic supply and sedentary demand along a relatively

[35] The Chinese situation is mainly based on: S. Jagchid and C.R. Bawden. 'Some Notes on the Horse-Policy of the Yüan Dynasty', *Central Asiatic Journal* 10 (1965), 246-69; M. Rossabi, 'The Tea and Horse Trade with Inner Asia during the Ming', *Journal of Asian History* 4 (1970), 136-69; Tani, 'A Study on Horse Administration'; Smith, *Taxing Heaven's Storehouse*; Perdue, *China Marches West*.

well protected border. Most of the breeding and trade of warhorses was concentrated at the very edge of empire, girdling the perimeter of the realm.[36] As a consequence, in China there was much less of a chance that horse-traders would turn into mounted warlords, infiltrate the empire and take power in Beijing from inside.

In sum, comparing the various horse-economies, we first encountered the Central Eurasian, Iranian and Middle Eastern situation where states were faced with a practice of internal, mostly nomadic horse-breeding. Although it made the states of this so-called Saharasia extremely dynamic and powerful, it also made them extremely unstable as it remained very difficult to keep the military power of the horse-breeding tribes at bay. This situation contrasts sharply with that of western and central Europe, where horse-breeding is equally internal but also much more integrated into the sedentary world which allows neither much agency nor political clout to breeders and traders. Again different, we came across Russia, India and China, all of which imported huge numbers of warhorses from Central Eurasia and, in the case of India, to a lesser extent, from Iran and the Middle East. Despite this resemblance, the differences stand out more clearly. China confined its horse-economy entirely to the frontier, Russia closely supervised supplies into the interior, India, finally, allowed a great deal of leeway to commercial intermediaries.

EURASIAN HORSE-INSTITUTIONS: MUGHAL INDIA AND MANCHU CHINA

Up until the eighteenth century, the Mughals in India and the Qing Manchus in China stand out as the two most important superpowers of their day. Much of this was based on their successful exploitation of both nomadic and sedentary resources, both of which had considerably increased during the previous centuries. From this perspective, both empires improved on their predecessors and, as such, they mark the highpoint of a long-lasting medieval tradition of frontier exploitation. Although both used gunpowder from their inception, these empires remained strongly dedicated to and dependent

---

[36] Smith, *Taxing Heaven's Storehouse*, 291.

on horses. Hence, in my view, both should not be considered as early-modern gunpowder empires but rather as very successful late-medieval frontier empires. In fact, both empires were founded by conquerors, who, although no nomads themselves, boasted a strong nomadic background and legacy that went back as far as Chinggis Khan. As so-called bicultural transfrontiersmen, i.e. settled warriors who remained committed to camp and horse, both the Mughals and the Manchus turned out to be extremely well qualified to rule predominantly sedentary societies without having to leave their cherished saddles. Horses were the crucial linchpin in a not too complicated tripartite relationship, well summarized by the early Manchu chief Hong Taiji: 'With booty from China and goods bought from Korea we buy horses from the Mongols and set out against China.'[37] Of course, to really administer their empires both dynasties had to become deeply entangled in all sorts of regional and local affairs. This not only tended to compromise their imperial authority but also tended to undermine their military superiority as based on extra-regional mobility and ongoing access to foreign horsepower. Consequently, there always was this Ibn Khaldunian fear of acculturation, softening and decline, most urgently felt in the most settled of societies in India and China. It made Mughals and Manchus alike repeatedly yearn for those early days of simple life and wandering adventure back 'home' in that unspoiled Central Eurasia, be it Turkistan in the case of the Mughals or Manchuria in the case of the Manchus.[38]

Apart from such post-nomadic nostalgia, in order to survive, both dynasties invented a whole range of pseudo-nomadic institutions, traditions and ritual that derived from their rich nomadic past in Central Eurasia. As suggested already by the late Joseph Fletcher, the

[37] G. Roth Li, 'State Building Before in 1644', in W.J. Peterson (ed.), *The Cambridge History of China*, vol. 9, part 1: *The Ch'ing Empire to 1800* (Cambridge: Cambridge University Press, 2002), 70.

[38] M.C. Elliott, *The Manchu Way: The Eight Banners and Ethnic Identity in Late Imperial China* (Stanford: Stanford University Press, 2001); Gommans, *Mughal Warfare*, 202. It goes without saying that the lifestyle of the early Mughals in Turkistan was not at all unspoiled or nomadic but already rather urban and post-nomadic.

Turks and Mongols – and here one could equally well read Mughals and Manchus – transplanted heir (imagined) nomadic traditions at various times in each of the agrarian societies, and in one form or another this legacy from the steppe persisted in each of them for an extended duration.[39] This was primarily done in recognition of the sheer power of nomadism, i.e. first and foremost its power to produce and employ large quantities of the most excellent warhorses. However, during the process of transplantation the nomadic traditions were thoroughly adjusted to the specific sedentary circumstances in India and China in order to make these more pliable and less disturbing instruments of state policy.[40]

Here I would like to discuss three such post-nomadic institutions that were highly significant as they were specifically geared to mobilize, organize and remunerate an army that remained based on the employment of massive numbers of Central Eurasian horse-warriors. It is my basic contention that it is really hard to imagine these institutions without the ongoing need to use Central Eurasian warhorses.

[39] J. Fletcher, 'Turco-Mongolian Monarchic Tradition in the Ottoman Empire', *Harvard Ukrainian Studies* 3-4 (1979-80), 242. See also J.A. Millward, 'The Qing Formation, the Mongol Legacy, and the "End of History" in Early Modern Central Eurasia', in L.A. Struve (ed.), *The Qing Formation in World Historical Time* (Cambridge Mass.-London: Harvard University Asia Center, 2004), 101, 110.

[40] Wink, 'India and the Turko-Mingol Frontier'; J. Waley-Cohen, 'Ritual and the Qing Empire', in N. Di Cosmo (ed.), *Warfare in Inner Asian History (500-1800)* (Leiden: E.J. Brill, 2002), 405-45; J. Waley-Cohen, 'Changing Spaces of Empire in Eighteenth-Century Qing China', in N. di Cosmo and D.J. Wyatt (eds), *Political Frontiers, Ethnic Boundaries, and Human Geographies in Chinese History* (London: Routledge Curzon, 2003), 324-50. To a lesser extent, this is even relevant for seventeenth and eighteenth-century Europe (think about the Huzars and other pseudo-tribal contingents) and Southeast Asia, regions that were far less conducive to the breeding and employment of strong warhorses in great numbers but also seem to have experienced a marked revival of pseudo-nomadic, equine rituals and apparel both in the armies and at the courts. For Southeast Asia, see J.J.L. Gommans, 'Trade and Civilization around the Bay of Bengal, *c.* 1650-1800', *Itinerario* 19 (1995), 82-108 and W.G. Clarence-Smith, 'Elephants, Horses, and the Coming of Islam to Northern Sumatra', *Indonesia and the Malay World* 32 (2004), 271-85.

The three institutions I have in mind are: (1) the moving camp or *ordo*, (2) the personal following or *nökör* of the leader separated from the bureaucracy, (3) the principle of sharing-out (*yurt*) as based on the equal distribution of spoils. The terms *ordo*, *nökör* and *yurt* are used since they refer to the original Eurasian background of these institutions and, as such, were recognizable to the Mughals, to the Manchus or to any other post-nomadic dynasty that claimed a nomadic tradition, be it true or invented. As much as present-day constitutions, parliaments or elections are part of the paraphernalia of the modern nation state based on the European model, in our period, *ordo*, *nökör* and *yurt* were part of the basic accoutrement of medieval frontier empires as based on the Central Eurasian paradigm. This does not mean, of course, that we cannot find elements of these three institutions in other parts of the world. For example, most of the following may remind one of such, much older, notions of feudalism, patrimonialism or sultanism but we tend to forget that such more general, often Weberian categories, have a concrete historic background which is different from modern ideal-types of sovereign states as based on the western experience. In our period, it was neither Europe, nor China, nor India but Central Eurasia that provided the most powerful model of successful state-formation. As such, the three institutions *ordo*, *nökör* and *yurt* were part of a post-nomadic repertoire that every ambitious new ruler had to adopt both to gain status and in order to survive. The following discussion will focus on these three institutions and discuss the ways in which this common Central Eurasian legacy necessitated different adaptations in accordance with the specific (mainly ecological) conditions in Mughal India and Manchu China, all three in the context of the persistent military significance of the warhorse.

ORDO

The pastoral economy of the nomad required the permanent movement of his camp (*ordo*) that included his extended household and his cattle. In military terms, his very mobility was the nomad's most vital strategic and tactical asset in his confrontations with sedentary armies. To maintain this advantage, nomadic conquerors of sedentary worlds, also after the conquest, tended to continue their itinerant ways of life often provoking grumbling of their new, sedentary advisors. Of course, they knew that a landscape of grain or paddy fields was not conducive for cavalry marches. Actually, it made the early Mongols consider radical plans to depopulate north China and convert its agricultural lands into pasture.[41] For the same reasons, their Khitan predecessors had already banned irrigating the land surrounding the capital. Permission was finally given, but not along the routes used by the army.[42] But whereas in the steppes and deserts of Central Asia and Iran, the degree and the direction of movement were indeed determined by the availability of grazing, in the sedentary surroundings of India and China, movement hinged on a sophisticated logistical network consisting of numerous itinerant transporters, bankers and merchants providing cash and fodder to the ever-moving camp. Not surprisingly also, the Mughal and Manchu rulers had a very strong awareness that travelling across their realm with the entire court and army was somehow crucial to their survival.[43]

Interestingly, the emperors who appreciated the need of the camp-cum-court in *perpetuum mobile* most were two contemporaries: Aurangzeb (r. 1658-1707) and the Kangxi emperor (r. 1662-1722). Both of them envisaged the danger of an empire rusting away by the increasing lethargy of their *mansabdari* and banner nobility; both of them personally commanded long-distance campaigns, chasing their political rivals (Shivaji and Galdan respectively); both of them clearly enjoyed the austere life in tents far away from the capital. Aurangzeb's restless behaviour was in line with at least some of the rhetoric in Indo-Islamic chronicles in which the king is recurrently warned that he 'ought not to keep his tent pitched at one place for two days even with a view to enjoying ease and comfort'. It was also clearly propagated in one of Aurangzeb's own counsels for kings:

---

[41] Th.T. Allsen, 'The Rise of the Mongolian Empire and Mongolian Rule in North China', in H. Franke and D. Twitchett (eds), *The Cambridge History of China*, vol. 6: *Alien Regimes and Border States, 907-1368* (Cambridge: Cambridge University Press, 1994), 376.

[42] D. Twitchet and K.P. Tietze, 'The Liao', in Franke and Twitchett (eds), *Alien Regimes*, 95.

[43] Gommans, *Mughal Warfare*, 100-11; Perdue, *China Marches West*, 409-29.

Next this, an emperor should never allow himself to be fond of ease and inclined to retirement, because the most fatal cause of the decline of kingdoms and the destruction of royal power is this undesirable habit. Always be moving about, as much as possible (*verse*): 'It is bad for both emperors and water to remain at the same place; the water grows putrid and the king's power slips out of his control. In touring lie the honour, ease, and splendour of kings; the desire of comfort and happiness makes them untrustworthy.'

Aurangzeb's perception that political fitness required movement was based on the experience his great-grandfather Akbar who had been able 'to capture the country by means of travelling through it'. The official chronicle depicts an ever-marching Akbar who was known to be fond of 'travelling and hunting', and who does never 'fix his heart to one place', gathering 'new affluence from every quarter'.[44] But Aurangzeb's advice that kings should never rest was also used to charge his father with political negligence as he chose to stay at Delhi and Agra, instead of being constantly on the march – a verdict that a later Manchu ruler, the Qianlong emperor (r. 1736-95), would certainly have recognized as he also was inspired by the politics of his grandfather, the Kangxi emperor (r. 1662-1722), in reversing the too stationary policy of his father, the Yongzheng emperor (r. 1723-35). In general, one may say that despite the advise of various Indian and Chinese ministers that kings should not leave the capital for too long, both Mughals and Manchus held a strong belief, based on long experience, that ongoing movement was not only vital for their own survival but also healthy for the body politic in general.

From my own calculations it appears that the four Mughal emperors Akbar (r. 1556-1605), Jahangir (r. 1605-27), Shah Jahan (r. 1628-57) and Aurangzeb (r. 1658-1707) were for about 65 per cent of their reign sedentary, i.e. staying for more than six months in one place, short trips not being counted. For about 35 per cent of their rule, they were migratory, be it on tour, on military campaigns or long hunting expeditions. Although based on different criteria and being extremely approximate, it strikes one that these calculations come very close to the Iranian figures for Shah Abbas I (r. 1587-1629). The Safavid king was travelling about one-third of his reign, resident

[44] Gommans, *Mughal Warfare*, 100-11.

in a capital for one-third, and static in other locations for the rest. The Safavid and Mughal figures are certainly much higher than in the case of the Ottoman rulers who stayed for much longer periods in the capital.[45]

What about Manchu movement? Apart from the two founding fathers Nurhaci (1559-1626) and Hong Taiji (1592-1643), also the later emperors also spent a considerable time moving from place to place. As indicated already, this was particularly true for the Kangxi and Qianlong emperors. The latter, for example, went on about 150 trips, and was on the road an average of three to four month a year, which comes close to the Mughal and Safavid figures. As in the Mughal case, the Manchu tours were a constant re-enactment of a narrative of nomadic conquest and military superiority. As such, it was very much part of the Central Eurasian heritage: to rule by personal presence of the man who marked his domains on horseback.[46] Apart from this, though, marching also implied surveillance, required road-building and enabled the emperor to keep in physical touch with his many regional and local co-sharers of the realm and to forge bonds of loyalty with them. With all the dust, smell and clamour surrounding it, the moving imperial camp was a permanent reminder of Mughal and Manchu sovereignty and, as such, a constant threat to any obstreperous governor or landlord considering disobedience or revolt. Since the camp showed the imperial grandeur on permanent display all over the empire, actual fighting could often be avoided. The pending arrival of the court was usually more than enough to bring people into submission.

Apart from these administrative advantages, another fairly obvious reason for keeping the court moving was to keep the cavalry army in permanent mobilization and readiness. In all the imperial tours, both Mughals and Manchus constantly evoked the theme of military discipline. Both insisted that their people ride on horseback, not on

[45] C. Melville, 'From Qars to Qandahar: The Itineraries of Shah Abbas I (995-1038/1587-1629)', in J. Calmard (ed.), *Études Safavides* (Paris Teheran: Institut Français de Recherche en Iran, 1993), 195-225; Gommans, *Mughal Warfare*, 101-3.

[46] Waley-Cohen, 'Changing Spaces of Empire'; Perdue, *China Marches West*, 409-29.

sedan chairs or palanquins, and both conducted riding and shooting exercises to test their skills. Both knew perfectly well that as much as political survival hinged on permanent movement, permanent movement hinged on good horsemanship. But although Mughals and Manchus shared the ideal of the *ordo*, it is also true that – despite the Southern Tours of the Kangxi and Qianlong emperors – the Manchus tended to move mainly along and across their northern borders whereas the Mughals tended to be more omnipresent throughout the entire realm.

The military and the administrative objectives of imperial mobility converged most visibly in the practice of hunting.[47] Hunts could be organized when the court was temporary settled in the capital but also when the army was on the move. Often we find, the Mughal emperor hunting or hawking 'as he went', 'on the way' or 'along the road'. In all these cases, the bulk of the imperial troops was left behind and continued its own route in its own pace. The hunting party accompanying the emperor usually involved a few personal retainers and never exceeded the number of 1,000 horse and foot. At every stage of the hunt, special friends could be selected to join the emperor. Away from the tight ritual of the court, the excitement of hunting created a more convivial atmosphere, reminiscent of the erstwhile warband of the Central Eurasian founding fathers, in which new relations of friendship and loyalty could be forged.

But apart from being an enjoyable pastime, hunting was an essential instrument of government. Under the veil of hunting, the Mughals both rallied and suppressed the enormous military potential of the country surrounding the imperial hunting grounds. Hunting expeditions were often organized to inconspicuously mobilize troops or inspect the contingents of the nobility. For example, Akbar's campaign against Chitor in 1567 started with a hunt in the Bari district, 'in order that the loyal and devoted leaders might come without the notoriety of being sent for', while others, either their servants or not, would, seeing that there was no prohibition, readily assemble in order to pay their respects. Similar to what we have seen

---

[47] Gommans, *Mughal Warfare*, 110-11; Th.T. Allsen, *The Royal Hunt in Eurasian History* (Philadelphia: University of Pennsylvania Press, 2006).

for marching in general, hunts were also useful for making surveys, for gathering local intelligence as well as for subjecting unruly areas, which, at least in the Mughal case, could be very near to the imperial centre; be it capital or hunting ground. Even more important, though, hunting expeditions were crucial for practising cavalry manoeuvres. Most prominent in this respect was the so-called *qamarghah*. From an area that could be as large as about 100 km in circumference all the wild game was driven together into a small, enclosed circle where the king started the actual hunt. Since the rounding up of wild animals involved similar manoeuvres as on the battlefield, i.e. skirmishers in front of the centre with two flanks wheeling around the enemy, hunting served as a most useful training camp.

Similar reasons made the Manchus relish the hunt. In 1684 the Kangxi emperor ordered the garrison generals to organize local hunts since 'if the officers and soldiers at the provincial garrisons are not made every year to go hunting to practice their martial skills, they will eventually become lazy'. Later, under the Qianlong emperor, hunting became an important ingredient of the 'Manchu way' as only the hunt could guarantee the maintenance of the art of shooting from horseback, so essential to the Manchu military repertoire. But in contrast to the Mughal hunting expeditions, which were conducted in the various hunting reserves throughout the empire, most Manchu hunts were organized in the laid-out hunting grounds beyond the Great Wall.[48]

NÖKÖR

Nökör is the second nomadic concept that was thoroughly transformed but remained in full operation after the nomadic conquest of sedentary lands. *Nökör* was the Mongolian term for the personal following or *comitatus* of the ruler. As such it was a means to create an institutional framework to move beyond the ephemeral tribal polities so characteristic of the political history of Central Eurasia. Hence this personal elite was supposed to overrule the tribal structure by creating a more selective, more meritocratic, artificial 'tribe' that was fully attached

---

[48] Elliott, *The Manchu Way*, 182-91.

to and often named after the new ruler who became the focus of obedience and allegiance.[49]

This supra-tribal principle of military service and loyalty remained the basic building block of the military organization after conquest but in order to rule an extensive sedentary realm also had to be tremendously expanded. In my view, in all their variety, both the *mamluk* system of the Islamic world, the *tümen* of the Timurids, the *mansabdari* system under the Mughals, even the *angaraksa* system of southern India,[50] and the *banner*-system of China, may all be considered as manifold adaptations of the nomadic concept of *nökör*. Like *nökör*, these sedentary inventions aimed at the incorporation of the existing tribal and ethnic leadership, in command of most of the existing horsepower, into non-tribal, or more appropriately, *new* tribal and ethnic organisations based on the loyal service to the emperor and as linked, in various degrees, to his personal bodyguard. But apart from being an instrument of incorporation through imperial service, *nökör* was also a means of exclusion since it tended to articulate a new imperial hierarchy based on personal loyalties and imperial service, often underscored by new ethnic and social categories that were to prevent the subversion of the imperial *esprit de corps* by random assimilation. This created a new elite of post-nomads, or surrogate nomads, as Joseph Fletcher has labelled them:[51] they were primarily recruited from the agrarian population but as cosmopolitan, often highly mobile men in service to the emperor also emphatically detached from it.

One proven way to keep the military from striking local and

[49] J. Nemeth, 'Wanderungen des Mongolischen Wortes *Nökür* "Genosse",' *Acta Orientalia* 3 (1953), 1-24; P.B. Golden, '"I Will Give the People unto Thee": The Cinggisid Conquests and Their Aftermath in the Turkic World', *Journal of the Royal Asiatic Society, 3rd Series* 10 (2000), 21-41; N. Berend, *At the Gate of Christendom: Jews, Muslims and 'Pagans' in Medieval Hungary, c. 1000-c. 1300* (Cambridge: Cambridge University Press, 2001) 145-6. Cf. D.H.A. Kolff, *Naukar, Rajput and Sepoy: The Ethnohistory of the Military Labour Market in Hindustan, 1450-1850* (Cambridge: Cambridge University Press, 1990).

[50] C. Talbot, *Precolonial India in Practice: Society, Region, and Identity in Medieval Andhra* (Delhi: Oxford University Press, 2001), 165.

[51] Fletcher, 'Turco-Mongolian Monarchic Tradition', 243.

regional roots was by keeping its organization apart from the sedentary bureaucracy. Hence, both Mughals and Manchus attempted to maintain a dual civil-military structure, which (whether or not it originated in ancient China), became a tremendous success-story all across those parts of Eurasia conquered by horse-based nomads.[52] In my view, this policy of military apartheid was not instigated by some 'modern' urgency to distinguish between ever more professional armies and ever more rational bureaucracies but by the need of post-nomadic rulers to keep the cavalry core of the army as loyal, fit and ready as possible. In this sense, *nökör* and the dual civil-military administration are two sides of the same coin.

To further prevent the sinification of their *nökör* the Manchus organized them into artificially constructed tribal groups or banners, still based on the ethnic distinction of Manchu, Mongol, and Chinese, but within these three categories, in new, mixed compositions. Consequently, the banner leaders were turned into the new, pseudo-tribal companions of the emperor. They also opted for a policy of rigid segregation in which the banners were quartered in a relatively isolated, strategically well-situated network of garrison towns. Although the Mughals would certainly have recognized the danger of the assimilation of their armed retainers, the more open and fluid frontier circumstances of the Indian subcontinent made segregation nonviable. Instead, to avoid the indianization of their rank-holding elites or *mansabdars*, the Mughals attempted to keep them in permanent circulation – alternating both offices and revenue-lands (*jagirs*).[53] But in the longer term, neither segregation nor circulation could prevent military elites in China and India from sinking roots into the indigenous society, or a regionally rooted gentry from making their way into the imperial ranks. Hence, after about three or four generations of Mughal and Manchu rule, we come across emperors – Aurangzeb and Qianlong – who attempted in a kind of rescue mission

[52] H. Franke and D. Twitchett, 'Introduction', in Franke and Twitchett (eds), *Alien Regimes*, 21-3; D. Ostrowski, *Muscovy and the Mongols: Cross-Cultural Influences on the Steppe Frontier, 1304-1589* (Cambridge: Cambridge University Press, 1998), 36-44.

[53] For Manchu apartheid and Mughal circulation, see resp. Elliott, *The Manchu Way* and Gommans, *Mughal Warfare*.

to counter the bloating of their *mansabdari* and banner systems.[54] In both cases they stressed moral rearmament and a return to the fundamental principles – the Islamic and Manchu ways respectively – that had always propped up their rule.

At first sight, *nökör* has nothing to do with the warhorse. Indeed, it is like the European *Gefolgschaft* or Ibn Khaldun's inner circle (*bitana*) of *mawali*.[55] In my view, though, the idea served best in the Central Eurasian context and under the Mongols gained a new lease of life.[56] Here the function of the chief's bodyguard (*keshig*) was to look after the horses of the Great Khan.[57] Hence, the selection of the Great Khan's companions was primarily based on good horsemanship. Not surprisingly, the Jurchen banners originated from war and hunting contingents (*niru*), which only later were turned into the more rigid ethnic garrison armies of the Manchus.[58] Mughal *mansabdari* rank was based on the quantity and quality of horses to be kept in service. Apart from these considerations, in both Manchu and Mughal cases, Central Eurasian horse warriors – Turks, Mongols or Manchus – received the highest salaries. This may have been based on ethnic proclivities but it was also well informed by actual performance on the battlefield. Finally and most importantly, the need for permanent grazing and exercise not only made the nomadic warrior but more in particular his warhorse extremely vulnerable to the debilitating effects of sedentarization. In this light the dual civil-military structure was essential to maintain at (literally) its grassroots the readiness of the cavalry and to keep it as much as possible detached from the highly sedentary Indian and Chinese bureaucracies.

YURT

To establish military superiority with the help of nomadic cavalry was one thing; but paying for it after the conquest was another. The

[54] Elliott, *The Manchu Way*, 306.

[55] D. Ayalon, 'Mamlukiyyat', *Jerusalem Studies in Arabic and Islam* 2 (1980), 340-50.

[56] Golden, 'I Will Give', 32.

[57] Jagchid and Bawden 'Some Notes', 247.

[58] Elliott, *The Manchu Way*, 56-63.

classical nomadic way – living off the country by raiding and distributing the spoils of war – was no longer a viable option. Besides, compared with the relatively cheap methods of the nomads, to maintain horses in the sedentary world was extremely expensive. Apart from the problems of climate and lack of grazing opportunities, it is also very labour intensive. Given the limited availability of both pasturage and fodder, sustaining large numbers of horses required that horses be distributed equally, not only among the conquerors, but across all the available pastures. Especially in the absence of a developed central economic-bureaucratic infrastructure the best option was to plug the horsemen directly into revenue sources in the countryside.[59] Hence the obvious solution had been to share out the available pastures as so-called *yurts* among the tribal companions. Similar to pastures and booty, territories and conquered peoples were also distributed among their retainers and followers, also often giving rise to extensive census operations after the initial conquest. Under the Mongols, the system could involve huge territories that were made personal appanages of the Great Khan's relatives. In the first half of the thirteenth century, under Ögödei and Möngke, the system was implemented in both China and Iran.[60] Since the grantees' own dependents played a prominent part in the administration of these shares, the sharing-out principle gradually trickled down toward the regional and local level.

In the Middle East the Mongol conquerors and their dependents were soon faced with the more or less well-established revenue system based on the so-called *iqta* which had emerged under the Buyids in the eleventh century. It implied that every holder of an *iqta* had the right to collect the revenue of an assigned piece of land in exchange for certain military or administrative services. As such, it involved, at least in theory, conditional and temporary rights to accurately assessed revenue proceeds.[61] Although having two entirely different

[59] A. Gat, *War in Human Civilization* (Oxford: Oxford University Press, 2006), 323-401.

[60] Th.T. Allsen, 'Sharing out the Empire: Apportioned Lands under the Mongols', in A.M. Khazanov and A. Wink (eds), *Nomads in the Sedentary World*, Richmond (Surrey: Curzon, 2001), 172-91.

[61] C. Cahen, 'L'évolution de l'iqta du IXe au XIIIe siècle: Contribution à une

antecedents, what we see from the thirteenth century onward is the gradual application of the idea of *yurt* as based not on the equal distribution of spoils but on *iqta*. In due course, *iqta* immensely facilitated the accommodation of the huge nomadic cavalry armies into the sedentary worlds of Iran and India. In all its later guises – as *pomest'ia* under the Russians,[62] as *timar* under the Ottomans, as *tiyul* under the Safavids, as *jagir* under the Mughals, and even as *nayankaramu* in Vijayanagara[63] – it was meant to support a certain number of warhorses with the centrally calculated revenue of a particular district. Hence *iqta* was the logistical and financial solution of the frontier empires to the problem of maintaining a strong cavalry army in a sedentary world. It also proved a most convenient instrument to peacefully accommodate the necessary influx of new nomadic horsepower.

Considering the breathtaking spread of *iqta*, one may wonder why such a successful institution did not strike root in China, a sedentary realm that was several times conquered by Central Eurasians. One obvious reason for this is China's sharp ecological frontier. As we discussed already with regard to the different horse-economies, immediately south of the Great Wall wild steppes gave way to fields of intensive agriculture that left no room whatsoever for supporting large numbers of warhorses. As a consequence, the Central Eurasian conquerors of China experimented with sharing-out but it was never combined with a revenue system that was geared to maintain cavalry units inside the realm. On the contrary, China under the Ming strongly disapproved of a militarized society based on the sharing-out principle and, instead favoured a centrally paid army under civilian control.

It is only in the most northern, more arid parts that we come across fiscal practices that seem somewhat similar to *iqta* but, on closer

---

histoire comparée des sociétés médiévales', *Annales: Économioes, Sociétés, Civilisations* 8 (1953), 25-52.

[62] Ostrowski, *Muscovy and the Mongols*, 48-9.

[63] Ph.B. Wagoner, 'Harihara, Bukka and the Sultan: The Delhi Sultanate in the Political Imagination of Vijayanagara', in D. Gilmartin and B.B. Lawrence (eds.), *Beyond Turk and Hindu: Rethinking Religious Identities in Islamicate South Asia* (Gainsville Fl.: University Press of Florida, 2000), 300-26.

inspection, was rather different. In this case, each soldier of the banner companies was to be assigned with a tract of land to be farmed by his household or his serfs. As such, it was not much different from the old policy to establish self-supporting military colonies along the border. But these soldiers did not own revenue rights. The relative success of the system in northern China under the Jin and Yuan explains why the Manchus initially wanted to implement a similar system of banner landholding around the capital of Beijing. In the first years of the dynasty, each male on the banner rolls arriving in Beijing received a tract of land depending on rank. Banner land was tax-exempt and was supposed to provide a permanent, inalienable source of income. The farming was left to Chinese tenants or serfs, the management to Chinese overseers.[64] Simultaneously, the Manchus started to experiment with stationing their troops among the Han-Chinese, to share houses, land and food.[65] After only twenty years this policy of integration and decentralized payments broke down, however, and, as mentioned already, bannermen were entirely segregated from Han society to become increasingly dependent on the court's steady supply of grain and cash. Maybe the nomadic idea of sharing-out continued under the Manchus when territories were granted to the Han generals who had supported them but Kangxi's repression of the Three Feudatories marked the final rejection of the military service state and the re-establishment of centralized power.[66]

Despite the common Central Eurasian mindset of the *yurt*, the differences in its adaptation and implementation between Mughal India and Manchu China stand out more clearly than in the case of the *ordo* and the *nökör*. The Mughal case appears to be much closer to the Safavid and Ottoman methods of military remuneration and stands somewhere midway between European feudalism – according to Azar Gat the most 'primitive' option with the least administrative supervision[67] – and the bureaucratic management under the Manchus. In this regard, both Mughals and Manchus created not something out of the blue but merely improved on the already proven ways of

---

[64] Elliott, *The Manchu Way*, 193.
[65] Elliott, *The Manchu Way*; Roth Li, 'State Building', 46-8.
[66] Perdue, *China Marches West*, 560.
[67] Gat, *War*, 343.

their predecessors to cooperate closely with a highly professional bureaucratic organization.

CONCLUSION

This unabashedly sweeping essay is to remind historians that as late as the eighteenth century the warhorse played a central role in the organization of frontier empires bordering on the Central Eurasian heartlands of nomadic power. The survival of the (semi-)nomadic conquerors after the conquest of sedentary societies continued to hinge on the production, trade and use of Central Eurasian warhorses. This helped to forestall the full sedentarization of these conquerors and conditioned the emergence of a post-nomadic political culture and organization in which Central Eurasian institutions like *ordo*, *nökör* and *yurt* continued to provide a forceful paradigm to mobilize, organize and enumerate cavalry armies. But as the specific ecological conditions supported different horse-economies, they also gave rise to different adaptations of the nomadic paradigm. This is demonstrated in the case of the Mughals in India and the Manchus in China. Both shared a common Central Eurasian heritage and both ruled the richest sedentary economies of their time. Nonetheless, considering the varying ecological circumstances – basically penetrating *inner* frontiers in India, circumventing *outer* frontiers in China – Mughals and Manchus had to seek different solutions to the post-nomadic predicament they had so much in common.[68] As a prelude to what should be a much more detailed comparative investigation, this essay merely aims to stress the basic ecological context of the warhorse as the *sine qua non* of Mughal and Manchu empire building.

---

[68] The same ecological circumstances may have contributed to the more monolithic stature of China against the far more fragmented image of India which apparently finds its one and only unity in its diversity. Although revisionist historiography tends to play down the centralizing capacities of both empires the differences between the two remain striking.

CHAPTER 5

The Embarrassment of Political Violence in Europe and South Asia, *c.* 1100-1800

THE SULTAN'S ANXIETY

THE VENETIAN TRAVELLER Nicolao Manucci (1653-1708) accounts the following anecdote of Abul Hasan (1672-87), the last ruler of Golkonda. In 1672 this young prince unexpectedly found himself installed as the new sultan of Golkonda. Feeling overwhelmed by this heavy task and failing the necessary political experience he decided to consult the factor of the Dutch East India Company in Golkonda about European statecraft and warfare. He was particularly interested whether there existed any horses in Europe since he had heard that the Europeans fought only at sea. The Dutchman answered that there were not only horses in his home country but also valiant warriors, who could even fight without their aid. The sultan, however, wanted some proof and requested a picture representing a European battle with cavalry. Some time later, the Dutch governor of Nagapatnam presented the sultan with a large painting in which a battle was depicted, fought by the Dutch against the Spaniards. It showed squadrons of cavalry and infantry contending most fiercely with one another, also various bloody and dying figures and objects. The sultan was truly impressed but after one night already, he told the Dutch to carry away the painting since all night long he had been unable to sleep merely by thinking of it. According to Manucci, these fears were typical for those who have never been in battle and have no experience of the world and it was from this cause that the sultan was to lose his throne to the Mughals only fourteen years later.[1]

Reading Manucci's story one might wonder whether the Sultan's behaviour was only a consequence of his supposed meekness or that their where other motives involved. Obviously, European artists held no monopoly on the depiction of gory violence. On the contrary, Indian paintings and sculpture display countless scenes full of conspicuous violence with all the conventional acts of killing, dismembering, disembowelling and other forms of gruesome destruction. Nevertheless, it remains to be seen to what extent these scenes really represent the actual state of affairs in the here and now of this world. Not used to the violent effects of real warfare, one may speculate that the sultan might have preferred the more figurative and symbolic images of Indian pictures to the more realistic and direct representation in the European painting. It is very well possible that the sultan had no problem with excessive violence as such but that he merely shrunk to see it implemented on the actual battlefield, as implied by the Dutch painting. This would suggest that the sultan and the Dutch were at odds with each other about the proper arena in which the use of excessive violence was permitted. In other words, it appears they had different feelings of embarrassment regarding political violence. Therefore, one way to understand the sultan's sudden anxiety is to compare the historical development of political violence in Europe with that in South Asia.[2] In this essay we will broadly follow this line of thinking by taking Norbert Elias' famous 'civilizing process' as a point of departure. From this we will explore how European polities came to terms with all kinds of erratic violence in

[1] N. Manucci, *Storia do Mogor*, trans. W. Irvine, 4 vols (New Delhi: Oriental Reprints, 1981), 126.

[2] The periodization of South Asian history is highly problematic. Here the term 'medieval' loosely refers to its conventional usage marking the period between AD 500 and 1500. For the present purpose it also includes the early-modern phase which usually involves the sixteenth to the eighteenth century. By doing this for mere convenience, I claim no parallel whatsoever with European periodization.

the public sphere. Turning towards South Asia, as an experiment we will follow Elias' controversial Freudian approach by paying attention to Sudhir Kakar's well-known but equally controversial description of the 'Indian' way of raising children to maturity and the way this involves different attitudes towards man's so-called inner drives of violence. After highlighting the most eye-catching differences with Europe, it will be attempted to relate the micro-level to the macro process of state formation in South Asia from about 1100 to 1800.[3] Or to put it briefly, in order to understand the sultan's anxiety the present essay will broadly compare European and South Asian civilizing processes of political violence.

CIVILIZING THE WARRIOR IN EUROPE:
FROM FREUD TO ELIAS

In Europe, the scholarly debate on the role of violence in public society is still largely informed by Norbert Elias' seminal work *Über den Prozess der Zivilisation*.[4] This civilizing process involves man's increasing tendency to restraint his or her affects (*Affectbeherrschung*). As such Elias takes resource to Freud's notion of the repression of instincts (*Triebverzicht*). Freud argued that growing social pressure created its own watchman in the psychic structure of individuals: the superego (*Über-Ich*). In this context, Elias felt that the civilizing process ran parallel to the development of any individual from un-restrained childhood to maturity. He gave this general idea a more specific historical and sociological turn by presenting the European absolutist state and the demands of its royal court as being the prime

---

[3] By using terms like Europe and South Asia it is not my intention to claim uniform historical developments or to deny considerable diversities in terms of sub-regions or religions. Europe mostly refers to Latin Europe which originated in the north-west but gradually spread towards the rest of the continent except for its Greek and Russian east. Although the term South Asia refers to the Indian subcontinent at large, in terms of values, it mainly involves Hindu traditions. Nevertheless, the main thrust of the essay is also relevant for the Indo-Islamic experience.

[4] N. Elias, *Über den Prozess der Zivilisation: Soziogenetische Untersuchungen*, 2 vols (Frankfurt am Main: Suhrkamp Verlag, 1991).

mover of this process. From the *ancien régime* courts the new style trickled down on society at large, and eventually, it even spread from western Europe to other parts of the world. The first victims of this taming process were the once joyfully aggressive but now 'uncivilized' knights of the bygone Middle Ages. Gradually they were disarmed and coerced into courtly behaviour in which sudden outbursts of violence were not tolerated any longer. In due course, they internalized the imposed style of comportment and even felt it to be a new achievement. Thus, although the polished code of behaviour was initially forced upon them, it quickly became part and parcel of their new noble identity. In the end, Elias claims, the social constraints of court life (being *Fremdzwang*) had turned into the courtier's self constraint (*Selbstzwang*). In other words, it appeared as if the warrior-turned-courtiers had shifted their former battlefields into their inner selves. Similar to what happened in the upbringing of children, there was a general transition from conduct motivated by fear of punishment to that based on a sense of shame or guilt. For Elias, this 'courtization of the warriors' (*Verhöflichung der Krieger*) was mainly brought about by intensifying state formation and of the increasing interdependence of society at large.[5]

On several scores the civilizing process of Elias has been heavily criticized. For example, following Johan Huizinga, he appears to have exaggerated the 'uncivilized' nature of the early Middle Ages.[6] Proceeding from his preoccupation with the Freudian *Fremdzwang-Selbstzwang* mechanism Elias also appeared to have overrated the power of the royal court and hence the victimization of the nobility. Not surprisingly, the restraint and fine-tuning of affects was soon to be found in numerous, small-scale, rather 'primitive' societies without strong political structures and without a large amount of inter-dependence. In addition, it was observed that internalization of emotions does not necessarily result in less violence. On the contrary,

[5] J. Duindam, *Myth of Power: Norbert Elias and the Early Modern European Court* (Amsterdam: Amsterdam University Press, 1995), 159-80.

[6] J. Huizinga, *Herfstij der middeleeuwen: Studie over levens- en gedachtenvormen der veertiende en vijftiende eeuw in Frankrijk en de Nederlanden* (Haarlem: H.D. Tjeenk Willink and Zoon, 1947).

the use of violence can often be highly controlled and rational. Finally, Elias, not succeeding in dissociating the particular European developments from the normative connotations of the term 'civilization', was attacked for being ethnocentric.[7]

Nevertheless, I feel that for the present comparative purpose Elias made some observations which are still instructive. Firstly, he was right by pointing out the changing attitude towards the intensity of violence itself. Public opinion increasingly disapproved the use of excessive violence, for example, against criminals or children. The former fondness of public mutilation and murder was replaced by a public embarrassment of violent behaviour against the weak. Michel Foucault, referring to Rusche and Kirchheimer, reminds us of the fact that public torture of criminals, even by the authority of the state, died out in Europe in the late eighteenth century. However, the way to this sudden humanitarianism was paved by the shift from the values of a feudal to those of a bourgeois society. In the former, honour and status were the most prized attributes, and violence was therefore directed against the person; in the latter, money and market relationships formed the basis of social organization, and violence was therefore directed against property. Equally, for the political authorities, property and rights were not sufficiently individualized to make its seizure a common form of punishment which made the body in most cases the only property accessible.[8]

Secondly, apart from the degree of violence, Elias also rightfully stressed the shifting legitimacy of political violence. The use of violence by the free individual as an instrument so settle open scores or to solve conflicts became less and less tolerated within the growing 'public

[7] N. Wilterdink, 'Die Zivilisationstheorie im Kreuzfeuer der Diskusssion: Ein Bericht vom Kongress über Zivilisationsprozesse in Amsterdam', in P. Gleichmann, J. Goudsblom and H. Korte (eds), *Macht und Zivilisation: Materialien zu Norbert Elias Zivilisationstheorie* 2 (Frankfurt am Main: Suhrkamp, 1984); H.P. Duerr, *Der Mythos vom Zivilisationsprozess*, 3 vols (Franfurt am Main: Suhrkamp Verlag, 1988-93); Duindam, *Myth of Power;* C. Marx, 'Staat und Zivilisation: Zu Hans Peter Duerrs Kritik an Norbert Elias', *Saeculum* 47, 2 (1996), 282-99.

[8] M. Foucault, *Surveiller et punir: Naissance de la prison* (Paris: Gallimard, 1975), 29-30 ; L. Stone, 'Interpersonal Violence in English Society 1300-1800', *Past and Present* 101 (1983), 22-33.

order' of the European states. Indeed, western Europe witnessed a general tendency, albeit by fits and starts, towards, what Max Weber labelled, the monopolization of violence by the sovereign state which allowed no violent self-help on the part of its 'citizens'. Hence, the civilizing process in Europe was closely tied in to the gradual monopolization of legitimate violence by the kings of the Latin West. Although the latter found its culmination in the nineteenth-century nation state, it started already much earlier during the eleventh and twelfth centuries. *Pace* those who hold this to be an ethnocentric position, it appears to have been a uniquely European experience conditioned by a number of particular, historical circumstances of which by far the most important were Europe's almost complete immunity for nomadic disturbances and an exceptionally rigid parcelization of rights and lands. Before taking a closer look at this European *Sonderweg*, let me first briefly describe the early-medieval situation in which the use of violence was not yet the prerogative of the state or its prime representative, the king.

CRIMINALIZING FEUDS

During the early Middle Ages the use of violence – Elias would say *Angriffslust* – was intimately related to the regular raiding practices of free warriors. Criss-crossing the natural frontiers marked out by marchland, forest and other wilderness, they would roam round the countryside and lay their hands on everything they could carry off from the settled lands such as ornaments, weapons, cattle, and if possible men, women and children. The legitimizing principle behind this raiding was the idea of the *feud*. Although it refers to all kinds of warlike activity, a rightful feud must be understood as a fight for one's rights. A feud started when a presumably wronged prince, nobleman, city or the like, found no redress for grievances. Then the injured party went to war for the sake of vengeance or redress, damaging the adversary until the latter was willing to make good for the presumed violation of rights. In addition, not only those directly involved but also the subjects, servants and helpers of the feuding parties became mutually enemies to each other. Whatever the ultimate purpose of the particular feud, inflicting the maximum of damage on these enemies was the immediate purpose. The usual mode of feuding was to march into the opponent's realm and to lay waste the land by plundering and burning. Hence, looting was a rightful weapon, while its material result, booty, was used for exacting reparation and attracting support. To put it differently, there was no fundamental difference between warlike activity and pillaging; both were perfectly normal means of collecting tribute or obtaining supplies. It was only much later during the sixteenth and seventeenth centuries that the juridical concept of war emerged victorious, to cover external conflicts between sovereign states. Within these states feuds had only been gradually suppressed. Earlier though, the notion of the rightful feud reveals the fact that there was no monopoly of the state on the legitimate exercise of violence, and that there was such a thing as rightful force below the level of state power.[9]

Although feuding was pervasive, it did not mean that there were no countervailing ideas that aimed to limit such uncontrolled eruptions of violence and destruction. First of all, there existed some restraints against excessive violence in the idea of the feud itself. For example, the killing of one's opponent was something rather to be avoided. Of course, this could be out of fear of supernatural punishment but, more importantly, killing would spoil the lucrative ransom and evoke the blood wrath of the victim's relations. Thus, even though it was legally recognized, as long as enmity existed between the parties, it went against the purpose of the feud – i.e. rectifying an injustice – to destroy the other party. Hence, initially, kings and Church leaders not so much agitated against the feud itself but, more in particular, against its erratic nature. During the eleventh and twelfth centuries several so-called Peace of God movements and other peace associations emerged which aimed to counter this very

---

[9] F. Redlich, 'The Praeda Militari: Looting and Booty 1500-1815', *Vierteljahrschrift für Sozial- und Wirtschaftgeschichte*, Beiheft, 39 (1956), 1-79; O. Brunner, *Land and Lordship: Structures of Governance in Medieval Austria* (Philadelphia: University of Pennsylvania Press, 1992), 1-95; M. Bloch, *Feudal Society*, 2 vols (London-New York: Routledge and Kegan Paul, 1989), 125-30; G. Duby, *The Early Growth of the European Economy: Warriors and Peasants from the Seventh to the Twelfth Century*, trans. H.B. Clarke (Ithaca: Cornell University Press, 1992), 48-9.

irregularity of the feud and its dreadful side-effects. These produced no general ban on feuds but only periods of general abstention in which one attempted to suppress the daily violence, exactions and unlawful fines exacted by plunderers and other troublemakers. Of course, this is still a long way from the state's monopoly of violence, but the peace movements clearly represent the first attempts to establish some kind of public order in medieval society.[10]

In attempting to limit the violent side-effects of the feuds, kings and Church leaders had at their disposal a long tradition of Roman and scholastic thinking on the legitimacy of political violence. St Augustine, already, proclaimed that any just war should be declared and waged on the authority of a prince. From the twelfth century onwards, scholastic teaching on war began to distinguish between war and other forms of violence like brawls, sedition and the exercise of judicial power. Thomas Aquinas claimed that war, properly speaking, was against an external enemy, one nation as it were against another. Brawls were between individuals, one against one or a few against a few. Sedition in its proper sense was between mutually dissident sections of the same people, when, for example, one part of the city rebelled against another. In a similar spirit, Gratian spelled out the four conditions for a just war: it should be ordered by a prince, without the participation of clerics, for the defence of the country attacked or for the recovery of despoiled goods, and free from violent, unlimited passion. Hence, unlike the feud, war should not be waged out of hatred nor insatiable cupidity. Reflections of this type led to the contrasting of two kinds of war, according to the outward bearing of the combatants. In opposition to 'mortal' war, waged with fire and blood, where all sorts of cruelties, killings and inhumanities were tolerated, or even systematically prescribed, there was that form of war described as *guerroyable*: regular war, honourable, *bonne guerre*, fought by good fighters in conformity with the law of arms or according to the discipline of chivalry.[11]

---

[10] R.W. Kaeuper, *War, Justice, and Public Order: England and France in the Later Middle Ages* (Oxford: Clarendon Press, 1988), 134-84; Ph. Contamine, *War in the Middle Ages*, trans. M. Jones (Oxford: Blackwell , 1984), 270-80.

[11] Contamine, *War in the Middle Ages,* 280-4.

Both this kind of scholastic thought and the Peace of God movements linked up well with the ideology of the Crusades. In all these cases, the state and Church authorities sought to mobilize warriors into the fight for a higher goal, to liberate Latin Christendom from the 'internal' feud and the 'external' infidel. On several occasions 'knights of peace' were instituted, and benefited, in exactly the same way as the crusaders, from the privilege of sacred immunity. Inspired by the example of the crusading orders, this kind of martial activity combined aggression with self-abnegation and, as such, acquired a spiritual and redemptive value. As a consequence, both the Peace Movements and the Crusades favoured the definition and promotion of a new chivalric ideal that gradually looked down upon the now rather pusillanimous violence of feuding. Finally, the latter became criminalized as the excess of the lawless, as a form of 'civil' disobedience. The crucial point is that by both taming and ejecting its inner 'barbaric' drives, Latin Europe succeeded in taking its first steps towards what later came to be the monopolization of legitimate violence by the state. However, as we will see, domesticating the restless barbarian could not succeed without simultaneously taming the inner and outer frontiers of Europe's remaining natural wildernesses.

BESTOWING BOUNDARIES

During the European High Middle Ages the Latin West went through a phase of rapid economic expansion. From the eleventh until the fourteenth century, it increasingly participated in a growing and converging global economy. Through an extensive web of commercial and financial networks it was linked up with the still far more advanced and monetized economies of the Islamic Middle East, South Asia and China.[12] This commercial build-up was closely tied in to a phase of rapid agrarian expansion. Everywhere in Europe, both internally and externally, there was a massive thrust to bring uncultivated forests – and even seas – under the newly developed heavy plough. Most spectacular was the extension of the agrarian frontier towards the

[12] J. Abu-Lughod, *Before European Hegemony: The World System AD 1250-1350* (Oxford: Oxford University Press, 1989).

East. This *Ostsiedlung* partly replaced earlier Slav settlements but also effectively reclaimed land which formerly had been either waste or pasture. The same was true of the Anglo-Saxon and Norman expansion into the Celtic West and the Spanish *reconquista* of the Muslim South. Simultaneously, within the existing outer frontiers of Latin Europe, more and more forests were reclaimed. This increased cerealization also meant that the fortified castles of obstreperous medieval warriors which had once dotted the peripheries of the cultivated territories became increasingly incorporated into the realm of the sedentary order.[13]

The rapid extension of the agrarian frontier in Europe during the High Middle Ages was far from unique. Simultaneously, other areas, especially South and East Asia, went through a phase of agrarian expansion. Nevertheless, it was only Latin Europe which from about 1100 enjoyed an extraordinary immunity from disruptions caused by external or internal nomads. As a matter of fact, Europe had always been relatively immune for nomadic inroads since its climate favoured relatively thick forests instead of semi-arid steppes and open savannahs which were more suitable for pasture. Of course, previously, it had been recurrently infested by the inroads of Vikings, Muslims and Magyars, partly through the Hungarian extension of the Central Asian steppe, partly through the inland seas of the Mediterranean and the Baltic. In general, though, the European climate and terrain were not conducive for wandering groups of pastoral nomads or bands of nomadic warriors. Hence, Latin Europe lacked a significant nomadic population of its own and its mobile warriors had a natural tendency to move to its outer peripheries, in particular in eastern and southern directions. Not surprisingly, the European mounted warrior was the first to lose ground to infantry units of archers and pikemen. Hence, from the eleventh century onwards, Latin Europe could consolidate its extended agrarian frontier without having to accommodate incoming nomads.[14]

---

[13] R. Bartlett, *The Making of Europe: Conquest, Colonization and Cultural Change 950-1350* (London: Penguin Books, 1994), 133-67.

[14] For the present purpose, the term nomad differs from its conventional meaning. Here it refers to all groups which have nomadic or semi-nomadic

Being exempt from nomadic inroads, European rulers were able to set up fixed and closed territorial boundaries within and outside their own realm. Agrarian expansion was characterized by a ruler bestowing 'boundaries and rights' to all those who came to settle new land. This mostly involved well-defined freedoms and immunities from external tolls and taxes. Potential conflicts between feuding parties were preferably solved by systematically defining and demarcating their respective rights. This parcelization and privileging of space and its inhabitants more and more produced self-contained, mutually exclusive, territorial units that came to be supervised by more permanent and more uniform political and judicial institutions, propped up by Roman law and the extraordinary legal authority of the European town. Gradually not only bishoprics, *villae* and towns, but the European state itself, became characterized by a well-defined, contiguous territory with a relatively centralized government differentiated from other organizations. Obviously, this systematic parcelization of rights and lands would have been unthinkable in a situation where internal or external nomads could easily cross an always open and fluid frontier in order to claim their share of the sedentary wealth.

How this relative stability of the sedentary realm relates to the notion of legitimate violence? As we have seen already, feuding had been a structural feature of early-medieval society. It was not the occasional excess of the bandit but the foremost activity of every free adult male. As such it was not a cause for embarrassment but a source of pride. As we have seen already, from the eleventh century onwards, the boundless violence of the feud was increasingly criminalized. Conflicts were not to be solved by feuding, but by juristic procedures which further defined and fixed the rights of the parties involved. The violent capabilities of the free European warrior were gradually monopolized by the state and, at the same time, externalized towards its own rigidly-defined demarcation lines. Not the open frontier connecting the cultivated *ager* with the wild *saltus* – the usual passage

---

life-styles whether or not combining their mobility with fixed dwellings. Hence, it includes all itinerant pastoralists, traders, warriors, ascetics and other regular wanderers.

way for the feudal raid – but the closed boundary of the sovereign state came to be the prime location of political violence. Not surprisingly, it was this very kind of legitimate inter-state violence which has proven to be by far the most excessive and destructive.

TRANSFORMING THE INNER DRIVES IN SOUTH ASIA: FROM KAKAR TO KAUTILYA

Before turning towards medieval South Asia, we should return to Norbert Elias and ask ourselves the question: was there a civilizing process in South Asia? In other words, to what extent South Asia came to terms with man's inner drives towards violence and aggression? This important question may be approached from several different perspectives but since Elias himself made a compelling comparison between the civilizing process proper and the upbringing of children, it might be fruitful to look briefly at the South Asian, that is 'Hindu', way of raising children to maturity; the more so since the latter has been comprehensively analysed by the Indian psychiatrist Sudhir Kakar.[15] Although his work lacks a historical and social dimension – it describes the situation during this century among upper-class Hindus – it still serves as an interesting South Asian example of civilizing the self which compares well with the Freudian model employed by Elias. The more so since Kakar appears to agree with Elias that the cultural traditions of society are internalized during childhood in the individual's superego, that categorical conscience which represents the rights and wrongs, the prohibitions and mores, of a given social milieu. Nevertheless, Kakar observes some major distinctions. For example, whereas Western education attempts to nurture ego boundaries – for example between inside and outside and between 'I' and others – the Hindu ideal seeks to undo the process of ego development which in the West takes place during early childhood. Here, the development of a superego entails the constraining of the primitive id, which is the mental representative of the organism's instinctual drives. The ego must be ever watchful and strengthened

[15] S. Kakar, *The Inner World: A Psycho-Analytic Study of Childhood and Society in India* (Delhi: Oxford University Press, 1981).

to control the eruptions of the 'ever seething cauldron' of the id. According to the Hindu ideal, though, the elemental and instinctual drives of the organism – called *chitta* or 'mind' in Raja Yoga – should be given free rein as they are capable of transformation, through 'gathered' and concentrated forms, into the final one-pointed form in which the unique aim is the union of 'I' and the 'not-I'. The central aim is to quiet the turmoil of *chitta* and bring it nearer to its perfect state of pure calm. So the ego is not in opposition to the id, as in the West, but merged with it.[16]

According to Kakar, this permissive attitude towards the human inner drives is also reflected in the upbringing of children. In this respect he stresses the sharp distinction between the individual human being and the social human being: until his fourth to fifth age the child remains a complete, innocent being who is considered to be a gift of the gods, to be welcomed and appreciated for the first few years of life. At this prolonged infancy the mother is inclined towards a total indulgence of her infant's wishes and demands. The Indian mother is predisposed to follow rather than lead in dealing with her child's inclinations and with his tempo of development. This contrasts sharply with modern Western attempts to encourage the child's individuation and autonomy at a very early age. Hence, the Hindu child's differentiation of himself from his mother – and consequently of the ego from id – is structurally weaker and comes chronologically later than in the West. After about the age of five, all of a sudden, the world of Indian boyhood widens from the intimate cocoon of maternal protection to the unfamiliar masculine network woven by the demands and tensions, the comings and goings, of the men of the extended family. The abruptness of the separation from his mother and the virtual reversal of everything that is expected of him may have traumatic developmental consequences. All in all, the conflict generated by this 'second birth' also leads to a relatively weaker differentiation and idealization of the Indian superego. By this Kakar means that the categorical conscience, as a representative of the rights and wrongs, the prescriptions and prohibitions of Indian *dharma*, does not exist as a psychic structure sharply differentiated from the

[16] Kakar, *The Inner World*, 15-52.

id and the ego, nor are its parts idealized as they tend to be in Western cultures. Much of the individual behaviour and adaptation to the environment that in westerners is regulated or coerced by the demands of the superego, is taken care of in Indians by a communal conscience. In contrast with the western superego, the communal conscience is a social rather than an individual formation: it is not 'inside' the psyche. To state it differently, instead of having one internal sentinel, Hindus rely on many external 'watchmen' to patrol their activities which all relate to social status.[17]

The weaker differentiation of the superego has several important consequences involving Elias' taming of man's inner aggression. One major difference with the European civilizing process is that Kakar's Hindu generally demonstrates a greater indulgence towards acts of violence in situations where these are not explicitly forbidden by the communal conscience or the hierarchical imperative. Although the social taboos on the expression of overt aggression are often very strong, they are not matched by complementary superego controls. When these taboos break down or where they do not apply, these relaxed controls permit a volatile aggressiveness which can quickly flare up and as suddenly die down. This in contrast to early-modern Europe where, after the medieval rejection of children, the ideological emphasis had decisively shifted towards inner suppression, stressing the child's training and inner discipline. Without claiming Kakar's findings to be representative of the South Asian attitude towards violence in general, these certainly suggest that the civilizing process as described by Elias should not be considered universal. On the contrary, India appears to have experienced a civilizing process of its own in which legitimate violence took a different direction, in another context.

India has produced more than a fair amount of texts and images which indeed demonstrate a remarkable lack of embarrassment towards extreme violence. Actually, in ways that are hardly conceivable to the modern eye, these exhibit a conspicuous predilection for gory scenes of killing and utter destruction. It appears, though, that a great deal of this excessive cruelty takes place in a religious context, more

[17] Kakar, *The Inner World*, 52-140.

specifically, they often seem to exhibit man's unbounded love and devotion (*bhakti*) to god. As in the well-known south-Indian story of the *Periya Puranam*, parents slaughter their children and offer them as curries in order to satisfy their 'hungry' gods.[18] By killing, hurting, abusing and destroying, devotees could not only express their love but also their steadfast allegiance to their particular branch of religion. Although at times they seem equally at ease hurting others, the main thrust of violence was directed against the self, either directly or through seemingly others. Hence, these devotees are often depicted as committing acts of self-mutilation or even suicide. This paradoxical conjunction of love and slaughter promised them recreation and nearness, or even unity, with their gods.[19] We should keep in mind, though, that this kind of devotional violence was not entirely devoid of political consequences. Actually, it could easily be transformed in some kind of magical power and even lead to political dominion, often depicted as being created out of blood sacrifice and violent dismemberment.[20]

The tradition of violent devotion became particularly pronounced from about the eleventh century onward. This not only coincides with the Islamic inroads, but also with the emergence of numerous *bhakti* sects throughout the Indian subcontinent.[21] These devotional

[18] G.L. Hart, 'The Little Devotee: Cekkilar's Story of Ciruttontar', in M. Nagatomi et al. (eds), *Sanskrit and Indian Studies* (New York: D. Reidel Publishing Company, 1979) 217-36; D.D. Hudson, 'Violent and Fanatical Devotion among the Nayanars: A Study in the Periya Puranam of Cekkilar', in A. Hiltebeitel (ed.), *Criminal Gods and Demon Devotees* (Albany: State University of New York, 1989), 373-404; D.D. Shulman, *The King and the Clown in South Indian Myth and Poetry* (Princeton: Princeton University Press, 1985), 102-29.

[19] D.D. Shulman, 'Battle as Metaphor in Tamil Folk and Classical Tradition', in S.H. Blackburn and A.K. Ramanujan (eds), *Another Harmony: New Essays on the Folklore of India* (Berkeley: University of California Press, 1986); Narayana V. Rao, *Shiva's Warriors: The Basava Purana of Palkuriki Somanatha* (Princeton: Princeton University Press, 1990), 3-31; C. Vamadeva, *The Concept of 'Vannanpu' 'Violent Love' in Tamil Saivism with Special Reference to the Periyapuranam, Uppsala Studies in the History of Religions* 1 (Uppsala: Uppsala University, 1995).

[20] S. Bayly, *Saints, Goddesses and Kings, Muslims and Christians in South Indian Society 1700-1900* (Cambridge: Cambridge University Press, 1989), 19-71.

[21] Rao, *Shiva's Warriors*, 3-31; D.D. Shulman, 'Die Dynamik der Sektenbildung

movements aggressively dissociated themselves from the Brahminical or Buddhist establishment. Not unlike the fanatic *mujahedin* of the time, they combined their uncompromising ways to worship their gods with an iconoclast zeal against the settled order of temples and *viharas*. The rise of these more aggressive sectarian movements coincided with the spread of the royal cult of Rama which also contributed in making Indian religious expression as a whole more assertive and aggressive.[22]

Countless other examples might be gathered which may illustrate the point that South Asia developed a certain acceptance of man's violent categories. As for example in the case of violent devotion, the aim was not to suppress but to gather and to transform one's violent drives towards the higher goal of salvation through love and devotion. In this case, a different civilizing process had taken place in which violence had been 'civilized' by *bhakti*, again not by being checked or denied, but by being internalized and redirected against the self. This should bear out the fact that South Asia developed a civilizing process of its own which was, however, different from the one experienced in Europe, where the concept of the unitary, sovereign state had prescribed its course. Leaving all transcendental aspirations of violence aside, the following will attempt to amplify this more general observation by focusing specifically on the way medieval South Asia came to terms with violence aimed at the here and now of the immanent sphere.

BRIDGING THE FRONTIER

The rise of new, more assertive sects and cults during the eleventh and twelfth centuries coincided with some breath-taking other changes which affected the Indian subcontinent at large. These changes primarily involved large-scale migrations of Muslim, Telugu and other warrior groups and the emergence of previously marginal tribal

---

im mittelalterlichen Südindien', in S.N. Eisenstadt (ed.), *Kulturen der Achsenzeit II: Ihre institutionelle und kulturelle Dynamik,* Teil 2: *Indien* (Frankfurt am Main: Suhrkamp Verlag, 1992), 102-29.

[22] S. Pollock, 'Ramayana and Political Imagination in India', *The Journal of the Asian Society* 52, 2 (1993), 261-97.

elements. Actually, the emergence of the Ghaznavids and Ghurids in northern India, the Yadavas in Maharashtra, the Kakatiyas in Andhra and the Hoysalas in Mysore all mark the growing importance of the arid and semi-arid marchlands of the South Asian interior. At the time, these jungles formed a vast, contiguous zone of savannah-like waste which served well as extensive pastures for herds of pastoralists, caravans of long-distance traders and armies of mobile warriors. This so-called Arid Zone stretched from the north-western deserts of Sind, Rajasthan and the Punjab to the dry Rayalaseema in the south. As such it served as a wide transitional area connecting the more humid, agrarian centres along the main river valleys. Its population consisted primarily of semi-nomadic pastoralists who earned a living by providing the sedentary world with their animals, animal products, transport facilities and, last but not least, their military expertise.[23]

The new rulers were well placed to tap the moveable resources of the arid zones; actually many of them had started their careers from a nomadic background. But in order to consolidate their power in the long term they were forced to bridge the frontier between the jungle where they came from and the arable where they settled. In other words, their challenge was to invest the movable wealth produced by pastoralism, trade and warfare into the newly conquered lands or into lands to be reclaimed from the jungle. Consequently, new capitals such as Delhi, Devagiri, Warangal and Dvarasamudram, and, at a later stage, similar headquarters like Bijapur, Golkonda and Vijayanagara had a peripheral location where not only warriors and pastoralists preferred to display their services, but also, long-distance traders could find wholesale bulking and transhipment facilities. The modern onlooker might regard these eccentric new capitals as extraordinary misfits in the middle of relatively poor and precarious countryside but, actually, these were ideally situated at the interface of unsettled marchlands and more settled agrarian fields, delivering its rulers the best of both worlds: movable wealth and secure investment.

---

[23] D. Ludden, 'History outside Civilisation and the Mobility of South Asia', *South Asia* 17, 3 (1994), 1-23; J. Gommans, 'The Silent Frontier of South Asia, c. 1100-1800 AD', *Journal of World History*, 9, 1 (1998), 1-25.

Clearly, the upsurge of mobile elements along the arid frontier could produce sudden outbursts of violence. Naturally, as they were closely tied to the jungle, mobile warriors could dispose of a plethora of violent power. Their nomadic lifestyle gave more space to militant heroism and provided them with a permanent training in the exercise of violence and in the resistance to the violence of others.[24] It would be wrong, though, to attribute a more violent characteristic to nomadic than to settled society. On the contrary, as a result of its extensive radius of action, nomadism was in fact far less destructive, the more so, since plundering other nomads was not only difficult but also not a durable means for enrichment. It was not nomadism itself, but rather its confrontation with the completely opposite way of life of the settled world, which delivered nomadism its violent reputation. There was a natural tension between the world of the nomad, always on the lookout for pasture and plunder, and that of the peasant, always eager to bring his land to the plough, yielding the richest harvest under the most peaceful conditions. Hence, the most violent acts of conquest and revolt usually occurred at the agrarian-cum-nomadic frontier between arable and jungle. Not surprisingly, many Indian epics, including the *Mahabharata* and *Ramayana*, are staged in a tripartite setting in which the violent battlefield connects the juxtaposed world of civilized court and wild forest.[25] But both in epic and history, violent activities were not the exclusive privilege of the nomadic warrior. In fact the nearness of the frontier facilitated the continued practice of seasonal raiding by settled chiefs and peasants; at this time to be increasingly regulated by a huge military labour market.[26] Actually, harvesting and raiding alternated with each other on the rhythm of the monsoon. The most violent season was that after the monsoon when peasants-turned-warriors first crossed the frontier into the wilderness and from there marched into the settled

[24] E. Gellner, *Anthropology and Politics: Revolutions in the Sacred Grove* (Oxford: Blackwell, 1995), 160-80.

[25] T. Parkhill, *The Forest Setting in Hindu Epics: Princes, Sages, Demons* (Lewiston: Mellen University Press, 1995).

[26] D.H.A. Kolff, *Naukar, Rajput and Sepoy. The Ethnohistory of the Military Labour Market in Hindustan 1450-1850* (Cambridge: Cambridge University Press, 1990).

fields of the enemy. In this way the nearness of the frontier also facilitated an ongoing switching of roles which could even turn Brahmins into warriors or kings into ascetics.[27]

Thus just at the time Latin Europe gained immunity from nomadic inroads, South Asia experienced a renewed upsurge of mobile elements. This upsurge was not only characterized by the recurrent inroads of external and internal nomads but, as we have seen, also by alternating cycles of raiding and harvesting on the part of the settled population itself. Under these turbulent and volatile circumstances it was nearly impossible for rulers to establish long-term internal stability and public order, let alone to impose a full monopoly on the use of violence. Of course, in this case, the civilizing process as described by Elias could never take root. Even the more powerful rulers could hardly be expected to suppress the inner aggression of warriors who continuously shifted their loyalties by recurrently marching out for new, lucrative *mulkgiris*, either within or outside the rulers' own realm. In terms of political stability, the South Asian experience stands much nearer to the Middle East and Central Asia than to Latin Europe. It was only in the nineteenth century that the modern means of the Pax Britannica succeeded in criminalizing and containing the always destabilizing mobility of pastoralists, traders, warriors and other wanderers.

Taking the political flux for granted, we should keep in mind that for the settled population interaction with nomads held many blessings in disguise. Apart from raiding and plundering, which in itself mobilized resources and increased its velocity of circulation, nomads could equally inject enormous amounts of movable wealth into the sedentary world. By being situated at the frontier, the Indian ruler always had to strike a shaky balance between tapping the movable wealth of the jungle and managing the exploitation of the arable. His main challenge was not to tame or to expel, but to incorporate the nomadic frontier into his realm without, however, releasing its huge potential for violence. Therefore, South Asia had to invent a whole range of political mechanisms which aimed at accommodating the recurrent break-ins from the frontier. Although relatively small-scale violence, such as plundering or thieving, was publicly tolerated, it

[27] Shulman, *The King and the Clown*, 149-51.

was never allowed to get out of hand and to undermine the delicate political balance. It is in this context that South Asia developed a political culture in which large-scale political violence was deemed highly imprudent and as much as possible to be avoided.

RULING THROUGH CONFLICT

Indian state building never really involved the intense conflict between the 'secular' and the 'sacred', as embodied in the historical struggle between State and Church in Latin Europe. The latter dichotomy clearly stems from the Christian separation between the earthly domain of man and the heavenly domain of God. Being cut off from the heavenly kingdom, both the state and the Church were convicted to vie with each other for earthly power and resources.[28] In the end, this historical struggle gradually produced rigid demarcation lines between their respective jurisdictions. In South Asia, however, the situation was different since the transcendental world continued to play an active part in this world which deprived earthly power from possessing an ultimate meaning in itself. The ultimate aim for both Brahmins and kings was not earthly power but eternal salvation; the way thither was not conquering the world but renouncing it. His pursuit also involved the well-known Indian ideal of *ahimsa* or non-injury to life. As it was only the ascetic who could really live up to it, world-renouncement and *ahimsa* had to be internalized by those whose *dharmic* role could not escape the evil consequences of worldly action. As such, this second-best option was most fitting for kings and warriors wielding earthly power. Hence, their *himsa* had to be detached, without self-interest; *ahimsa* only serving to calm their inner world. This also explains why even martial sects of ascetic warriors saw no problem at all in subscribing to the vow of *ahimsa*.[29] In any case, it is clear that *ahimsa* was not meant to establish peace in the public realm as had been the prime objective of the European peace movements.

The ideal of renunciation made the position of the worldly king

---

[28] F.W. Buckler, 'Regnum et Ecclesia', *Church History* 3, 1 (1934), 31-2.
[29] J.N. Farquhar, 'Fighting Ascetics in India', *Bulletin of the Rylands Library* 9 (1925), 431-52.

extremely problematic. The king could not be prevented from being contaminated by evil for the very essence of his activity as a ruler. As we have seen already in the case of the ascetic devotee and warrior, one way to limit the contamination by *himsa* was by acquiring an inner state of detachment. But this was no final way-out. Another option was to renounce all power and riches produced by worldly activities. In other words, in order to gain legitimate power, the king had to renounce it. Hence, we witness the king driven, as if by inner obsession, to distribute his power to Brahmins or 'gods' in the form of *brahmadeyas* and *devadeyas*.[30] Interestingly, while the Western state and Church stood face to face competing for earthly power, the Indian king and Brahmin had turned their backs against each other competing for being relieved of it. Nonetheless, gift-giving (*dana*) to Brahmins was only one aspect of royalty. *Rajadharma* clearly prescribed the king to wield his staff (*danda*) in order to protect his subjects and to punish evildoers. For the king there was no escape: he was forced to be who he was created to be. This confronted the king with a huge dilemma: how he was supposed to combine wielding with renouncing power? As we have noticed already, both roles could neatly alternate each other in the annual cycle of raiding and dispensing. But even then the situation could easily get out of hand. With a looming frontier always near at hand, the crucial question remained: how to accommodate warrior groups recurrently released and recruited from the frontier without being overwhelmed by them?

Faced with this dilemma, the king admirably managed to make *dana* itself part of the solution. Keeping a public image of disinterestedness, endowments of land to Brahmins and temples enabled the king to direct the moveable wealth of the frontier, both cash and cattle, into the productive channels of agrarian investment. In this way, the king not only redistributed economic resources but also alleviated potentially disruptive inequities in his realm. In other words, by redistributing plunder and other movable wealth, temples and

---

[30] R. Inden, 'The Ceremony of the Great Gift (*Mahadana*): Structure and Historical Context in Indian Ritual and Society', in M. Gaborieau and A. Thorner (eds), *Asie du sud: Traditions et changements* (Paris: Centre national de la recherche scientifique, 1979), 131-36; Shulman, *The King and the Clown*, 3-47.

brahmadeyas served as major connecting links between the unruly, martial elements of the frontier on the one hand, and the settled agricultural society, on the other.[31] We should keep in mind, though, that *brahmadeyas* – like the medieval European landed rights – mostly involved well-defined territorial units marked by solemn ceremonies of demarcation and consecration.[32]

Another way-out to the king's predicament was the time-honoured bestowing of temporary and concurrent, that is shared, rights in the proceeds of the agrarian exploitation. In this way, the king gave away much of his realm to his followers or other warrior groups without definitely losing his stake in it, the more so since these rights often involved rebellious, unconquered or uncultivated lands. Although being just one of many shareholders, the king managed to keep his stake in each field and every threshing floor in order to ensure the collection of his share.[33] Or to put it differently, by distributing over-lapping rights the king not only incorporated incoming frontiers-men, but also controlled them by exploiting their ongoing rivalries. Actually, ruling in South Asia often came down to managing conflict, not eliminating it.

Hence, both through *dana* and co-sharing, South Asian rulers attempted to channel the mobile resources of the frontier towards agrarian expansion. But as we have stressed already, the recurrent input of new people and resources made the position of the king also extremely vulnerable. He was only one player in an intricate web of mutual dependence between numerous share-holders. His ruling was in fact a dangerous balancing exercise in which any rash action could easily result in endless *renversements des alliances* which could well undermine the king's powerbase itself. Given the restricted checks and balances, the king always had to operate with utmost care. Equally,

[31] G.W. Spencer, 'Temple Money-Lending and Livestock Redistribution in Early Tanjore', *The Indian Economic and Social History Review* 5, 3 (1968), 277-93.

[32] B. Stein, 'All the Kings' Mana: Perspectives on Kingship in Medieval South India', in J.F. Richards (ed.), *Kingship and Authority in South Asia* (Madison: University of Wisconsin-Madison Publication Series, 1978), 145.

[33] J. Heesterman, 'Traditional Empire and Modern State', unpublished paper IDPAD symposium on State and Society, New Delhi, 5-9 March 1990.

for any shareholder in such a conflict-ridden system, to act with sudden, large-scale violence was highly imprudent.

Thus, in a remarkable concurrence of circumstances, both frontier and *dharma* demanded from the king an extremely cautious policy of giving and taking. In this respect, the policy recommendations of the *Arthashastra* proved perfectly adequate. These amount to no more than stratagems which the king should employ to keep the delicate political balance intact. As concisely presented by André Wink, the well-known *Arthashastra* of Kautilya draws up a list of four general political 'means' (*upaya*), with the following order of precedence: conciliation (*santva, sama*), gift-giving (*dana*), sedition and winning over (*bheda*), and force (*danda*).[34] The application of force comes in the last place and is also explicitly discredited in favour of sedition (*bheda*) in all those situations which involve multitudes of people. Such depreciation of force as a means of politics and conquest is not a peculiarity of Kautilya. It recurs in Indian political writings until as late as the eighteenth century as the scheme of the four *upaya*. It is likewise found in Manu that a king should try to conquer his enemies by conciliation (*sama*), by gifts (*dana*), and by sedition (*bheda*), used either separately or conjointly, never by fighting.

So far, we have mainly examined the position of the Hindu king. Islamic rulers had their own set of ambivalences regarding power politics. As most comprehensibly analysed by the Arab historian Ibn Khaldun (1332-1406), Islam had a long history of close interaction with the nomadic frontier.[35] Once and again, the settled Islamic world had to withstand and incorporate nomadic warriors who otherwise would have plundered it. Not much different from their Hindu colleagues, Muslim rulers attempted to accommodate these nomadic intruders by distributing concurrent tax rights among them. They also attempted to enhance their power by forging close personal bonds with them, mainly through the institution of elite slavery (*mamlukiyat*).

---

[34] A. Wink, *Land and Sovereignty in India: Agrarian Society and Politics under the Eighteenth-century Maratha Svarajya* (Cambridge: Cambridge University Press, 1986), 9-51.

[35] Ibn Khaldun, *The Muqaddimah: An Introduction to History*, trans. F. Rosenthal (London: Routledge and Kegan Paul, 1967).

But by bringing in the particularism deriving from tribal cohesion (*asabiya*) and alliance, rulers could not but disregard Islamic universalism as symbolized by the unitary *umma* and the prescriptions of the sharia. Like the Hindu *dharma*, the Islamic law of sharia could hardly be implemented in the daily routine of ruling, especially when Muslims were only a small minority such as in South Asia. Although the Islamic ideal could prescribe *jihad* or holy war against infidels, in practical terms ruling came down to what Wink and Chamberlain labelled *fitna*: sowing dissension and forging alliances.[36] This policy was not much different from the already mentioned recommendations of the *Arthashastra*.[37] Indeed, thanks to the persistence of the nomadic frontier both the Hindu and Muslim king ruled their realm by managing conflict on the basis of overlapping rights, and not, as in Latin Europe, by solving conflicts through demarcating rights. More than in Europe, diplomacy and intrigue were essential elements of the political game. To this should be added that this was further stimulated by the high degree of monetization in the Middle East but in South Asia in particular. Money could not only serve as a movable means to buy off conflicting parties but could also be used as a measurement of shifting hierarchies, as it did so well, for example, in the *mansabdari* system of the Mughals.

Under these circumstances, warfare in South Asia usually involved the slow and showy manoeuvring of armies, preferably with all the superpower panoply of elephants or heavy artillery, but merely aimed at the strengthening of positions at or under the negotiating table. This theatrical aspect of violence also served to demonstrate the warrior's military abilities which might interest new customers. Hence, communications were always kept open on all sides and all parties involved continued to be on speaking terms with each other. We should keep in mind that, as long as there were no fanatic spoilsports from outside, these armies could still be successful in accommodating

[36] Wink, *Land and Sovereignty in India*, 21-34; M. Chamberlain, *Knowledge and Social Practice in Medieval Damascus, 1190-1350* (Cambridge: Cambridge University Press, 1994), 45-7.

[37] A. Wink, *Al-Hind: The Making of the Indo-Islamic World,* vol. 1: *Early Medieval India and the Expansion of Islam, 7th-11th Centuries* (Leiden: E.J. Brill, 1990), 200.

the enemy. In this respect the whole pomp and circumstance of courts and bazaars that surrounded the large Indian armies could be very instrumental in winning a battle or taking a fortress.[38] This tallies well with the findings of the well-known military historian John Keegan. He characterized all oriental warfare by a general predilection for evasion, delay and indirectness of confrontation. Actually this was the very tactic of the mounted archer. He chose to fight at a distance, to use missiles rather than edged weapons, to withdraw when confronted with determination, and to count upon wearing down an enemy to defeat rather than by overthrowing him in a single test of arms. Unlike the mounted archer, the western knight preferred to press home his charge against the main body of the enemy, rather than skirmishing against it at a distance.[39] By referring to the mounted archer, we find ourselves in the midst of the nomadic frontier again. Indeed, thanks to the persistence of the frontier, fighting on horseback remained the heroic model and rule in South Asia until as late as the nineteenth century.

THE SULTAN'S ANXIETY REVISITED

From the foregoing it appears that Europe and South Asia both experienced civilizing processes of their own. Latin Europe witnessed a uniform process in which the violence of the feud and the barbarian were externalized. This facilitated the rigid definition and demarcation of rights, arbitrated by a state informed by Roman law and equipped with a monopoly on the use of legitimate violence. The latter had shifted from the domain of the free male individual to the sovereign nation state. In South Asia, the ongoing impact of the open, nomadic frontier made it impossible and even imprudent to draw rigid boundaries of rights and jurisdictions. Actually, through the institution of *dana* to Brahmins the movable wealth of the frontier was channelled into agrarian investment whereas mobile groups of warriors, traders and pastoralists were accommodated by a flexible system of concurrent

---

[38] J. Gommans, 'Indian Warfare and Afghan Innovation during the Eighteenth Century', *Studies in History* 11, 2 (1995), 261-80.

[39] J. Keegan, *A History of Warfare* (London: Pimlico, 1994), 386-92.

rights. In this system, large-scale violence was as much as possible to be avoided since this could easily unsettle the existing political balance. So here we have a dual tendency. On the one hand, the persistence of the frontier ensured that the use of violence remained the legitimate domain for each and every warrior. On the other hand, this violence was to be contained by a settled society eager to accommodate the frontier and informed by the ideals of *rajadharma*. As a result, Indian rulers took recourse to the well-tested stratagems of the *Arthashastra* which envisaged violence to be used only in the last resort. In other words, in a system of concurrent rights the king had to tolerate the use of small-scale, casual violence both within and outside his realm but could not allow for large-scale political violence which was considered imprudent and extremely dangerous. Such an embarrassing measure of violence could easily endanger the proper operation of the system, the latter of which was not only the responsibility of the king but of all co-sharers of the realm. Returning to Elias, the civilizing process of violence in South Asia was not imposed by the state but was mainly springing from the working of the nomadic frontier and, partly, from the king's *dharma*. Hence, not the state but the nomad and the Brahmin were its driving forces in South Asia.

This brings us back to the riddle of the sultan's sudden anxiety. Although as a Muslim Abul Hasan could hardly be concerned about his *dharma*, he could not escape from the usual policy of distributing his realm to numerous co-sharers. Of course, under Muslim rule the latter involved less temples and Brahmins and more *khanaqahs* and Sufis. As we have stressed already, the prevailing system of concurrent rights primarily responded to the exigencies of the nomadic frontier, also in Golkonda. Even when the sharia itself was at stake, this precluded the use of exceptional violence. But it hardly could have been violence as such which disturbed the sultan, the more so, since he had specifically asked for it himself. He also must have been familiar with the idea that battles could be tragically depicted as metaphors or demonstrations of cataclysmic violence. What may have been bewildering to him, however, was the glorification of violence in the context of a real battle, taking place in this world, as suggested by the direct and realistic scenery of the Dutch picture. In this case, the sultan and the Dutch may have had different expectations about the where and how of political violence. For the first this would involve small-scale violence, for example, aimed at punishment, looting or, as we have seen, devotion through self-injury. It was also characteristic of a regime that was hard with the soft and soft with the hard. Maybe for the Dutch political violence was primarily about large-scale battles and wars in loyal service to the sovereign state. But what may have been brave and logical to the Dutch could have been rather absurd and irrational to the sultan. Of course, all this remains a matter of wild speculation but what seems clear is that the sultan suddenly realized that his initial play with European violence had nightmarish consequences.

CHAPTER 6

Indian Warfare and Afghan Innovation during the Eighteenth Century

INTRODUCTION

IN 1757 THE English East India Company took its first step towards empire. At Plassey the highly professional European army under Colonel Robert Clive defeated the larger Indian forces under the reigning Bengal Nawab Siraj al-Daula. The Company's victory was partly the result of its more advanced firepower and discipline, partly of its superior skill in diplomacy. Before the advent of the battle the British had successfully seduced the backbone of the army into revolt against the Nawab. Hence, at the time of battle the great body of the Indian army stood aloof watching while the remainder was butchered by the controlled salvoes of the British sepoys. Moreover, though, the best troops of the Nawab were not at all present at the battle scene. They were deployed on the Bihar frontier to meet a possible attack of the Afghan invader Ahmad Shah Durrani (1747-73) who had sacked Delhi, Agra and Mathura and had proclaimed himself emperor and *shahanshah* of India. Apparently, Siraj al-Daula considered the Afghan threat more dangerous than the British one and it is only with the benefit of hindsight that we may judge him wrong.[1]

The battle of Plassey neatly fits the all too familiar idea that Indian 'tradition' was overwhelmed by European 'modernity'. In other words, the European armies based on disciplined firepower of light artillery and well-drilled infantry, equipped with flintlock muskets with

[1] B.J. Gupta, *Sirajuddaullah and the East India Company, 1756-1757: Background to the Foundation of British Power in India* (Leiden: E.J. Brill, 1966), 116-26.

bayonets, could easily spread discord into the open ranks of their Indian enemies, mostly mounted on horses, sometimes willing to fight à l'arme blanche, but, mostly on the lookout for individual gains. Hence, it is generally claimed that after Plassey Indian armies gradually adapted themselves to European drill, tactics and technology. At the end of the eighteenth century Europe's art of war had clearly gained momentum in India.[2]

Much earlier, from the twelfth to the sixteenth centuries, not European but Turkish and Iranian expertize was setting the military trend and had yielded the Muslims their Indian empires.[3] Two centuries later history still seemed to repeat itself. At almost each major engagement the invading Afghan armies proved again superior to their Indian counterparts. In spite of these awe-inspiring successes, Afghan history is far too often placed within a narrow regional perspective. Hence its military past is consistently associated with tribal or, at best, feudal backwardness. Here, it is my intention to probe the still dominant picture of a persisting Afghan backwater. Therefore this article will highlight some neglected military innovations adopted by the eighteenth-century Afghan invaders. Since it is only in the wider context of early-modern Indian warfare that the achievements of the Afghan military will gain some perspective, I will first consider some eye-catching characteristics of the prevailing military tradition in early-modern India.

THE INDIAN WAY

During the first three quarters of the eighteenth century, Indian armies were still essentially armies of horsemen. Even at the end of the

[2] For a discussion of the so-called European military revolution, see M. Roberts, *The Military Revolution, 1560-1660* (Belfast: M. Boyd, 1956); G. Parker, *The Military Revolution: Military Innovation and the Rise of the West 1500-1800* (Cambridge: Cambridge University Press, 1989); J. Black, *A Military Revolution? Military Change and European Society 1550-1800* (London: MacMillan Education, 1991).

[3] One of the first to stress the importance of Middle-Eastern expertize was H. Goetz, 'Das Aufkommen der Feuerwaffen in Indien', *Ostasiatische Zeitschrift* 12 (1925), 226-9; Cf. B. Rathgen, 'Die Pulverwaffe in Indien', *Ostasiatische Zeitschrift* 12 (1925), 11-30, 196-217.

century, the East India Company's officials and army commanders were fully aware of the fact that, in the long run, they could not hold or expand their newly acquired territories without a substantial enlargement of their cavalry establishment. As a result, the Bengal cavalry grew from a mere 500 horse in 1793 to 6,000 horse in 1809. Despite this increase, the British cavalry force was still very much inferior to the massive cavalry contingents of the eighteenth-century Indian states. Of course we know that, after about 1760, well-drilled and trained infantry units clearly started to make headway, albeit in the territories under European influence. But even at the middle of the century the bulk of the indigenous infantry still consisted of a multitude of people assembled together without regard to rank; some with swords and spears, who could never stand the shock of a body of horse; some bearing matchlocks, which without order, could produce but a very uncertain fire. This was the practice in early-modern India, whereas in Europe disciplined infantry of pikemen and fusiliers had already considerably reduced the impact of the massive cavalry charge.

Nevertheless, as argued by Dirk Kolff, one of the main features of the Indian military scene was an almost limitless availability of armed peasants who presented themselves constantly on the Indian military labour market.[4] As a consequence, the bulk of the imperial army of the Mughals consisted of all kinds of matchlockmen, servants, sutlers, tradesmen, travelling musicians and all those numerous individuals who belonged to the bazaars, kitchens and harems which accompanied the army. Obviously, all these people with all their appropriate rings and bells considerably hampered the free and quick manoeuvring of the army. The head of the French factory at Kasimbazar, Jean Law de Lauriston, drew a very telling picture of such an eighteenth-century military troupe. Interestingly, he contrasted the typical Indian army of his days with that of the Afghan Durranis which he considered far more efficient. To quote him in full:

There is none of that ostentation to be found amongst most of the Muslim or even the Hindu princes, which serves only to dazzle but in the end has

[4] See D.H.A. Kolff, *Naukar, Rajput and Sepoy: The Ethnohistory of the Military Labour Market in Hindustan 1450-1850* (Cambridge: Cambridge Univeristy Press, 1990).

no substance. In the armies of India one merely sees colours, flags and standards, one is continuously overwhelmed by the noise of timbals, trumpets, oboes and fifes and so on. A unit of five thousand horsemen assuredly makes more noise than an army of 100,000 men in Europe. There are nearly as many officers, or those who fancy themselves to be such, as there are soldiers, five times as many servants or traders, and at least ten times more women of all persuasions. As soon as a chieftain who, from the humble status of attendant, through the protection of eunuchs secures from the prince a title or a commission to command a small cavalry corps, he can no longer move without a great following. He has scarce enough to live upon, but you will see him doing without necessities and selling or pawning all his valuables, and all this merely for the sake of being preceded by a horse, two large standards, even small flags, timbals, fifes, flutes and trumpets as well as twelve attendants who, lacking muskets, carry sticks wrapped in red cloth. Even those who have muskets usually have neither powder nor shot.[5]

About a Mughal noble (*mansabdar*) with a rank of 3000 he remarked disdainfully:

His camping quarters are first perceived by a large flag where one would normally expect to find the three thousand horses. On approaching, there is nothing to be seen. Five or six poor gun carriages, on some of which women display themselves, some twenty beasts of burden that are to carry the baggage of the mansabdar, complete with all his bazaar, in other words the market without which his troop could not survive.[6]

During the seventeenth and eighteenth centuries there had been an increased tendency to recruit soldiers, not directly by the state authorities, but through numerous small military brokers and mercenary captains (*jamadar*) who always kept an eye on their own self-interests. As a result, the proportion of cavalry paid and mounted by the state on state-owned horses had dropped significantly and the bulk of the Indian armies were turned into full mercenary forces under the control of small military entrepreneurs.[7] The recruits of these armies consisted partly of professional specialists but mostly of

[5] Jean Law de Lauriston, *Mémoires sur quelques affaires de l'Empire Mogol 1756-1761* (Paris: Alfred Martineau, 1913), 191-2 (my translation).

[6] Law de Lauriston, *Mémoires*, 193 (my translation).

[7] Cf. P. Mason, *A Matter of Honour: An Account of the Indian Army, its Officers and Men* (London: Jonathan Cape, 1974), 52-3.

part-time landlords and cultivators for whom warfare was part and parcel of the yearly routine. These men, high and low, served only for what they could get, and were ready at any moment, when things went badly, to desert or transfer themselves to the highest bidder;[8] or to continue with Law de Lauriston:

They gather as many horsemen as they can muster, make a few advances and occasionally fight out of conviction, although most do so only in the hope of a fortunate event that might bring them compensation.[9]

From this and other instances Law concluded that:

It is evident from what I have recounted how ready a nabob, or any other very rich man, might be to revolt; the richest party always prevails, whilst loyalty and patriotism are virtues unknown in India.[10]

Obviously, their commercial outlook often functioned as a brake on their valour and courage at the battlefield. Only when they found themselves clearly in a position of superior strength were they prepared to actually engage the enemy. Otherwise, as a British officer noted in 1772:

The cavalry are not backward to engage with sabres but are extremely unwilling to bring their horses within the reach of our guns, not so much through fear of their lives as for their fortunes which are all laid out in the horse they ride.[11]

Under these circumstances, diplomacy and intrigue were essential elements of the whole concept of war. The slow and showy manoeuvring of an army, preferably with all the superpower panoply of heavy artillery and elephants, merely aimed at the strengthening of positions at or under the negotiating table. Communications were always kept open on all sides and all parties involved continued to be on speaking terms with each other. During the months preceding the famous

[8] W. Irvine, *The Army of the Indian Mughals* (London: Luzac and Co., 1903), 297.

[9] Law de Lauriston, *Mémoires*, 194 (my translation).

[10] Law de Lauriston in G. Deleury (ed.), *Les Indes florissantes: Anthologie des voyageurs francais (1750-1820)* (Paris: R. Laffont, 1991), 351 (my translation).

[11] R.O. Cambridge, *Account of the War in India 1750-1760* (London: T. Jefferys, 1772), viii.

battle of Panipat, where the combined Afghan troops defeated the Marathas, even the orthodox Sunni Afghan chiefs of Rohilkhand wrote numerous apologizing letters for joining the Durrani side to the Hindu Marathas in order to keep all their options open.[12]

Although European observers are quite unanimous in their disqualifications of Indian armies, we should keep in mind that, as long as there were no fanatic spoilsports from outside, these armies could still be successful in accommodating the enemy. In this respect the whole pomp and circumstance of courts and bazaars that surrounded the large Indian armies could be very instrumental in winning a battle or taking a fortress.[13]

Under these conditions, the numerous stone and mud fortresses of Hindustan served perfectly well as, what Streusand called, 'units of political bargaining power'.[14] Bombardment by artillery and other siege operations only served to strengthen the attacker's bargaining position and to erode that of the defender. Like the Ottomans, the Mughals had such huge guns that only their noise was sufficient to tear down the battlements.[15] Nevertheless, during the eighteenth century, large-scale sieges with heavy artillery occurred only incidentally. The lack of mobility of heavy guns was a major problem. Besides, the location of most of the Indian fortresses made the siting of heavy guns against them very difficult. In Bundelkhand, forts were protected by a wide belt of thorny jungle. Similarly, the forts in Rohilkhand

[12] See the somewhat biased but very detailed account, mainly based on Marathi manuscripts, by T.S. Shejwalkar, *Panipat: 1761* (Poona: Deccan College, 1946), 45.

[13] Many illuminating examples of the prevailing 'Indian' ways of warfare, including critical European comments, are found in C. Compton, *A Particular Account of the European Military Adventurers of Hindustan from 1784 to 1803* (Oxford: Oxford University Press, 1976). Very interesting is also the anthology of French travellers, recently edited by G. Deleury, *Les Indes*.

[14] Term adopted from D.E. Streusand, *The Formation of the Mughal Empire* (Delhi: Oxford University Press, 1989), 65.

[15] C.M. Cipolla, *Guns, Sails and Empires: Technological Innovation and the Early Phases of European Expansion 1400-1700* (Yuma: Sunflower Univeristy Press, 1992), 95. For an interesting example of this, see the Durrani siege of Nishapur in 1751 as described in Ganda Singh, *Ahmed Khan Abdali* (London: Asia Publishing House, 1959), 94.

were often surrounded by bamboo hedges which a cannon ball was unable to penetrate.[16] According to Wendel, even the mud forts of the Jats were too thick to be penetrated by artillery.[17] With the decline of Mughal power, fortresses became new rallying points for the rising local magnates who used them as safe havens, storehouses and granaries during the war season.

Apart from small fortresses, the continued use of the elephant in India also fits very well into the general atmosphere in which Indian battles took place. They were not so much about crushing, but more about outbluffing and impressing the enemy. From the twelfth century onwards elephants lost much of their military effectiveness as they were no match to the mounted archers from Central Asia.[18] Even so, in the eighteenth century we still find them in great numbers among the emerging Mughal successor states. They could serve well for the transport of artillery but because of the expenses oxen and camels were generally preferred.[19] Moreover, a large elephant establishment (*pil-khana*) was one of the requirements to fulfil the royal pretensions of the aspiring new courts. In Hyderabad and Awadh, for example, more than 1,000 elephants were held by the reigning Nawab.[20] Probably, elephants were still available from mainland Hindustan and the northern Deccan as large parts of these areas still consisted of uncultivated forests. Also the Afghan Nawabs in northern India used to send elephant hunting expeditions (*kheda*) to the jungles of the

[16] Irvine, *Army*, 260-9.

[17] Francois-Xavier Wendel, *Les mémoires de Wendel sur les Jats, les Pathan et les Sikhs*, ed. J. Deloche (Paris: Ecole Française d'Extrême-Orient, 1979), 95.

[18] Streusand, *Formation*, 55-6. For pre and early Islamic India, see A. Wink, *Al-Hind: The Making of the Indo-Islamic World*, vol. 2: *The Slave Kings and the Islamic Conquest of India 11th-13th Centuries* (Leiden: E.J. Brill, 1997); for Delhi Sultanate, see S. Digby, *War-horse and Elephant in the Delhi Sultanate* (Karachi: Oxford Orient Monographs, 1971); for Mughals, see P. Horn, *Das Heer- und Kriegswesen der Grossmoghuls* (Leiden, E.J. Brill, 1894), 51-6; Irvine, *Army*, 174-81 and R.J. Phul, *Armies of the Great Mughals* (Delhi: Oriental Publishers and Distributors, 1978), 64-6.

[19] See the accounts of Dubois, Desvaulx and Maitre de La Tour in Deleury, *Les Indes*, 551, 553-4, 579.

[20] Irvine, *Army*, 180.

Tarai at the fringes of the Himalayas.[21] The elephant was still considered a symbol of authority, but was not any longer an indication of real military power. Consequently, the elephant figures prominently in the official local chronicles in which Nawabs and Rajahs fight each other from the top of their *haudas*. Nevertheless, only very incidentally were elephants used on a large scale in warfare. In real battles they mostly served as high platforms from where the commanding generals could watch and be watched from the battle scene. The visibility of the leader was of the utmost importance for his troops. If he was killed or, more frequently, just disappeared, the army had lost its employer and, in a wink of time, all fighting men had left the battle scene and the battle was lost. Although the *hauda* could be made of high and strong metal plates, the leader was always extremely vulnerable by being a clear and visible target for the enemy's archery or musketry. Thus, there was always a kind of ambivalence on the part of the commanding generals: although riding an elephant in battle offered them status and authority, it also very much exposed them to unfriendly fire. The Durranis were astonished to find their Indian counterparts still making such fuss over their elephants. For example, in response to remarks that elephants were specially used by the emperors and nobles of Hindustan, Ahmad Shah Durrani said that:

Elephants were admirable means of baggage transport. But a mount, the control of which is not in the hands of the rider, and it can carry him whither it wills, should not be resorted to; while a litter is only suitable for a sick man.[22]

Indeed, the grand scale Indian armies, including their elephants and heavy artillery, were no match at all for the Afghan invaders, with their more disciplined and mobile firepower.

Although lances and swords continued to play their part in the Indian war scene, guns and firearms were produced all over the Indian

[21] *Gazetteer of the Rampur State* (Allahabad: Government Press, United Provinces, 1911), 11-12.

[22] Ghulam Hasan Samin Bilgrami, 'Ahmad Shah Abdali and the Indian Wazir Imad-al-Mulk (1756-1757) (Bazi az ahwal-i Ahmad Shah Badshah Abdali)', trans. W. Irvine, *Indian Antiquary* 36 (1907), 60.

subcontinent. Wherever possible, military entrepreneurs had set up their own ironworks and gunneries.[23] Before the spread of European expertize in the late eighteenth century, metallurgical tradition was modelled on the fine Persian damascened barrel. The gunsmiths of north-western India were reputed for the fabrication of gun and musket barrels. The wide distribution of gunneries and the relative low costs involved made firearms widely available.[24] The use of firearms was further stimulated by the widespread supply of saltpetre which facilitated the production of gunpowder. Nevertheless, compared with Europe, indigenous production was not standardized and weighted heavily towards the dispersed production of a limited number of fine matchlocks or muskets. The effective use of matchlocks required a much greater degree of drill and discipline than the fluid Indian armies could provide. Initially, even the Indian sepoys under the command of European generals continued to serve in their traditional way: without much order, preferably behind a wall or in a trench. It appears that they were fully submitted to European drill and training only after the more convenient flintlock musket with bayonet made its appearance in India.[25] Indeed, I would argue that the introduction of the lighter, faster-firing and more reliable flintlock musket, whose powder was ignited by a spark produced through the action of flint on steel, in combination with the ring or socket bayonet, greatly enhanced the effectiveness of European methods of drill and discipline.[26] This entailed not only an increased demand for European

[23] See for example Deleury, *Les Indes*, 145 (Lahore) and 159 (Chhatarpur in Bundelkhand).

[24] B. Lenman, 'The Weapons of War in 18th-Century India', *Journal of the Society for Army Historical Research* 46 (1968), 33-43; Z. Zygulski, 'Oriental and Levantine Firearms', in C. Blair (ed.), *Pollard's History of Firearms* (London: Country Life Books, 1983), 444-54. For a detailed account of the metallurgical technology in early nineteenth-century Kashmir, see W. Moorcroft and G. Trebeck, *Travels in the Himalayan Provinces of Hindustan and the Panjab; in Ladakh and Kashmir; in Peshawar, Kabul, Kunduz and Bokhara*, vol. 2 (London: J. Murray, 1841), 195-213.

[25] S. Bidwell, *Swords for Hire: European Mercenaries in Eighteenth-Century India* (London: J. Murray, 1971), 56-7.

[26] Interestingly, one of the first to adopt flintlocks were the Iroquois Indians whereas the European armies generally adopted the flintlock as late as the second

officers but also for engineers who could maintain and repair the new locks. As in Europe, the combination of drill and the new firelocks doubled the infantry's firing power and increased its flexibility. Not surprisingly, after the middle of the eighteenth century, its success triggered off a wave of modernization across the Indian subcontinent and within a few decades all Indian armies relied heavily on disciplined units of infantry staffed with European officers and equipped with modern flintlocks and bayonets.[27] But, although the Indian rulers were very keen to catch up with the European developments, and at times proved even very successful against their British adversaries, the usual Indian politics of permanent sedition and ever shifting alliances remained completely at odds with the new military principles. In the end, technological and tactical adaptation only served to accentuate the persisting cultural and institutional dichotomy between India and Europe.[28]

THE DURRANI EMPIRE

Ahmad Shah Durrani is often depicted as the founder of modern Afghanistan. Being a leading member of the Durranis, one of the most powerful Afghan tribes, he was able to establish a new imperial

---

half of the seventeenth century: J.P. Puype, *Dutch and Other Flintlocks from Seventeenth Century Iroquois Sites* (Rochester: Rochester Museum and Science Center, 1985). In Europe the adoption of the flintlock required profound adaptions in the drill and organization of armies which were still used to the disciplined employment of matchlocks and pikes. Nevertheless, after their adoption and the accompanying changes in training and supply, they were employed with maximum effectiveness both in Europe and in India. Indeed, it is hard to imagine how unwieldy European *tercios* of drilled and disciplined matchlock- and pikemen would have overawed the mobile cavalry units of their Indian counterparts.

[27] Though gunflints were relatively rare in India, agates and other quartzose minerals could be used as effective alternatives: G. Watt (ed.), *A Dictionary of the Economic Products of India* (Calcutta: Superintendent of Government Printing, 1889-93): 'Flint' and 'Carnelian').

[28] For a good analysis of the deep cultural gulf between Indian and European warfare, see D.H.A. Kolff, 'The End of an *Ancien Régime*: Colonial War in India 1798-1818', in J.A. de Moor and H.L. Wesseling (eds), *Imperialism and War: Essays on Colonial Wars in Asia and Africa* (Leiden: E.J. Brill, 1989), 22-49.

configuration in the area which had its core in present-day Afghanistan but which included very extensive territories to both its west (eastern part of Khorasan), east (Punjab, Sind, Kashmir, Sirhind) and south (Baluchistan and the Makran coast). *De jure* the Durrani emperor followed in the footsteps of Nadir Shah by claiming to be the sovereign over all Iran, Turan and Hind (India). *De facto*, though, he took over only the most lucrative eastern parts of his patron's former empire while the largest part of Iran was left to several other break-away generals of Nadir Shah.[29] Effective control of this extensive area, full of impenetrable mountains and deserts which were full of obstreperous nomads and tribes, was only exercised along the major trade routes and cities. Beyond, Durrani rule could only be maintained from the saddle. This implied permanent campaigning in order to suppress revolts, to enforce financial contributions, to claim revenue payments and to enter into new alliances. But even more important, to keep the wheels of empire moving the Durranis had to exploit the still enormous riches of India.

Apart from any plunder, the regular income from the Indian provinces accounted for at least 70 per cent of the total Durrani revenue. Thus, in imitation of their Ghaznavid and Ghurid predecessors, the Durrani Empire was fully geared to the exploitation of India. Invading India had the advantage that one could easily recruit irregular, but zealous, companies of *ghazis*, always eager to take part in an enterprise which gained both the treasures of the present world as eternal merit in the world to come. In fact, the greedy Durrani inclination towards India was not much different from that of Nadir Shah who had only once invaded India in 1739. It is claimed that the latter took with him Rs. 150,000,000 in cash besides jewellery, rich clothing and furniture worth Rs. 500,000,000 more.[30] The paucity of sources makes it difficult to assess the relative weight of Indian plunder and other kinds of forced contributions for the financial situation of the Durranis. The total revenue of the Durrani territories is variously estimated but probably ranges somewhere

[29] See my *The Rise of the Indo-Afghan Empire, c.1710-1780* (Leiden: E.J. Brill, 1994), 45-68.

[30] Sarkar, *Fall of the Mughal Empire*, vol. 1 (Calcutta: Sarkar and Sons, 1932-50), 3.

between Rs. 20 and 30 million.[31] Unfortunately, nothing specific is known about the total plunder that was taken by the Durranis on their Indian campaigns. But if we take the already mentioned amount of Nadir Shah as a broad indication then it appears that the total booty taken from India between the years 1750-70 must have been at least equal to the total assessed (*jama*) land revenue for these years. All this remains of course very much a matter of speculation but, in any case, we should not deem the Durrani invasions to be mere chaotic tribal outbreaks. On the contrary, these campaigns, planned by a strong coalition of Afghan leaders and Hindu bankers, were designed to keep trade running and cash flowing.

But how should we assess the military means these enterprising men had at their disposal? In other words, what do we, actually, know about the structure and strength of the Durrani army?

THE DURRANI ARMY

Ahmad Shah had served so long under Nadir Shah that it should be of no surprise that he organized his army on similar lines as the Persian one. Hence, the backbone of the Durrani army consisted of an elite corps of about 10,000 royal slaves (*ghulam-shahis*), of one third being former Qizilbashes who had served in the Nadirid army.[32] The

---

[31] Based on the following estimates: Ghulam Sarwar's report at the end of the century, in *Oriental and India Office Collection (OIOC)*, London, Bengal Proceedings, Secret and Political Consultations, P/Ben/Sec/41, 7-7-1797 (Rs. 27,000,000); Comte Louis de Féderbe de Modave, *Voyage en Inde du Comte de Modave 1773-1776*, ed. J. Deloche (Paris, 1971), 369 (20,000,000); Strachey, 'Memoir on the Revenue and Trade of Caubul', in *Oriental and India Office Collection*, London, Home Miscellaneous, H/659, fols. 1-133 (23,000,000); 'Abd al-Karim Nadim b. Isma'il Bukhari, *Histoire de l'Asie Centrale par Mir Abdoul Kerim Boukhari (Ahwal-i Kabul wa Bukhara)*, trans. J. Schefer (Paris: E. Leroux, 1876), 6-7 (more than 13,000,000).

[32] Kashi Raj, 'Narrative of the Battle of Panipat (Karzar-i Sadashiv Rao Bhau wa Shah Ahmad Shah Abdali)', *Asiatic(k) Researches*, 3 (1792), 104-5. Cf. *OIOC*, Bengal Proceedings, Secret and Political Consultations, P/Ben/Sec/41, 7-7-1797, 'Intelligence Report Ghulam Sarwar'; Elphinstone, *An Account of the Kingdom of Caubul*, vol. 2 (Oxford: Oxford University Press, 1972), 267.

remainder was also recruited from the old troops of Nadir Shah or other non-Afghan groups such as Tajiks, Abyssinians, Kurds and Kalmuks.[33] Later he made an arrangement with the chiefs around Kabul and Peshawar to supply him with new young recruits in return for assignments in land.[34] The *ghulam* corps was organized in regiments (*dastas*) of 1,200 each under the command of *qullar-aqasis*. These were selected from amongst royal bodyguard of 500 eunuchs and personal attendants (*pishkhidmatgar*). The *pishkhidmatgars* – under the Safavids called *qurchis* – were the only soldiers who were mounted on the Shah's special *Turki* horses, procured from the Uzbek and Turkoman territories along the river Oxus, whereas the mounted *ghulams* and the other regiments owned their own horses.[35]

The guard of *ghulams* in the Durrani army served particularly well during Ahmad Shah's Indian campaigns. They were mostly kept in reserve behind the main lines of artillery and cavalry in order to reinforce weak spots in the defence or to charge into the already broken or dispersed lines of the enemy. Additionally, in a retreated position they were able to keep the men in place and check possible desertions and unnecessary flights from the battle field.[36] As such, especially in the short run, military slaves proved a splendid antidote to the huge gravitational forces of permanent sedition and shifting alliances. Even the ever sceptical but sharp-witted Law de Lauriston was impressed as he claimed that: 'That which renders the Pathans superior is primarily the discipline and subordination so strictly observed in the Abdali [Durrani] army.'[37]

Part of the *ghulam* troops were equipped and trained with short

[33] *OIOC*, Bengal Proceedings, Secret and Political Consultations, P/Ben/ Sec/41, 7-7-1797, 'Intelligence Report Ghulam Sarwar.' The majority, however, were Persians and Tajiks see *OIOC,* London, Elphinstone Mss, Eur. F.88, Box 13, H: 'Sketch of the Dooraunee History', fol. 590.

[34] Elphinstone, *Account*, vol. 2, 267.

[35] Elphinstone, *Account*, vol. 2, 272-3. Cf. Ganda Singh, *Ahmed Khan*, 357-64.

[36] Abu al-Hasan b. Muammad Amin Gulistana, *Mujmal al-Tarikh-i Ba'd-Nadiriya*, vol. 2, ed. O. Mann (Leiden, 1891-6), 130-1.

[37] Law de Lauriston, *Mémoires*, 191 (my translation).

light blunderbusses with a heavy calibre called *sher-bachas*.[38] Since these firearms were effectively used by mounted troops, they most probably consisted of flintlocks or of Mediterranean *miquelet* locks, which were far more manageable and more adaptable for cavalry purposes than matchlocks.[39] In imitation of the mounted Janissaries of the Ottomans, the Safavid emperor Shah Abbas (1588-1629) had already set up a corps of mounted infantry armed with muskets, swords and daggers.[40] The eighteenth-century Persian army of the Zands also consisted of an elite corps of 1400 mounted *ghulams* armed with flintlocks (hence *ghilman-i chaqmaqi*).[41] Thus, it seems that in India the phenomenon of the mounted musketeer armed with modern flintlock firearms was not exclusively introduced through European channels – this occurred during the mid-eighteenth century – but also through the Durrani invasions and subsequently adopted by the indigenous Indian states, most notably by the Jats of northern India.[42]

A contemporary Indian account presents us with a vivid picture

---

[38] Cf. Irvine, *Army*, 112; W. Egerton, *Indian and Oriental Armour* (London: Arms and Armour, 1968), 28-9; Ghulam Ali Khan Naqawi, 'Nigar-Nama-yi Hind', trans. extracts A.R. Fuller in H.M. Elliot and J. Dowson, *The History of India as told by its Own Historians,* vol. 8 (London: Trübner and Co., 1867-77), 398-9. An interesting collection of Persian *sherbachas* (in German: *Wallbüchsen*) is found in the Historical Museum of Bern. For a description, see R. Zeller and E.F. Rohrer, *Orientalische Sammlung Henri Moser-Charlottenfels* (Bern: Bernisches Historisches Museum, 1955), 267-72.

[39] For the implications of mechanical ignited firearms on cavalry tactics, see: C. Blair, 'The Sixteenth Century', in Blair, *Pollard's History*, 60-1; on the *miquelet* in Ottoman and Persian armies, see Zygulski, 'Oriental and Levantine Firearms', 431, 446.

[40] *Tadhkirat al-Muluk*, 32-5; H. Inalcik, 'The Socio-Political Effects of the Diffusion of Firearms in the Middle East', in V.J. Parry and M.E. Yapp (eds), *War, Technology and Society in the Middle East* (London: Oxford University Press, 1975), 208; 'Barud', in *Encyclopaedia of Islam*, 2nd ed. (Leiden: E.J. Brill, 1960-2005), 1068.

[41] J.R. Perry, *Karim Khan Zand: A History of Iran, 1747-1779* (Chicago and London: University of Chicago Press, 1979), 280.

[42] Ghulam Husain Khan Tabatana'i, *Seir Mutaqherin* (Siyar al-Muta'khkhirin), vol. 4, trans. Haji Mustafa (M. Raymond) (Calcutta: T.D. Chatterjee, 1902), 28.

of how the mounted musketeers were actually employed on the battlefield of Panipat:

At noon, the Bhau on horseback and Wiswas Rao, the Nana's son, on an elephant, delivered a charge and engaged in fighting at close quarters with spear and musket and sword. Mir Atai Khan was slain. Ahmad Shah saw that his troops were now very hard pressed; he summoned the *Bash Ghul* squadrons – which means his slaves who numbered 6000 men – and cried out: 'my boys! This is the time. Encircle these men.' The three squadrons of slaves moved from three sides and brought the vanguard of the Bhau's army under musket fire all at once, and swept away their firm stand. The Maratha vanguard retreated and mixed with the division under the Bhau himself. A great tumult arose; men turned their faces to flight. The Bhau's personal guards showed some firmness and kept standing at some places. One squadron of slaves, numbering 2000 men, came from the right and after firing off their muskets went away to the left. Another squadron which came from the left, after emptying their muskets, went away to the right. The third squadron which came from the front, discharged their muskets at the Bhau's vanguard and then turned to the rear. Before the enemy could recover, these men had loaded their muskets again and arrived, the left squadron on the right wing and the right squadron on the left wing, while the squadron that had been originally in front fell on the rear. During this circular manoeuvre, they quickly discharged their muskets from one side and went away to the other. It looked as if on all four sides troops were attacking the Marathas simultaneously. The fighting went on in this manner. The Maratha soldiers who had been spread over the field drew together into a knot at their centre. It came to such a pass that these three squadrons enveloped that lakh of troopers and revolved around them.[43]

As a follow up of this wheeling around, the Shah ordered the heavy cavalry, armed with swords and spears, to charge massively into the shaken lines of the enemy. Similar tactics had already been highly successful against the Rajputs at the battle of Manipur in 1748. Here the *ghulam* cavalry corps had galloped up within easy range of the Rajputs. After they made a volley of fire, they galloped back as swiftly

---

[43] Nur ud-Din Husain Khan Fakhri, 'An Original Account of Ahmad Shah Durrani's Campaigns in India and the Battle of Panipat' (*Tawarikh-i Najib al-Daula*), trans. J. Sarkar, *Islamic Culture* 7, 3 (1933), 452-3.

as possible thereby completely surprising the Rajput cavalry who had prepared for a hand to hand fight.[44]

Although contemporary Indian observers seem to have been completely stunned by it, the manoeuvre of wheeling around the enemy's units was of course nothing new as it was an essential part of the characteristic tactics of the Mongol and Turkish mounted archers of Central Asia.[45] Indeed, mounted archers were still present in the eighteenth-century Afghan and Indian armies but this was only on a relatively small scale, as they required an inordinate amount of training and experience, both in the art of horsemanship and archery. In this respect the handling of a flintlock or miquelet musket was undoubtably easier and much more convenient. Although having certain similarities with traditional Central Asian warfare, it is plausible that the Durrani cavalry tactic should be related to the sixteenth-century European *caracole*. In this manoeuvre successive ranks of cavalry charged toward the enemy, discharged their pistols or carbines and wheeled off to the sides.[46] In Europe this tactic was soon outmoded

[44] H.C. Tibbiwal, *Jaipur and the Later Mughals (1707-1803): A Study in Political Relations* (Jaipur: Hema Printers, 1974), 112-13.

[45] Cf. D. Martin, 'The Mongol Army', *Journal of the Royal Asiatic Society* 49 (1943), 69-76; 'Harb', in *Encyclopaedia of Islam*, 2nd ed. (Leiden: E.J. Brill, 1960-2005), 179 ff.

[46] Much earlier the lances of the feudal knights had already fallen before the concentrated defence of massed pikemen and the increased firepower of archers and gunners. Another major factor that led to the disappearance of the fully armoured horseman was The invention of the wheellock ignition (from circa 1540) produced a marked change in cavalry tactics by facilitating the production of short firearms like pistols and carbines which could be used most effectively on horseback. The result was that the heavily-armoured knights who charged with their lances were rapidly replaced by mounted pistoleers. Although the *caracole* continued to be practised into the seventeenth century – it remained very effective against the Ottoman infantry which did not use pikes and were less drilled – the 'new' European armies again resorted to the simpler tactic of the direct charge with the naked sword and lance into the enemy troop formation. (D. Chandler, *The Art of Warfare in the Age of Marlborough* (London: B.T. Batsford, 1976), 27-61; E. Von Frauenholz, *Das Heerwesen in der Zeit des dreissigjährigen Krieges,* vol. 1: *Das Söldnertum* (Munich: Beck, 1938), 38-42, 58-62).

although the Habsburg cavalry employed it with much success against the Ottomans.[47] Most probably, the Ottomans and in their wake the Safavids and Afghans, adopted the *caracole* and started to equip and train their mounted slave corps with *miquelet* or flintlock firearms.[48] At the same time, though, the remainder of the cavalry stuck to their traditional way of fighting *à l'arme blanche* which was, anyway, not much different from the prevailing cavalry tactics in contemporary Europe.[49]

Apart from the elite corps of bodyguards and *ghulams*, the bulk of the Durrani army was composed of cavalry recruited from the Afghan tribes, lightly armored and armed with lances and broadswords. These tribal units consisted of both nomads, semi-nomads and peasants. During the eighteenth century, many of the nomadic Durrani leaders had gained extensive landed interest in the Kandahar area. Nadir Shah had assigned lands in *tiyul* (assignment of land-revenues, the Persian counterpart of the Mughal *jagir*) to the Durrani chiefs in remission of the crown revenue but subject to the supply of 6,000 horse, which was at a rate of one horseman for every plough. Ahmad Shah continued this arrangement but not without increasing the rate to two horses for each plough; upon the understanding that the two were not to be employed simultaneously, but to relieve each other. In this way, he assured himself of a ready and fresh supply of both horses and soldiers. In addition to their *tiyul* rights the Durrani tribal contingents were paid according to the time they actually spent in service but always receiving a minimum of a three-month pay, either in cash from the royal treasury or by *barat* (written assignment) upon the produce of certain districts. Apart from their *tiyul* lands, which was reckoned to be 6 tomans (*c.* 70 Company Rs.), this amounted to about 19 tomans (*c.* 210 Company Rs.) which resulted in a total annual military salary of about Rs. 280 to which has to be added plunder and booty. Although I will not go into the details of

---

[47] V.J. Parry, 'La manière de combattre', in Parry, *War*, 225, 231. There was, however, a change from pistols to short firearms which had a longer range. It is interesting to note that in Poland the mounted musketeer also remained very popular (Roberts, *Military Revolution*, 8).

[48] Parry, 'La manière', 244-5.

[49] Inalcik, 'Socio-Political Effects', 199-201.

internal Afghan developments, it should be kept in mind that the income and plunder from military service, especially from the many campaigns in India, was flowing directly into the pockets of the Durrani landlords or their financiers who had provided them with financial advances on the future receipt of plunder or revenue. This enabled them to make considerable investments in agricultural development and, especially in the Kandahar area, extensive areas were taken under the plough.[50]

Next to the slave and tribal cavalry, infantry and heavy artillery were almost irrelevant in the Durrani military. At Panipat the Durranis had some 20 to 40 heavy cannon which was much less than the Maratha ordnance of about 200 pieces. Neither heavy artillery nor slow-moving linear infantry were appropriate for long-distance operations. Especially in rough areas like Iran and Central Asia, their immobility rendered them ineffective. Here, the use of artillery was mainly confined to siege warfare and it is not too much to say that the Durranis never really made any effective use of artillery in the field.[51] In fact, the Durranis from Herat and Kandahar, inherited the Turko-Persian tradition which was based on mobility and long-distance marches in the extensive spaces and steppes of the arid zones of the Middle East and Central Asia.[52]

Under these circumstances, it was the one-humped camel or

[50] For information on tribal cavalry in Kandahar, see H. Rawlinson, 'Report on the Dooranee Tribes', in L.W. Adamec (ed.), *Historical and Political Gazetteer of Afghanistan,* vol. 6: *Kandahar* (Graz: Akademische Druck und Verlagsanstalt, 1985), 509-77.

[51] Take for example the battle of Gulnabad (1722) where the Persian artillery was completely overrun by the swift Afghan cavalry advance (L. Lockhart, *The Fall of the Safavid Dynasty and the Afghan Occupation of Persia* (Cambridge: Cambridge University Press, 1958), 130-43. Even at sieges the Durrani artillery was not very successful. One exception was the capture of the Jat fort of Ballabhgarh. According to the official Durrani chronicle, European artillery played a decisive role at the siege of Nishapur (O. Mann, 'Quellenstudien zur Geschichte des Ahmad Sah Durrani', *Zeitschrift der Deutschen Morgenländischen Gesellschaft* 52 (1898), 331) but, probably, reality was too embarrassing to be openly admitted (Ganda Singh, *Ahmed Khan*, 94).

[52] Cf. the dual organization of the Russian army: European on its western border but sticking to mobile cavalry on their eastern and southern borders (T.

dromedary which fared extremely well as it was the animal most suited for long-distance travel in the Middle East and Iran. The camel may be considered by far the most efficient beast of burden, since it was not only capable of going on for several days without drinking, it could also carry relatively heavy loads (150-200 kg.) at a speed of about 30 km. a day. Unloaded it could travel for many days at a double that speed. Unlike cattle, camels are economical feeders that never overgraze the vegetation and keep on moving while feeding. Moreover, as a beast of burden it was capable of keeping pace with the rapid movements and long marches of cavalry units. Camels were, however, only fit for relatively flat and dry terrain and in general not able to survive in the very humid conditions of the Indian monsoon climate. For this reason, their radius of action was confined to the north-western parts of India. Here the best camel breeds were the small but strong Afghan or Pahari camels which were also fit for the cold and hilly circumstances of Central Asia. In fact, the best of these Afghan camels were produced from inter-breeding one-humped dromedaries – predominant in the Middle East – with two-humped Bactrian camels – predominant in Central Asia. Not surprisingly, the initial result was a one and a half humped camel which, by careful breeding, could be developed into a very strong stock of dromedaries. In India proper, excellent camel breeds were found around the Thar Desert in Rajasthan and in Sind, Kutch and Kathiawar.[53] More to the south and east the semi-arid terrain of the camel gradually gave way to the more humid terrain where the draught and carriage bullock dominated the scene. Nevertheless, according to Modave, in the four

---

Esper, 'Military Self-Sufficiency and Weapons Technology in Muscovite Russia', *Slavic Review* 28 (1969), 185-208.

[53] On the camel in general, see R.W. Bulliet, *The Camel and the Wheel* (Cambridge, Mass.: Harvard University Press, 1975) and H. Gauthiers-Pilters and A.I. Dagg, *The Camel: Its Evolution, Ecology, Behavior, and Relationship to Man* (Chicago: University of Chicago Press, 1981). For India, among others see J. Deloche, *La circulation en Inde avant la révolution des transports,* vol. 1: *La voie de terre* (Paris: Ecole Française d'Extrême-Orient, 1980), 229-50; J.H. Steel, 'Notes on the Camel as an Animal of Transport', *Quaterly Journal of the Veterinary Society* 4 (1886), 155-71, 244-58. See also the remarks of Maitre de la Tour in Deleury, *Les Indes,* 573.

cities of Faizabad, Lucknow, Agra and Delhi there were still at least as many as 100,000 camels.[54] Turning towards the Deccan or Bengal, though, camels became increasingly rare. Obviously, for any expanding military power in India, be it the horse and camel based army of the Afghans or the infantry and bullock based army of the British, crossing this ecological transition zone – let say from camel to bullock terrain – posed serious logistical challenges.[55]

In India camels were only rarely used in direct military engagements because they lacked the spirit and manoeuvrability of horses. Of course, like elephants, camels were used for transport of heavy guns. During the seventeenth century, however, a light two-pounder swivel-gun was fixed to the saddle of the camel which could be fired from the camel's back. It appears though that only in the eighteenth century the Afghan armies began to employ the camel-gun (*shutur-nal*, *zamburak*, *shahin*) on a larger scale and in a more systematic way than before.[56] In any case, the successful Afghan invasions of the maid-century, further stimulated the use of camel-guns among their counterparts in India – especially among the Rajputs, Rohillas and Sikhs – as well as in Iran. Most probably, the use of mobile camel guns was introduced by the Ghilzai-Afghans as early as 1722 at the battle of Gulnabad and, subsequently, adopted by the Persian armies.[57]

Similar to the mounted musketeers, the camel-gun was aimed to adapt the newly increased firepower to the mobility of Middle Eastern and Indian warfare. The swivel ensured an increased flexibility in

[54] Modave in Deleury, *Les Indes*, 510-11.

[55] See for example the ongoing debates on the use of camels and bullocks in the Bengal Military Consultation for the years 1794-1821 or the Board's Collections 1801-1806: F/4/208, nos. 4632-4 (*OIOC*).

[56] Already during the seventeenth century Bernier describes a camel-gun which could be fired from the camel's back. But during the second half of the 18th century references to camel-guns become much more frequent (Irvine, *Army*, 136-7; Horn, *Heer*, 28) Cf. R.K. Gupta, 'The Military System of the Jodhpur State, *c.*1212 to 1947 AD', *Asiatische Studien* 45 (1991), 85.

[57] F. Colombari, *Les zemboureks: Artillerie de campagne a dromedaire, employée dans l'armée persane* (Paris: Martinet, 1853), 292. The very effective use of the *zamburak* at Gulnabad is confirmed by the contemporary sources (L. Lockhart, *The Fall of the Safavi Dynasty and the Afghan Occupation of Persia* (Cambridge: Cambridge University Press, 1958), 131, 141.

firing. In order to increase the speed of fire the Durranis attached two gunmen to each camel. To control the fire and to protect the animal against the shock of discharge the camel was usually made to kneel on the ground. To prevent its rising and make it immoveable each leg was fastened with a cord. But even under attack, camels proved extremely stolid and relatively unperturbed by gunfire.[58] Obviously, different kinds of guns could be used. Whenever lighter guns were employed the gunman could also fire directly from the back of the camel without feeling the need to kneel down. Sometimes, even a second gun was attached to the saddle.[59] Obviously, on the part of the camel-drivers all this required a great deal of training and drill.

It appears that the word *shutarnal* (lit. camel-gun barrel) denotes a relatively long swivelgun similar to the heavy Afghan rifle or jezail. The *zamburak* (lit. 'a little bee', i.e. 'like a bee') was a much shorter specimen. *Shahin* (lit. falcon; hence falconet) was used by the Durranis themselves and under this name we find the camel-gun also used, for example, in eighteenth-century Damascus.[60] On the whole, however, there did not exist a clearly recognizable distinction between them, and during the eighteenth century all three generally denoted a camel-gun. On the Indian scene the swivel-guns mounted on camels proved spectacularly effective. In fact, it was a development toward more mobile, faster firing artillery which was smoothly integrated into the cavalry-oriented army of the Durranis.[61]

The Durrani *shahin* corps amounted to a total of about 5,000 camels. At the battle of Panipat, Ahmad Shah was reported to have employed 2,000 of them as a kind of mobile artillery directly behind the heavy artillery but in front of the infantry and the cavalry. Unlike the heavy ordnance, which could at best be useful during the earliest

---

[58] D.B. Burn, 'Camel Corps', *Quaterly Journal of Veterinary Studies* 7 (1889), 98.

[59] R.K. Saxena, *The Army of the Rajputs* (Udaipur: Saroj Prakashan, 1989), 203.

[60] A.K. Rafeq, 'The Local Forces in Syria in the Seventeenth and Eighteenth Centuries', in Parry, *War*, 295.

[61] The chronicles give many examples of the effectiveness of camel-guns. See e.g. Irvine, *Army*, 135-7 and Maistre de la Tour in Deleury, *Les Indes*, 574.

stages of a battle, the camel artillery had the main advantage of being relatively flexible. Positioned in one line and under disciplined command the volleys of *shahin* shots proved to be quite effective against massive cavalry charges. In 1759, for example, the camelry corps successfully covered the massive crossing of the Durrani army of the Yamuna River against the heavy cavalry attacks of the Marathas.[62] Hence, it is of no surprise that, during the eighteenth century, the light camel artillery became a distinctive element of each serious army, both in the Middle East and in northern India.

Apart from the regular units of cavalry and camel corps, each Indian campaign saw a rapid increase of all kinds of irregulars, mostly adventurers and mercenaries who attached themselves to the Durrani train and who held various functions. A contemporary report claims that to each Durrani soldier four mounted skirmishers or *yatims* (lit. robbers) – elsewhere also referred to as *qarawuls* – were attached.[63] These forces were intended to harass and pillage the enemy and followed directly in the rear of the regulars. They were also employed as light cavalry which tried to cut off supplies and prevent the enemy from foraging the countryside. When the Durrani army was on the march they were sent in all directions to a distance of 5 *kos* (about 4 km) from the Shah's train and there they set up their encampments. Their movements and behaviour were controlled by a special unit of *nasaqshis*, a kind of military police which main duty was – like the *ghulam* corps – to stand in the rear of the army and to cut down every one who dared to flee. At Panipat, the *nasaqshis* were very successful in keeping the Afghan lines in place and, at times of utmost necessity, in mustering additional troops from amongst the multitudes of camp-followers. Most of the irregulars did not receive any pay but lived exclusively from the spoils they were able to collect in the countryside.[64]

What can finally be said about the total size of the Durrani forces?

[62] Shejwalkar, *Panipat*, 22.

[63] Ghulam Hasan Samin, 'Ahmad Shah Abdali', 16.

[64] Ghulam Hasan Samin, 'Ahmad Shah Abdali', 16-17. Cf. Irvine, *Army*, 227; Kashi Raj, 'Panipat: 1761 (Karzar-i Sadashiv Rao Bhau wa Shah Ahmad Abdali)', trans. J.N. Sarkar, *Indian Historical Quarterly* 10 (1934), 263-4.

According to a British intelligence report from 1793, the Afghan army consisted of 25,000 regular cavalry, exclusive of the corps of *ghulams*, directly around the person of the Shah. In addition, there was a local militia of about 35,000 men. The artillery amounted to 644 pieces of which 153 were on the ramparts of the various forts and cities, and 491 on the ground, 313 of these in Kandahar. In addition there were 5,800 light camel artillery of which only less than half was thought fit for service. According to the report, an additional 60,000 men had been serving in Ahmad Shah's army but since then the imperial control of the countryside had very much declined and many local chiefs were openly ignoring orders from court.[65]

Earlier reports about the Indian campaign seem to corroborate these figures. Combining the figures given by Kashi Raj, Muhammad Jafar Shamlu and Law de Lauriston, it seems safe to estimate the total number of Ahmad Shah's forces at about 100,000, mostly cavalry, and his expeditionary force in India at about 40,000.[66] But after his successors were forced to curtail their operations in India, the army quickly shrunk to about 50,000, a size which was more consistent with the limited state resources and land revenue in Afghanistan proper. Although these figures should be used with great care, it may nevertheless again underline the impressive size of the Durrani army under Ahmad Shah. Also in terms of costs, an army of about 50,000 to 100,000 soldiers, many of them horsemen, was very substantial indeed and must have absorbed almost the entire regular Durrani revenue.[67]

CONCLUSION

It is generally claimed that by the middle of the eighteenth century the European sepoy regiments demonstrated the vast superiority of

[65] 'Intelligence Report Ghulam Sarwar', in *OIOC*, Bengal Proceedings, Secret and Political Consultations, P/Ben/Sec/41, 7-7-1797.

[66] Kashi Raj, 'Panipat'; Muhammad Jafar Shamlu, Tarikh-i Manazil al-Futuh, trans. extracts A.R. Fuller in Elliot and Dowson, *History of India*, vol. 8, 398-9; Law de Lauriston in Deleury, *Les Indes*, 912.

[67] E.g. the Iranian army amounted to about 45,000 men (Perry, *Karim Khan*, 279). For some comparable European figures, see Black, *Military Revolution*, 6-7 (based on Parker).

disciplined armies based on a better drilled infantry and more mobile and lighter artillery. But, as we have seen, not only through European channels new military technology and tactics reached India. The eighteenth-century invading armies of Persians and Afghans introduced new modes of warfare as well. First of all, they employed new tactics and technology based on mounted musketeers. Furthermore, they introduced light camel artillery to Iran and made a more effective use of it in India. In both ways the Afghans successfully adopted the achievements of the military revolution of firepower and adapted it to the long-range mobility and speed of warfare in the Middle East and India. Under these conditions the continued preponderance of the cavalry tradition was ensured, whereas infantry and artillery remained very much underdeveloped until the more radical reforms along European lines.[68] Apart from technology and tactics, the Durrani armies were reputed for their improved discipline. Here again, tradition proved instructive. Discipline heavily depended on the genius of Ahmad Shah and the effectiveness of his personal slave corps, mounted on strong Turki horses and equipped with modern firearms. Besides, substantial material incentives further stimulated the Durrani aggressiveness and readiness for battle. Interestingly, all this fits very well into the pattern of the earlier Ghaznavid and Ghurid invasions also based on the superior abilities of military freemen and slaves on horses. Nevertheless, at the end of the eighteenth century, following the death of Ahmad Shah, the military strength of the Durranis rapidly declined as a result of repeated internal uprisings and dispersing allies. By that time, even the Durrani chiefs and their slaves, their pockets filled with Indian cash and other plunder, reverted to their usual ways of permanent sedition and intrigue.

But if eighteenth-century technological and tactical innovation was not necessarily a monopoly of the Europeans, how should we explain the general shift of the world military balance of power to

[68] It appears that the cavalry of the South-Indian state of Mysore was also more or less influenced by Iranian military tactics as its Nawab Hyder Ali extensively recruited both horses and officers from Iran (M. Wilks, *Historical Sketches of the South of India*, vol. 2 (Madras: Higginbotham, 1869), 718-19 and Ghulam Husain Khan, *Seir*, vol. 3, 123).

the West? Of course, this remains to be the perennial question which still awaits a convincing answer but leaves us ample room for speculation. Therefore, I would tentatively advocate a slightly different approach. Perhaps more important than technological and tactical alterations was the European introduction of a completely different attitude towards warfare as such.[69] During the seventeenth and eighteenth centuries, European warfare had gradually become the exclusive domain of the modern nation state, motivated by such familiar requisites like national honour and national interests. As a result, the European way of warfare became gradually less open to intrigue and sedition and, therefore, less open to peaceful accommodation and settlement.[70] For example, in the early European Middle Ages the phenomenom of looting and plunder was still considered a fully legitimate act on the part of the individual. By 1500 it was still taken as a matter of course, then it became more regulated, but by 1800 it became a practice to be fully condemned and eliminated. By that time general conscription implied that booty was no longer needed as a bait for enlistment in war. Simultaneously, the state had managed to take the organization of warfare out of the hands of the military contractors. From this 'modern' European perspective, the fluidity and openness of South-Asian warfare was increasingly scaled down to mere 'collaboration', 'treason' and 'corruption.' Simultaneously, for the Indians and Afghans, the British generals became increasingly 'inaccessible', 'ungrateful', 'rude' and 'violent'.[71] In the end, on the Indian battlefield these very characteristics of 'Europeaness' could

---

[69] For the early nineteenth-century colonial wars in India, Dirk Kolff, has similarly stressed and analyzed this point in his 'The End of an *Ancien Régime*'.

[70] For this gradual 'cultural' shift in European warfare, see F. Redlich, 'The Praeda Militari. Looting and Booty 1500-1815', *Vierteljahrschrift für Sozial- und Wirtschaftgeschichte*, Beiheft 39 (1956) and his 'The German Military Enterpriser and his Workforce, 13th to 17th Centuries', *Vierteljahrschrift für Sozial- und Wirtschaftgeschichte*, Beiheft 47-8 (1964-5). Also interesting in this respect are the recently rediscovered ideas of Otto Brunner on the medieval feud (O. Brunner, *Land and Lordship. Structures of Governance in Medieval Austria* (Philadelphia: University of Pennsylvania Press, 1992).

[71] See, for example, Ghulam Husain Khan, *Seir*, vol. 3, 191-213.

turn out to be at least as important as any technological or tactical innovations.

Notwithstanding these important European advantages, in the end the expansion of the British Raj stranded in the wide deserts and mountains of Iran and Afghanistan. Not surprisingly, this was the territory *par excellence* of the camel and the horse, leaving the pack and draft bullock, and with that the advanced panoply of the British army, way behind in the muddy and sultry plains of Hindustan.

CHAPTER 7

Warhorse and Gunpowder in India,
c. AD 1000-1850

INTRODUCTION

SINCE MICHAEL ROBERTS launched his thesis on the military revolution
it has been vehemently debated among the military historians of the
West. His revolution occurred in western Europe during the century
after 1560. Basically, it came down to the replacement of relatively
small, undisciplined, cavalry troops by huge, well disciplined and
drilled, gunpowder infantry armies.[1] More recently, Geoffrey Parker
has complemented Roberts' thesis by stressing the radical development
of European fortification as a response to the increased challenge of
artillery. Parker's main contribution, though, was his successful attempt
to extend the Roberts debate to Asia. Hence, it was Roberts' infantry,
the *tercio* or rather the latter's ever more linear adaptations, and Parker's
bastion, the *trace italienne*, which came to represent the European
military *sonderweg* that would ultimately pave the way for the rise of
the West. According to Parker's subtle formulation, the native peoples
of America, Siberia, Black Africa and Southeast Asia lost their
independence to the Europeans because they seemed unable to *adopt*
Western military technology, whereas those of the Muslim world
apparently succumbed because they could not successfully *adapt* it
to their military system.[2]

[1] M. Roberts, *The Military Revolution 1560-1660* (Belfast: M. Boyd, 1956).
See also J. Black, *A Military Revolution? Military Change and European Society*
(London: MacMillan Education, 1991).
[2] G. Parker, *The Military Revolution: Military Innovation and the Rise of the*

The impact of gunpowder technology in Asia had been discussed before. The Chicago historians Marshal Hodgson and William McNeill applied the term gunpowder empire to the large sixteenth- and seventeenth-century Muslim states of the Ottomans in the Middle East, the Safavids in Iran and the Mughals in India. These empires were supposed to owe their long-term stamina to their effective and exclusive use of firepower, employed by both infantry and artillery. Hence, as is often claimed for the European case as well, the coming of gunpowder in Muslim Asia blew up the old feudal order of forts and heavy cavalry.[3] This idea was also implicitly present in David Ayalon's pioneering work on Mamluk Egypt in which he claimed that the traditional, cavalry-oriented Mamluks were defeated by the more innovative Ottomans who made effective use of infantry equipped with modern matchlocks. Here again, modern gunpowder, in this case employed by a disciplined infantry of Janissaries, stood at the basis of rapid imperial expansion.[4]

Combining the debates on the military revolution and the gunpowder thesis raises the paradox that although a stagnant Muslim east was believed to have failed to adapt gunpowder, it was nevertheless able to exploit it for impressive imperial expansion. Of course, it may be argued that their gunpowder weaponry was good enough for their Asian but failed to impress their Western rivals, but this argument has never been explicitly made. Interestingly, a similar contradiction appears in the discussion on military fiscalism where historians have stressed the enormous fiscal pressures exerted by the state's urge to adopt modern gunpowder armies. Whereas in Europe this is claimed to have stimulated the emergence of highly centralized states, in Muslim Asia the results appear to have been more ambiguous. For

---

West 1500-1800 (Cambridge: Cambridge University Press, 1988), 136. Parker made an exception for the Far East.

[3] M.G.S. Hodgson, *The Venture of Islam,* vol. 3: *The Gunpowder Empires and Modern Times* (Chicago: University of Chicago Press, 1974); W.H. McNeill, *The Pursuit of Power: Technology, Armed Forces and Society Since AD1000* (Chicago: University of Chicago Press, 1982); 'Interview with William H. McNeill', *The Historian* 53 (1990), 1-16.

[4] D. Ayalon, *Gunpowder and Firearms in the Mamluk Kingdom: A Challenge to Medieval Society* (London: Valentine Mitchell, 1956).

example, in the case of the sixteenth-century Ottoman Empire, military expansion and state-wide decentralization went hand in hand. According to Halil Inalcik, the spread of modern firearms and the resulting increase in financial burdens provided the local elites and peasantry with the military and financial means to withstand the central authorities.[5] A similar dual tendency was noticed by Burton Stein in the case of fifteenth- and sixteenth-century south India. Although Stein stressed the prominent role of modern gunpowder in the making of the Vijayanagara Empire, he also rightfully observed a fundamental contradiction between, on the one hand, military fiscalism leading to centralization, and on the other, the ongoing existence of vigorous regional chieftaincies.[6]

This essay sets out to examine this apparent contradiction in the process of state-formation by focusing on the foremost military developments in medieval and early-modern India. After stressing the *longue durée* of India's geographical context and military culture, it will specifically address revolutionary changes in the use of the warhorse and gunpowder. It is only in the conclusion that I will briefly discuss the way these military developments may have brought about, simultaneously, both state-formation and deformation.

THE INNER FRONTIER

Warfare in pre-modern India was not a matter of sovereign states. Indeed, from a European perspective, Indian wars were often described as permanent 'civil' wars waged by inveterate rebels and traitors. All this is not very surprising since India lacked the idea of a closed, sovereign state, favoured with well-demarcated borders in which the king enjoyed a monopoly on the use of legitimate violence. As early as the eleventh century the Latin West started to live up to this ideal and came very near to achieving it in the eighteenth century. By contrast, Indian kingdoms remained open-ended entities without

---

[5] H. Inalcik, 'Military and Fiscal Transformation in the Ottoman Empire, 1600-1700', *Archivum Ottomanicum* 6 (1980), 283-337.

[6] B. Stein, 'State Formation and Economy Reconsidered', *Modern Asian Studies* 19 (1985), 387-413. Apart from gunpowder Stein also mentions the state's need to purchase warhorses.

fixed external borders but with numerous inner frontiers. These open marches marked the divide between a highly productive sedentary society mainly inhabited by peasants, and the still extensive space of arid and humid waste, traversed by highly mobile pastoralists, traders and all sorts of warrior bands. In other words, these transitional zones gave access to mobile resources such as cattle, cash and military recruits, all crucially needed for the exploitation of the sedentary, agrarian economy. Thus to wield power the Indian king had to have his stakes on both sides of the divide.[7] One way to achieve this was through warfare, or rather, through campaigning along the inner frontiers of empire. Any Indian king with imperial ambitions had to be constantly on the move, as may be viewed in the case of the more successful Indian rulers such as Akbar (r.1556-1605), Shivaji (1627-80) and Aurangzeb (r.1658-1707). Indeed, as in the case of Iran and the Middle East, warfare and state-formation often eccentrically gravitated towards the frontier.[8]

Apart from being focused on the peripheral zones, Indian campaigns danced to the tune of the monsoon. Any military expedition, or *mulkgiri*, had to wait for October, when the monsoon rains withered away, the roads became dry and the rivers fordable. As such, it also nicely tuned to the grazing season and the *kharif* or autumn harvest, both of which facilitated the supply of food and fodder to the army. Before the start of the hot Indian summer, at about March or April, in any case before the busy time of harvesting the winter crops and sowing the summer crops, the war season came to an end and most warriors returned to their urban quarters or to their villages. For many Indian rulers, this seasonal campaigning was part and parcel of the

[7] The idea of such an Indian inner frontier derives from Jan Heesterman, see e.g. his 'The "Hindu Frontier"', *Itinerario* 13, 1 (1989), 1-17 and more recently 'Warrior, Peasant and Brahmin', *Modern Asian Studies* 29, 3 (1995), 637-54. For further elaborations on the theme, see A. Wink, *Al-Hind: The Making of the Indo-Islamic World*, vol. 2: *The Slave Kings and the Islamic Conquest 11th-13th Centuries* (Leiden: E.J. Brill, 1997) and my 'The Silent Frontier of South Asia, c.1100-1800 AD', *Journal of World History* 9, 1 (1998), 1-25.

[8] Hence the eccentric location of imperial capitals such as Delhi, Vijayanagara, Bijapur in India; Tabriz, Qazwin and Teheran in Iran; and perhaps even Vienna and Berlin in Europe.

annual administrative routine. In the case of the Mughals, these campaigns should not be considered as exclusive military affairs. Their expeditions not only involved the movement of the army but of the entire royal court with all its entourage of thousands of officials, merchants, ladies, musicians, dancers, artists and other non-military personnel. Far from being swiftly moving forays, imperial campaigns often came down to dignified, slow-moving processions, with a usual speed of about 5 km a day, taking breath every one or two days. What was viewed was not an army but a moving *darbar* with all the proper offerings of homage and wealth to the emperor, the latter distributing numerous robes of honour, ranks and other gifts in return. Of course, in all this, the emperor was theatrically presented as the glorious and conquering warrior, the possessor of vast riches, all lavishly displayed to both his following and his enemies.[9] Even though these pompous and noisy campaigns may have been tactical disasters from the purely military point of view, for the Mughals these proved to be the sine qua non of empire.

FITNA

While gazing at these carnavalesque parades, we are faced with another structural feature of pre-modern Indian warfare: before, during and after a battle or siege, the enemy's loyalty was nearly always for sale. Military alliances were as easily forged as broken, taking no account of so-called ascribed affiliations of caste, religion or ethnicity. On the contrary, warriors usually affiliated along open, pragmatic loyalties, which could as rapidly change as their identities.[10] Although, this

[9] There are numerous descriptions of the pomp and circumstance of royal Indian armies. For an early seventh-century example, see E.B. Cowell and F.W. Thomas (trans.), *The Harsacarita of Bana* (London: Albemarle, 1897), 199-211. For the Mughal camp, see e.g. M.A. Ansari, 'The Encampment of the Great Mughals', *Islamic Culture* 37 (1963), 14-24. For an eighteenth-century example, see the account of Maistre de la Tour of the so-called 'grand safari' of Haider Ali of Mysore in G. Deleury (ed.), *Les indes florissantes: anthologie des voyageurs francais (1750-1820)* (Paris: R. Laffont, 1991), 568-72.

[10] Unfortunately, a great deal of energy has been wasted by stressing the political inhibitions of the Indian caste-system. It is most prominent in the more

kind of political arithmetic was not in accordance with the prescriptions of Indian *dharma*, it was in full agreement with the traditional Indian 'science of politics' of *arthashastra* which also recommended a policy of conciliation, gift-giving and sowing dissension. In the Islamic sources, this permanent tendency of sedition is often referred to as *fitna*, literally meaning rebellion but practically implying the manipulation of ever-crumbling alliances.[11] In this context, the un-embarassing exhibition of riches by the Mughal army merely served to entice potential allies into the imperial camp. After an often pre-arranged desertion or defeat, the former rebels were usually left unharmed and were symbolically incorporated into the imperial *mansabdari* system, a system which conveniently translated each warrior's honour in more quantifiable terms of rank and salary. Obviously, this ostentatious ritual of incorporation critically depended on the bold expenditure of enormous treasures. Hence, the imperial army was always accompanied by a huge military bazaar. Its merchants and bankers not only facilitated the transfer of cash and commodities back and forth to the imperial camp but also demonstrated the on-going creditworthiness of the army's commanders.[12]

In order to keep one's ground under such volatile conditions, every

---

argument supporting the 'divided we fell' (against the united Muslim and British forces) thesis. Political alliances were not about rigid caste loyalties but about the ever-shifting loyalties of *fitna*.

[11] For the role of *fitna* in Indian politics and warfare, see A. Wink, *Land and Sovereignty in India: Agrarian Society and Politics under the Eighteenth-Century Maratha Svarajya* (Cambridge: Cambridge University Press, 1986), 9-51. See also D.H.A. Kolff, 'The End of an Ancien Régime: Colonial War in India, 1798-1818', in J.A. de Moor and H.L. Wesseling (eds), *Imperialism and War: Essays on Colonial Wars in Asia and Africa* (Leiden: E.J. Brill, 1989), 22-49 and my 'Indian Warfare and Afghan Innovation during the Eighteenth Century', *Studies in History* 11, 2 (1995), 261-80. For southern India, see P. Price, *Kingship and Political Practice in Colonial India* (Cambridge: Cambridge University Press, 1996).

[12] Generally speaking, the court's almost permanent 'trade deficit' appears to have facilitated the drawing of cash (and thus revenue) by bills of exchange. For the logistics of bills of exchange in India, see S. Subrahmanyam's introduction to S. Subrahmanyam (ed.), *Money and the Market in India 1100-1700* (Delhi: Oxford University Press, 1994), 33.

Indian ruler was eager to create an elite corps of trusted warriors who were personally attached to the royal house. For this, the early Turkish rulers of India turned to the traditional Islamic expedient of purchasing military slaves or *mamluks*. At a later stage, the Afghan sultans of Delhi attempted to establish elite loyalties on the basis of shared ethnic identities. The Mughals forged bonds of loyalty through marital relations with the Rajputs gentry of northern India. In addition, relying on religious models of Sufism and bhakti, they created an esoteric circle of personal disciples who served both in the army and the bureaucracy.[13] In the end, though, even these *mamluks*, relatives, disciples and other 'sons' of the royal house could not withstand the huge incentives offered through *fitna*.

What does all this mean for the military historian? Does it imply that Indian rulers were not interested in effective instruments of war and that they were not sensitive to technological innovations? Partly this is true. One example is the curious Maratha behaviour during the early-eighteenth century, after their defeat of the Mughal army. Despite the fact that they had clearly demonstrated the superior logistics of their light cavalry against the slow-moving, heavy armies of the Mughals, they decided to adopt the pompous ways of their former overlords. They probably realized that these moving *darbars*-cum-*bazaars*, accompanied with all the superpower panoply of un-wieldy artillery and elephants, still played a significant role in prevent-ing violent and costly battles and sieges. Besides, considerations of purely military effectiveness gave way to their prime need to demon-strate their rise to political eminence and to administer their newly conquered territories. Hence, they established a full imperial court with all proper pomp and circumstance, geared to collect the revenue and to control the flow of trade. At the same time, though, the rise of the Marathas exemplifies that military power did make a difference. Like the early Turkish and Mughal invaders, their conquests were

---

[13] For this, see the excellent studies of J.F. Richards, 'The Formulation of Imperial Authority under Akbar and Jahangir', in J.F. Richards (ed.), *Kingship and Authority in South Asia* (Madison: University of Wisconsin, 1978) and 'Norms of Comportment among Imperial Mughal Officers', in B.D. Metcalf (ed.), *Moral Conduct and Authority: The Place of Adab in South Asian Islam* (Berkeley: University of California Press, 1984), 255-90.

launched from the arid and semi-arid zones of the subcontinent. Their early campaigns were swift moving operations with relatively small but disciplined armies which could produce staggering victories on the battlefield. Their initial successes were not achieved through ritual or monetary means, the blessings of the settled powers, but mainly through their military might.[14] Even in India, military power and technology could still achieve tremendous fortune and, only slightly less, fame. In the political game of *fitna*, military superiority always remained an important trumpcard. In order not to waste its costly use, once or twice it had to be majestically demonstrated. But after the initial show of force, the typical pattern of *fitna* usually reasserted itself: conquerors gradually compromised their early ways of violent raiding and turned to the more usual ways of peaceful accommodation and incorporation in which violence was only used in the last resort.[15] In the case of the Mughal army, it also explains the shift from Babur's small but effective warrior band to Aurangzeb's large and pompous court retinue.[16] Nevertheless, both were equally effective and both made extensive use of the military instrument *par excellence* in India: the warhorse.

THE HORSE-WARRIOR REVOLUTION (AD 1000-1200)

During the eleventh and twelfth centuries India experienced a sudden rise of formerly peripheral but highly mobile warrior tribes. Whereas the north witnessed the well-known inroads of Muslim horse-warriors from Central Asia, the south, saw the ascendancy of new dynasties, such as the Yadavas, Kakatiyas and Hoysalas, who issued from the dryer uplands of the Deccan and the Carnatic.[17] It appears that all

---

[14] For Maratha warfare, see S. Sen, *Military System of the Marathas* (Calcutta: The Book Company Ltd., 1928).

[15] In the case of the Delhi sultanate, the continued threat of a Mongol on-slaught from the northwest served as an important brake on this development and kept alive the nomadic traditions of mobile warfare.

[16] I am fully aware that this argument comes very near to that of the fourteenth-century Arabian historian Ibn Khaldun. See his *The Muqaddimah*, trans. F. Rosenthal (London: Routledge and Kegan Paul, 1967).

[17] It is very well possible that India, like the Middle East, witnessed a growing

these developments are more or less related to a more effective use of the warhorse. The south experienced some major technological innovations regarding the horse's harness. The saddle, for example, was turned into a large hollow seat giving more support to the rider. In addition, a breast plate was attached to the saddle which improved its steadiness. The introduction of the martingale permitted a better control of the horse whereas the old complicated headstall was superseded by a simpler one. Hence, except for the curb bit, the south-Indian rider could dispose of all the pieces constituting the modern harness. Together with the diffusion of nailed horseshoes and the stirrup, the latter being introduced already one or two centuries earlier, these improvements must have considerably enhanced the manoeuvrability and shock-power of the twelfth-century cavalry trooper in south India.[18]

Conspicuously absent at this stage in south India was the mounted archer who had become the prominent military figure in the northern plains of Hindustan. Although mounted archery was not entirely new to the subcontinent, with the coming of the Turks it experienced a marked revival during the eleventh and twelfth centuries.[19] Mounted archery became particularly effective in the hands of nomadic warriors who could build on the rich Central Asian legacy of equine technology and mobile tactics. In terms of technology, it appears that the

---

interest in technical treatises on statecraft (*nitishastra*) and warfare (*danurveda*) at about this time. This remains speculative, however, since the dating of these works remains extremely troublesome.

[18] Interestingly, the Turkish harnesses were provided with a curb bit which thus may have given the Turkish rider a more severe control of his horse. On the development of horse-equipment in southern India, see the pioneering studies of Jean Deloche, *Military Technology in Hoysala Sculpture (Twelfth and Thirteenth Century)* (New Delhi: Sitaram Bhartia Institute of Scientific Research, 1989) and *Horses and Riding Equipment in India Art* (Pondicherry: Indian Heritage Trust, 1990).

[19] There is a great deal of literature on the horse-archer. Latest is André Wink who takes issue with Simon Digby on the issue whether or not the horse-archer signalled the decline of the elephant on the Indian battlefield (*Slave Kings*, 79-110). According to Digby the elephant remained of prime importance to the Delhi sultanate (S. Digby, *War-Horse and Elephant in the Delhi Sultanate* (Karachi: Oxford Orient Monographs, 1971).

combined use of the stirrup and the composite bow gave the Turkish warrior a decisive edge against their Indian counterparts; the first relatively recent, the latter a relatively rare phenomenon in India. Although both were earlier inventions, it was only after the eleventh century that their full impact was felt throughout the Indian subcontinent. Apart from the well-known fact that the improved stability of the stirrup stimulated the rise of heavy cavalry, it also considerably enhanced the firing power of the mounted archer. Much older than the stirrup, the composite bow had already been the long-time weapon of the Central Asian horse-warrior. It had evolved in a Central Asian arms-race against the ever-improving quality of mail and plated body armour.[20] As a result, it had an effective range of over 250 yards in which the arrow retained much of its excellent penetrating qualities. However, its construction, in particular under the humid Indian conditions, required specific skills since it was glued together from a number of materials, mostly containing horn, wood and sinew.[21] Although India was already familiar with the composite bow, at the time of the Turkish invasions, many Indian warriors appear to have stuck to the simpler, single-curved bow made of a single piece of wood, mostly of bamboo. Most probably, these served perfectly well against warriors barely covered by body armour but made no

[20] C. Uray-Kühalmi, 'La périodisation de l'histoire des armements des nomades des steppes', *Études Mongoles* 5 (1974), 145-55.

[21] For the composite bow, see the studies of J.D. Latham and W.F. Paterson, *Saracen Archery* (London: Holland Press, 1970) and 'Archery in the Lands of Eastern Islam', in R. Elgood (ed.), *Islamic Arms and Armour* (London: Scolar Press, 1979), 78-88; J.D. Latham, 'Some Technical Aspects of Archery in the Islamic Miniature', *The Islamic Quarterly* 12, 4 (1968), 225-34; J.D. Latham, 'The Archers of the Middle East: The Turco-Iranian Background', *Iran* 8 (1970), 97-103; W.F. Paterson, 'The Archers of Islam', *Journal of the Economic and Social History of the Orient* 9 (1966), 69-87 and A. Boudot-Lamotte, *Contribution à l'étude de l'archerie musulmane* (Damascus: Institut français de Damas, 1968). For the composite bow in India, see the studies of W.F. Paterson, 'Archery in Moghul India', *The Islamic Quarterly* 16, 1-2 (1972), 81-95 and E. McEwan, 'Persian Archery Texts: Chapter Eleven of Fakhr-i Mudabbir's Adab al-Harb (Early Thirteenth Century)', *The Islamic Quarterly* 18, 3-4 (1974), 76-99; E. McEwan, 'The Chahar-Kham of "Four-Curved" Bow of India', in Elgood, *Islamic Arms and Armour*, 90-6.

impression at all against the mail and plated protection of Central Asian heavy cavalry.[22]

In terms of tactics, mounted archery came only fully into its own when employed on a massive scale. While wheeling around at a safe distance, large contingents of archers had to launch a relentless rain of arrows on the enemy. Thus after wearing out the enemy in a series of skirmishes and feigned retreats, the charge of the heavy cavalry often struck the final blow. Hence, next to mounted archery, the deployment of heavy armoured cavalry remained an important feature of Turkish battle tactics. Even more so, as late as the seventeenth century, heavy cavalry once more came to represent Mughal invincibility on the battlefield whereas the sultanates of the Deccan and Central Asia continued to rely more on mounted archery.[23]

Of course, apart from technology and tactics, mounted warfare required special training and strong warhorses. Both were major assets in the hands of the nomadic pastoralists of Central Asia. As a result, mounted archery became the prized specialty of the Turks, be it as nomads or as professional *mamluks* with a nomadic background. At the same time, the Turks were particularly well placed for drawing horses from Asia's best breeding grounds. However, there was no Turkish monopoly on the breeding of good warhorses. As long as Indian breeders were regularly supplied with new foreign stock, they were perfectly capable of producing excellent warhorses of their own. In fact, the location of the best Indian horse-breeding grounds often coincided with the inner marches along the drier zones of the subcontinent.[24] These extensive breeding zones not only brought

[22] For the use of the simple bow in India, see Deloche, *Military Technology*, 12; G.N. Pant, *Indian Archery* (Delhi: Agam Kala Prakashan, 1978), 264-301 and K.S. Lal, 'The Striking Power of the Army of the Sultanate', *Journal of Indian History* 55, 3 (1977), 98-9.

[23] This follows from the military exploits of Aurangzeb in both Central Asia and the Deccan. See the detailed accounts in the 4 volume study of J. Sarkar, *History of Aurangzeb* (Calcutta: M.C. Sarkar, 1925-30).

[24] J. Gommans, *The Rise of the Indo-Afghan Empire, c.1710-1780* (Leiden: E.J. Brill, 1995), 68-104. For example, between Lahore and Delhi there were long-time horse-breeding tracts which sustained the tenth-century Hindu Shahi state based at Bhatinda as well as the later Muslim states centred round Lahore

forth excellent warhorses but also the most important beasts of burden in India: dromedaries and bullocks. Any major military campaign was unthinkable without the help of an extensive service network of professional camel and bullock transporters. It generally appears that during the eleventh and twelfth centuries, these animals began to be employed on a wider scale than before. In any case, while the rapid developments in equine technology and tactics made the warhorse far more effective on the battlefield, the simultaneous spread of dromedaries and bullocks further extended its radius of action.[25] It was this horse-warrior revolution which gave rise to new state-formation and, at the same time, uprooted the more settled parts of Indian society and made these more open to outside influences.

Despite these equine developments, medieval Indian armies conti-nued to deploy India's most traditional war-animal, the elephant. Comparable to heavy cavalry, elephants were mostly used as massive breaking rams. They were, however, far less manageable than horses and great care had to be taken to avoid their trampling the own troops. At the same time, when well targeted, the elephant's frenzy could be extremely effective in raising panic in the enemy's ranks. The coming of the more mobile and evasive mounted archers made these tactics less effective but, as late as the eighteenth century, ele-phants continued to play a eye-catching element of Indian armies, albeit because they served so well in the grand style of Indian warfare.[26] Nevertheless, by about 1200 the warhorse had definitely replaced the elephant as the ruler of Indian battles.

---

and Delhi. For the Hindu Shahis, see B.K. Majumdar, *The Military System in Ancient India* (Calcutta: Firma K.L. Mukhopadhyay, 1960), 108.

[25] See Gommans, 'The Silent Frontier'.

[26] For the role of elephants in medieval Indian warfare, see the studies of J. Sarkar, *Military History of India* (Calcutta: M.C. Sarkar, 1970), 163-8; Digby, *Warhorse*, 50-83; Lal, *Power*, 97; J. Larus, *Political-Military Behaviour: The Hindus in Pre-Modern India* (Calcutta: Minerva, 1979); S.K. Bhakari, *Indian Warfare: An Appraisal of Strategy and Tactics of War in Early-Medieval Period* (Delhi: Munshiram Manoharlal, 1981), 61-85; Gommans, 'Indian Warfare', 266-7 and Wink, *Slave Kings*, 79-111.

SIEGES AND FORTS (AD 1200-1400)

The uprooting effects of the horse-warrior revolution during the eleventh and twelfth centuries began to be experienced even more thoroughly during the following two centuries, mainly as a result of a sudden advance in siege technology. This was also the first time that gunpowder came to play a more prominent role in oriental warfare; rising portions of saltpetre had turned it from a mere incendiary into a far more powerful explosive device. These gunpowder bombs and grenades proved particularly effective in sieges, both by having them thrown by traditional mangonels (*manjaniqs*) into the besieged fort or by having such dug under the rampart. In the latter case, sappers needed to operate directly under the fort's walls for which an underground tunnel (*naqb*) had to be dug, or, if the subsoil was too hard, a covered passage (*sabat*) had to be built. Under the rampart, a widened hole would be filled with gunpowder and other explosive and combustible material after which the whole construction would be set ablaze. Although there was always a danger that the explosion would fall on the onrushing attackers themselves, as a result of gunpowder, mining became increasingly effective in breaching a wall or, merely by its threat, in driving the besieged to an early surrender. Although, the rising potency of early gunpowder has mainly been studied for the Middle East and China, it generally appears that Indian armies speedily adopted this new technology. Apart from mines, this involved related devices such as explosive and fire-throwing bamboo tubes, most of which had come from China, reaching India through Mongol channels.[27]

As important as early gunpowder was the simultaneous introduction, from the Middle East, of the so-called counterweight mangonel or

---

[27] For the use of early gunpowder in China, see J. Needham et al., *Science and Civilization in China*, vol. 5: *Chemistry and Chemical Technology,* part 7: *Military Technology: The Gunpowder Epic* (Cambridge, 1986), 1-18, 161-92; for the Middle East, see A.Y al-Hassan and D.R. Hill, *Islamic Technology: An Illustrative History* (Cambridge: Cambridge University Press, 1986), 106-20; for India, see Iqtidar Alam Khan, 'Coming of Gunpowder to the Islamic World and North India: Spotlight on the Role of the Mongols', *Journal of Asian History* 30, 1 (1996), 27-45.

the trebuchet (*manjaniq maghribi*, lit. Western mangonel). The trebuchet was a great technological advance on earlier ballistic machines that were operated by crews of men pulling down on the arm. More specifically, the trebuchet could launch heavier missiles – from 130 by the mangonel to 560 pounds by the trebuchet – and from a greater distance – from 133 by the mangonel to 300 yards by the trebuchet. Clearly, these *maghribis* – as they are called in the Indo-Persian sources – posed an increased threat to most of the existing walls of the time. In contrast with later artillery, trebuchets did not hinder the army's mobility since they were mostly fabricated from the wood available near the beleaguered strongpoint. Not surprisingly, for a long time after the introduction of cannon the trebuchet remained an important siege weapon.[28] Although more research is certainly needed in this respect, it is highly plausible that early gunpowder and trebuchets provided a significant advantage to the besieging vis-à-vis the besieged party. It would explain, for example, the fact that during the early fourteenth-century virtually every city and fort of importance in central and southern India succumbed to the Muslim raids of Malik Kafur. More generally, one may conclude that, together with the earlier horse-warrior developments, improved siege technology clearly marked the heyday of offensive warfare in India. As mentioned already, it was also the time of new conquering elites who staged their careers from the fringes of settled society. By ruling from the saddle and by destroying ramparts, these new dynasties were perfectly situated to bridge India's inner frontiers and to bring about a renewed fusion between the mobile world of the horse and the sedentary world of the fort.[29]

Improved siege technology immediately raises the question of

---

[28] For the trebuchet in China, see J. Needham and R.D.S. Yates, *Science and Civilization in China*, vol. 5: *Chemistry and Chemical Technology,* part 4: *Military Technology: Missiles and Sieges* (Cambridge: Cambridge University Press, 1994), 203-40; for the Middle East, see D.R. Hill, 'Trebuchets', *Viator* 4 (1973), 115 and under the heading 'hisar', in *Encyclopaedia of Islam*, 2nd ed. (Leiden: E.J. Brill, 1960-2005); for India, see several reference in M. Habib, *The Campaigns of 'Ala'u'd-din Khilji Being the Khaza'inul Futuh (Treasures of Victory) of Hazrat Amir Khusrau of Delhi* (Madras: Taraporewala, 1931).

[29] Cf. Wink, *Slave Kings*, 3.

military architecture. How did gunpowder and trebuchets affect the building of forts in India?[30] From the Middle East we know that the more sophisticated siege techniques of the thirteenth century caused major adjustments in fortification. For example, to counter an easy approach of sappers and mangonels, walls were provided with towers which could give flanking fire whereas more loopholes and machicolations were used to throw down nafta and other anti-siege incendiaries.[31] Taking account of the scattered information on various Indian forts, it is very well possible that India went through a similar process. For example, the appearance of round towers may have been directed against sappers who would have found easy cover at the tower's corners. At some places, earlier massive walls were provided with inner vaulted chambers and depots which improved the ability to provide counterfire against attackers along the walls. To compensate for the building of these inner chambers and gangways walls became much thicker and more solidly built. At some places, such as at Tughluqabad in Delhi, the increased threat of mining and trebuchets may also have led to the building of projecting bastions, often provided with a base of scraped rock and sloping bolster plinths. Besides, more and more forts were built on steep, rocky hills and in the midst of dense thorny forests, hard to approach by sappers or other siege engineers. Hence it appears that thirteenth- and fourteenth-century siege technology did indeed give fresh impetus to the building and rebuilding of Indian forts. By the sixteenth century, most Indian forts were better equipped

[30] Information on Indian forts has been gathered from various studies on specific forts. Of a more general nature are: J. Burton-Page, 'A Study of Fortification in the Indian Subcontinent from the Thirteenth to the Eighteenth Century AD', *Bulletin of the School of Oriental and African Studies* 23, 3 (1960), 508-22; P.V. Begde, *Forts and Palaces of India* (New Delhi: Sagar Publications, 1982). Again paving the way is Jean Deloche in his 'Études sur les fortifications de l'Inde, I: les fortifications de l'Inde ancienne', *Bulletin de l'École française d'Extrême-Orient* 81 (1994), 89-131 and 'Études sur les fortifications de l'Inde, II: les monts fortifiés du Maisur méridional (1re partie)', *Bulletin de l'École française d'Extrême-Orient* 82 (1995), 219-66.

[31] See e.g. H. Kennedy, *Crusader Castles* (Cambridge: Cambridge University Press, 1994), 180-5.

than ever to sustain sieges, even if the besiegers employed the newest gunpowder device of the day: cannon.

THE FALSE DAWN OF GUNPOWDER (AD 1400-1750)

As mentioned already, the gunpowder thesis is mostly related to the sixteenth century when new armies equipped with cannon and matchlock firearms paved the way for the ascendancy of more centralized states. Before touching briefly upon the latter supposition in my conclusion, this paragraph will first broadly examine the thesis' relevance from a purely military perspective and, as such, will primarily focus on 'true' guns, i.e. guns in which a gunpowder charge exerted a propellant effect on a projectile launched from a metal barrel. Although a much earlier invention, it appears that it was mainly during the fifteenth century that cannon diffused more widely in India, most prominently in the Muslim sultanates of the Deccan. During the second half of the century their use spread towards the north while light firearms (*banduq, tufang, toredar*), equipped with matchlock trigger mechanisms, also became more widespread. Subsequently, both cannon and portable firearms figure prominently in the Mughal chronicles. In due course, gunpowder, cannon and firearms were manufactured in considerable numbers all over the Indian subcontinent. It remains to be seen, though, to what extent these new weapons were really effective on the actual battlefield or in sieges.

To begin with the latter, even for the new gunpowder armies, investing a fort remained a protracted and complicated affair. Mostly, the main problem was to find a proper staging ground from which the often extremely heavy siege guns could be effectively employed. Actually, from the fourteenth to the late-eighteenth century, there are only rare instances of artillery being successfully employed against forts. From time to time, walls were adjusted to accommodate artillery, but this primarily involved the levelling of towers and walls to make room for the installation of heavy guns, in other words, for defensive purposes. It appears that cannon became at least as important to the besieged as to the besiegers. The numerous failures of artillery to take forts suggests that the thirteenth- and fourteenth-century adjustments of fortification, together with the particular geographical conditions of the Indian subcontinent – remember its steep hills and inaccessible forests – made a later development following Parker's *trace italienne* rather superfluous.

As a consequence, there was nothing like the widespread decastellization – or as Joseph Needham would say, Sinification – witnessed in Europe.[32] On the contrary, forts re-emerged as the strongholds of a new martial elite, in the north often referred to as Rajputs, in the Deccan as Marathas, in the south as Nayakas. This was a self-conscious gentry comprising armed settlers of mostly marginal areas who had gained certain protection rights in return for military services to the conquering dynasties of the previous centuries. As these warriors sedentarized, their forts became the new centres of an expanding agrarian economy. However, these agrarian estates should not be equated with the more primitive, closed feudal fiefs of Europe since these were highly monetized economies that remained as open as ever to the outside influences of interregional trade and politics. This implied that the gentry could never fully settle down in their flourishing estates. As mentioned already, they were in fact permanently subjected to the uprooting forces of *fitna*. Although they could withstand the siege technology of the day, they could not reasonably be expected to reject endless offerings of cash and employment. Except for new conquerors who wanted to set an early example, taking a fort became less about shelling and more about bribing the besieged. Or to use the words of Douglas Streusand, even the strongest of Indian forts became ideal 'units of political bargaining power'.[33]

The crucial importance of *fitna* in taking forts reveals itself most explicitly in the numerous sieges of Aurangzeb in which his commanders outstripped each other in offering the highest bribes to the besieged Marathas. After every submission the latter were allowed to leave the fort unharmed and with all their personal belongings. Not surprisingly, within a few months after the Mughal train had taken up another siege, the fort was easily recaptured by their former occupiers.

---

[32] Needham and Yates, *Military Technology: Missiles and Sieges*, 240.

[33] D.E. Streusand, *The Formation of the Mughal Empire* (Delhi: Oxford University Press, 1989), 65.

Characteristically, all these sieges took place in a hostile environment in which the Mughals were mostly besieged themselves which forced them to spend considerable time and effort in building and digging extensive lines of defense. But this was nothing compared with the efforts they had to put into the maintenance of the numerous supply-lines to the north. In this respect, once more, their fate hinged on the mobile horse-warrior who foraged the countryside for food and supplies and protected the caravans bringing fresh supplies along the trade routes.[34]

Not much different from sieges, the tactical deployment of cannon at the battlefield proved extremely troublesome. Characteristically, the firing of guns, always stationed at the front rank, served solemnly to announce the start of battle after which the guns lay inactive on the ground to make room for the cavalry. Usually, the heaviness of most Indian guns made rapid manoeuvring in the field impossible. Although Indian armies disposed of various brass and bronze specimens, most of their guns were bound together from wrought iron bars which made them not only heavy but also unreliable and extremely inaccurate. While aiming such a heavy Indian bombard was already a major problem in itself, lack of standardization of both barrels and shot – the latter had often to be hammered in order to fit into the barrel – made it extremely difficult to co-ordinate the firing of a number of guns. Apart from these inconveniences of the gun itself, it also appears that Indian gunpowder was less powerful since it was not granulated.[35] Besides, mobility was hampered by very primitive gun carriages. Indeed, there is a general consensus among military observers that most Indian armies failed to introduce a light field artillery that could be employed in combination with cavalry. To repeat the words of Geoffrey Parker, they adopted cannon but failed to adapt it to their cavalry.[36]

[34] For a detailed account of these sieges, see J. Sarkar, *History of Aurangzeb*, vol. 4: *Southern India, 1645-1689* (Calcutta: M.C. Sarkar, 1930).

[35] See the accounts of Cossigny and Modave in Deleury, *Les indes florissantes*, 471-3.

[36] See e.g. Sarkar, *The Art of War in Medieval India* (Delhi: Munshiram Manoharlal, 1984), 324-5; and B. Lenman, 'The Weapons of War in 18th-Century India', *Journal of the Society for Army Historical Research* 46 (1968), 33-6.

To alleviate the problem of mobility, guns were at times produced at the very spot where they were needed. Attempts were also made to affix guns to elephants (*gajnal* or *hathnal*, both lit. elephant barrel) and dromedaries (*shutarnal, zamburak, shahin*; lit. resp. camel barrel, little wasp, falcon). During the eighteenth century light camel guns were successfully employed by the Durrani Afghans but even these rapidly moving units had only a limited range of action and could only serve in the drier parts of the Indian subcontinent.[37]

Coming to the use of firearms, the prime focus should be on the effectiveness of Indian infantry. We should keep in mind that according to received opinion, it were new infantry tactics based on rigid squares or *tercios* comprising drilled military professionals, equipped with standardized hand-held matchlocks and pikes, that during the sixteenth century announced a new gunpowder age in western Europe. As such, the rise of infantry also signalled the decline of heavy cavalry. Turning to sixteenth- and seventeenth-century India, the presence of infantry is certainly eye-catching. As argued by Dirk Kolff, there appears to have been an almost limitless availability of armed peasants who presented themselves constantly on an extensive military labour market.[38] For most of these armed peasants, military employment was only a part-time business, complemented by other activities such as sowing, harvesting and weaving. As mentioned already, military employment neatly followed the agrarian calendar. Especially in the semi-arid zones of India, such as in Rajasthan, Maharashtra and the Carnatic, which often depended on only one uncertain crop, military services could become a crucial part of the population's survival strategy.[39]

Despite the sheer masses of Indian foot soldiers, their role appears

---

[37] See my 'Indian Warfare', 275-7.

[38] D.H.A. Kolff, *Naukar, Rajput and Sepoy: The Ethnohistory of the Military Labour Market in Hindustan, 1450-1850* (Cambridge: Cambridge University Press, 1990), 1-31.

[39] Cf. S. Gordon, 'Zones of Military Entrepreneurship in India, 1500-1800', in his *Marathas, Marauders and State Formation in Eighteenth-Century India* (Delhi: Oxford University Press, 1994), 182-208. It was only during the nineteenth century that new crops began to change the existing agrarian calendar and to reduce the peasant's off-season. It is no coincidence that it was just at

as only marginal on the actual battlefield. Although the Indo-Persian chronicles give relatively high figures for infantry, the latter figures barely in the battle accounts which remain dedicated to the horse-warrior. The same goes for the many pictorial representations of miniatures. Although this may have been the result of a cultural bias on the part of the horse-loving artist, the minor role of infantry in battles is confirmed by the disqualifying comments of European contemporaries and modern military scholars alike.[40] Most of all, following the almost endless availability of superior cavalry, India lacked the inducement to develop disciplined and drilled infantry which could operate in the open field or could keep up permanent fire. As in the case of artillery, their firearms and shot lacked standardization. It further appears that the bulk of Indian infantry served off the battlefield, as local militia or as various attendants (*ahsham*) of court and cavalry. From a purely technological point of view, the foot soldier remained at a clear disadvantage vis-à-vis the horse-warrior. Mostly the first lacked any body armour and was only equipped with primitive bows and spears. Even when he could dispose of a matchlock arquebus, he was still at a loss against the mounted archer. For example, the firing rate of the matchlock could not compete with the composite bow. A trained archer could get off six aimed shots per minute without difficulty. According to the French observer Bernier this was in any case three times as often as a musketeer could shoot his matchlock. Besides, at long range, the bow retained more of its penetrating power and was definitely more accurate. Nevertheless, both matchlockmen and archers had great difficulty in penetrating the horseman's body-armour which could only be achieved from within 100 yards.[41] Hence the ongoing popularity of heavy Mughal armour, consisting of helmet, vambraces, mail shirt and trousers, and,

---

this time that the British authorities finally managed to disarm the Indian gentry and to impose a near monopoly on the use of violence.

[40] See e.g. G.L. Coeurdoux, *Moeurs at Coutumes des Indiens (1777)* (Paris: Ecole Française d'Extrême-Orient, 1987), 180; Sarkar, *The Art of War*, 320-1 and the still classical account of W. Irvine, *The Army of the Indian Mughals* (London: Luzac and Co., 1903), 161 fol.

[41] For this information on the bow and the matchlock, see R. Elgood, *Firearms of the Islamic World in the Tareq Rajab Museum, Kuwait* (London: I.B. Tauris,

in particular during the seventeenth and eighteenth century, the *chahar ayina* (lit. four mirrors), a plated cuirass in four sections.[42]

Notwithstanding the many defects of the matchlock gun, we should keep in mind that the use of the matchlock required considerably less training and less expenses than cavalry warfare. As such, it was indeed the weapon *par excellence* of the peasant who preferably employed it on an individual basis. In fact, there were only a few relatively well-trained, regular infantry units consisting mostly of peasants and petty landlords such as the Baksariyas and Bundelas. Other infantry troops were recruited from so-called savage tribesmen from the fringe of sedentary society, such as Kanarese Berads, who, apart from being notorious dacoits and cattle-lifters, had a reputation for being the best matchlockmen of the Deccan. These infantry units hired their services to the highest bidder and they were certainly an asset for any military commander. They failed, however, to revolutionize Indian warfare. Indian matchlockmen, usually operating their guns behind trenches or some other cover, should certainly not be confused with the Swiss *tercios* of Europe. But even if they had been organized in drilled formations, the slow-firing matchlockmen, either or not assisted by pikemen, would still have been at a loss against the massive deployment of the highly professional, well-armoured, and still much more mobile cavalry troops. *Tercios* could only be successfully deployed in western Europe where cavalry lacked both numbers and sophistication. Or to formulate it from the Indian perspective, the almost cornucopian availability of professional horse-warriors and, at the same time, the lack of rapidly-firing gunpowder arms, conditioned the ongoing supremacy of cavalry in India. The latest Turks in India, the Mughals, could only follow in the path of their predecessors. Even the Mughal's best-known gunpowder victory at the battle of Panipat (1526) was not the work of heavy artillery but of traditional Central Asian tactics combining mounted archery with the so-called *tabur* or

---

1995), 16. It also contains the best general survey on Oriental and Indian firearms to date.

[42] See e.g. D. Nicolle and A. McBride, *Men-at-Arms Series,* vol. 263: *Mughul India 1504-1761* (London: Osprey, 1993), 40, 43 and H. Russell Robinson, *Oriental Armour* (London: Jenkins, 1967), 107.

wagenburg. This consisted of a number of wagons chained together in order to form an effective barrier against cavalry charges and to give cover to matchlockmen and a few light guns. At the start of the sixteenth century the wagenburg made an initial impact but it certainly failed to herald a new gunpowder era in which artillery and infantry were to dominate Indian warfare.[43] Illustratively, almost 250 years later, at the same battlefield, the heavy artillery of the Marathas was again defenceless against the more mobile cavalry forces of the Afghans.[44]

THE GUNPOWDER REVOLUTION (AD 1750-1850)

It is my contention that it was only during the second half of the eighteenth century that gunpowder really revolutionized Indian warfare. Improved casting and boring techniques made artillery a more effective instrument of war, not only in Europe but also in India. Employed by European specialists, it proved more effective, both in sieges and – being lighter and thus more mobile – in the field. Next to artillery, the effectiveness of infantry also radically improved. First of all, the introduction of the flintlock mechanism considerably enhanced the firing power of European infantry units.[45] The new flintlock musket had many advantages against the older

---

[43] The same argument could be made for Central Asia and Iran. Even Ayalon's claims that Ottoman victories against the horse-loving Mamluks and Safavids were based on their superior command of gunpowder may be questioned. See e.g. the comments on the battle of Caldiran by D. Morgan, *Medieval Persia 1040-1797* (London: Longman, 1988), 117; and J.R. Walsh, 'Caldiran', in *Encyclopaedia of Islam*, 2nd ed. (Leiden: E.J. Brill, 1960-2005). For the Mamluks, see the review by Robert Irwin of Shai Har-El, *Struggle for Domination in the Middle East: The Ottoman-Mamluk War 1485-91* (Leiden: E.J. Brill, 1995) in the *Journal of the Royal Asiatic Society* 7, 1 (1996), 136.

[44] There are numerous descriptions on both battles. E.g. for Panipat 1526, see Streusand, *The Formation*, 52; and for Panipat 1761, see T.S. Shejwalkar, *Panipat 1761* (Poona, 1946).

[45] J. Black, *European Warfare 1660-1815* (London: UCL Press, 1994), 38-9. Black emphatically shifts the European military revolution to the eighteenth century.

matchlock arquebus. Basically, it was lighter, less unreliable and easier to fire and blessed with more hitting power. Thanks to the spread of pre-packaged paper cartridges the fire rate was almost doubled. Equipped with the new socket bayonet the defensive and offensive capacities of infantry against cavalry were increased. Moreover, it simplified the very complicated drill of the *tercio* since the new system could dispense with the pikemen. Although, flintlocks had been in use in both Europe and Asia before, it was only during the early-eighteenth century that flintlocks were generally adopted in Europe. India rapidly followed suit. Obviously, this did not mean that the traditional horse-warrior could be pensioned off immediately, but it certainly marked a general shift from cavalry to infantry, or rather, to well-drilled sepoy regiments equipped with flintlock muskets and socket bayonets.[46] Although Indian armies lacked experience with drilling infantry regiments, they could easily hire European officers who adjusted the Indian foot soldiers to the simplified European tactics.[47] Thanks to the flintlock, the period from 1750 to 1820 was to be the highpoint of the European military adventurer who gradually replaced the traditional Turkish and Afghan mercenary dealing primarily in cavalry and warhorses. It remained difficult, however, to integrate fully the infantry units into the larger Indian armies which were still emotionally attached to the horse. Now the dialectics of progress began to work to the advantage of Europe. For example, the new armies of the East India Company effectively combined the new infantry with a highly mobile field artillery. In terms of technology, the Company's army did not at all lag behind their European counterparts. Actually, the so-called India Pattern musket of the Company became the standard British infantry weapon throughout the Napoleonic wars.[48]

---

[46] See e.g. Lenman, 'Weapons', 37-9; Elgood, *Firearms*, 144ff and P.K. Datta, 'Guns in Mughal India', *Bulletin of the Victoria Memorial Hall* 2 (1968), 30-8.

[47] For the effectiveness of the late-eighteenth and early-nineteenth century Maratha infantry – only lacking bayonets – and artillery, see e.g. J. Pemble, 'Resources and Techniques in the Second Maratha War', *The Historical Journal* 19, 2 (1976), 375-404.

[48] D.W. Bailey and D. Harding, 'Form India to Waterloo: The "India Pattern"

Not surprisingly, the new flintlock mechanism also considerably effected cavalry tactics. First of all, Indian cavalry began to follow the Turkish and Afghan example of using short blunderbusses (*sherbashas* or lit. lion's whelps), preferably equipped with flintlock trigger mechanisms. The way they used these muskets were similar to the Central Asian tactics of the archer, wheeling around and wearing out the enemy by relentless fire.[49] Both in comparison with composite bows and matchlocks, flintlocks simplified the deployment of such tactics and opened up their feasibility to the non-specialist. There was also a gradual shift from heavy to light cavalry.[50] In India this was certainly the result of the improved firing and hitting power of infantry, but, as in Europe, it also marked a revival of nomadic tactics practised by tribal cavalry; Skinner's and other Indian light horse finding their counterparts in the hussar and cossack regiments of Europe. Off the battlefield, light cavalry still served well in all sorts of guerrilla tactics and, more importantly, in maintaining supply lines over long distances. Besides, the Indian nobility often stuck to their horses since horses offered superior opportunities for raiding and plundering. This is heard, for example, in the numerous complaints of European foot soldiers who were always too late to claim the war booty since all of it had long been carried off by their mounted Indian allies who had awaited the outcome of the battle from a safe distance. Hence, it would still take a century and a Mutiny before the proud mounted warrior of India finally succumbed to the more pliable sepoy. In the end, however, the combined use of more advanced field artillery and drilled infantry broke the long-standing supremacy of the Indian horse-warrior.

Musket', in A.J. Guy (ed.), *The Road to Waterloo* (London: National Army Museum, 1990), 48-57.

[49] See my 'Indian Warfare', 270-8. It remains unclear, however, whether the Afghans approached the enemy slowly, like the European *caracole*, or rapidly like the traditional archers (personal communication by J.P. Puype, Delft Army Museum).

[50] Interestingly, very heavily armoured cavalry came back into fashion in eighteenth-century Hyderabad which suggests that body-armour was still bullet proof (Nicolle and McBride, *Mughul India*, 43).

CONCLUSION

From about 1000 to 1850 India experienced two military break-throughs. The first occurred from about 1000 to 1200 and involved the ascendancy of the horse-warrior in general, and the mounted archer in particular. In the following two centuries this horse-warrior revolution gained its full weight due to the marked improvement in siege technology. For about 700 years, the effects of the revolution perpetuated in varying degrees all over the Indian subcontinent. The second breakthrough took place from about 1750 to 1850 and involved improved gunpowder technology in the hands of both artillery and infantry and should genuinely be called a gunpowder revolution. From the Indian perspective, both revolutions were intro-duced by outside 'barbarians', infiltrating the subcontinent along the long inner-frontier and littoral zones that connected India both with the Central Asian steppes and the Indian Ocean. Both started their Indian political career as relatively small, armed units which proved their military superiority on the battlefield against their outmoded Indian counterparts. Both successfully crossed India's inner frontier and brought about a renewed fusion of settled society with the more mobile world of pastoralism or trade. However, in the case of the Turks, their fusion could only be achieved through invigorating the disordering forces of the frontier. Hence, their rule remained wedded to the culture of the camp, the horse and the bow. This also explains the apparent contradiction, mentioned in the introduction between military expansion, on the one hand, and disintegration on the other. Here the military superiority of horse-warrior could never achieve fully centralized states. To state it differently, horse-warriors could never fully control a densely fortified countryside without losing the very base of their supremacy. Hence, their political organization stood somewhere midway between a sedentary and nomadic empire. Once and again, military chiefs broke away, built their own forts and started a princedom for themselves. Obviously, the traditional Indian politics of *fitna* served these volatile circumstances very well. Its work-ing was further stimulated by a highly monetized military labour market in which the military enterpriser had always played a prominent part. Perhaps one may even say that the easy availability of both horses and money precluded the rise of centralized medieval states in India.

Of course, military superiority could bring rapid conquests but its consolidation required the extremely malleable and uprooting politics of *fitna*. In other words, the continued vigour of the inner frontier, *fitna* and the warhorse conditioned both the rise and, almost simultaneously, the fragmentation of empire.[51]

In contrast to the Turkish and Afghan horse-warriors, the Indo-European armies of infantry and artillery succeeded in eliminating India's inner frontiers and, after more than a century, imposed a full monopoly on the use of legitimate violence. For the colonial government, the heroic frontiersmen became 'criminal tribes' whereas *fitna* was equated with corruption and treason. India at large was sealed off from the outside world and after about 1850 came very near to being a modern centralized state. Clearly, all of this would have been unthinkable without the help of gunpowder.

---

[51] In the case of *fitna* a similar argument has been made already by André Wink in his *Land and Sovereignty*, 34.

CHAPTER 8

Slavery and *Naukari* among the
Bangash Nawabs of Farrukhabad

THE FIRST BANGASH NAWAB of Farrukhabad, Muhammad Khan, was a son of Malik Ain Khan, descendant of an Afghan farmer in the Kabuli district of Bangash. Early during the reign of Aurangzeb (1658-1707) he had quitted his native country and settled in Mau-Rashidabad, 34 km. west of present-day Farrukhabad. Here he enrolled as an officer (*sarkar-i anfar*) in the cavalry of a fellow Afghan named Ain Khan Sarwani.[1] Both were in the service of the local *jagirdar*, Nawab Mirza Khan, who was a grandson of Khwaja Bayazid Ansari, famous founder of the Roshaniyya sect.[2] Although in the beginning this heterodox movement had been thoroughly anti-Mughal, Bayazid Ansari's descendants had, on the contrary, risen to prominence thanks to their military services to the Mughals. As a reward for their help during the Mughal campaigns in the Deccan, Mirza Khan and his brethren had received some *jagirs* in the area of Mau and Shamshabad. Still on the basis of their charismatic leadership and spiritual guidance, they had drawn many recruits from their tribal Roshaniyya following in Roh.

Malik Ain Khan married in Mau and when he died left two sons. The oldest, Himmat Khan died while on a military expedition in the

---

[1] Muhammad Wali al-Lah, *Tarikh-i Farrukhabad*, British Library, Oriental and India Office Collections (OIOC), Or. 1718, fol. 10b; M. Elphinstone, *An Account of the Kingdom of Caubul*, vol. 2 (Oxford: Oxford University Press, 1972), 51.

[2] For details vide: W. Irvine, 'The Bangash Nawabs of Farrukhabad', part 1, *Journal of the Asiatic Society of Bengal* 1, 47, 1 (1878), 357-64.

Deccan. His second son was Muhammad Khan who at the age of twenty took service with Yasin Khan Ustarzai Bangash, one of the leading Afghans of Mau. As a small mercenary jobber (*jamadar*) he and his gang were primarily engaged in local conflicts in the Deccan and Bundelkhand. After Yasin Khan was killed at a siege, Muhammad Khan started a *jamadari* business of his own. He started with seventeen followers but soon after his first successes, more and more Afghans were willing to join his standard. Although in numerous small campaigns Muhammad Khan was able to increase his wealth and power considerably, his status among his fellow Afghans of Mau remained highly controversial. For them, he was just one among equals and every sign of elitist pretensions, like riding an elephant or sitting on an elevated cushion amongst his brother Afghans, could only be met with their ridicule or open opposition. This strongly egalitarian atmosphere, reminiscent of similar conditions among his Afghan brethren in Rohilkhand, made claims to legitimate leadership always extremely complicated. In the case of the Bangash chiefs this problem was further aggravated by their small numbers in and around Mau.

In this essay I will discuss two aspects that played a crucial role in determining the power and authority of the Bangash Nawabs among the political elite in and beyond Farrukhabad, be it fellow Afghans or other co-sharers of the declining Mughal realm. The first aspect relates to their use of so-called *chelas*, elite slaves, mainly serving the Bangash military and the administration. These *chelas* were incorporated into the Nawab's household to prop up their masters' position among the other, mostly Afghan, chiefs of the little kingdom in and around Farrukhabad. In order to understand the internal context in which the *chelas* operated, I will start by describing the Nawab's Bangash descent (*nasab*) and their Farrukhabad household. The second aspect relates to the turbulent vicissitudes of the Farrukhabad state as engendered by the Bangash Nawabs' ongoing involvement in India's bustling mercenary trade. In the end, it was neither their *nasab* nor their *chelas* but their extensive military services (*naukari)* which brought them all-India fame and fortune. Although Farrukhabad slavery continued to serve its limited purpose in Farrukabad itself, it was soon drowned, so to speak, in the turbulence and fluidity of the huge military labour market that was India. At the end of the century, with the Nawab's *naukari* business on the decline, we once again find *chelas* active, this time not obeying but manipulating an increasingly isolated and 'decadent' Farrukhabad court.

SLAVERY

Bangash Nasab

The Bangash tribe derived its name from the hilly area, north of the Sulaiman Mountains, from Bannu to the Safed Koh, in between the Indus and the Kurram River. Its main centre, situated along the lower river valleys was the town of Kohat, also synonymous with the area called Lower Bangash or Pain Bangash. Under Mughal rule this area had been a nominal district (*tuman*) of the Kabul province.[3] The area was of some importance as it provided an alternative northern route to the Khyber Pass from Kabul to Peshawar and India. The Mughal Emperor Babur, however, describes it as rather peripheral and being infested with nothing but Afghan highwaymen and thiefs, such as Khugianis, Khirilchis, Turis and Landars; all of whom, obviously, declined to pay taxes. Nevertheless, in 1505 Babur decided to raid and plunder the district following some intelligence about its great riches.[4] Indeed, in Kohat the Mughals found cattle and corn in great abundance but they were not able to settle the area on a permanent basis since most of the tribes could temporarily retreat into the more impregnable upper hills called Bala Bangash or Kurram. For any central government, either from Delhi, Qandahar or Kabul, these hilly tracts of the Bangash *tuman* always remained an obstreperous area. As a result, Shia Islam and heterodox movements, such as that of the Roshaniyya, mostly combined with and fuelled by anti-Mughal resentment, remained very strong indeed in this area.[5]

[3] I. Habib, *An Atlas of Mughal India* (Delhi: Oxford University Press, 1982), 3.

[4] *Babur-Nama*, trans. A.S. Beveridge (Delhi: Penguin Books, 1970), 230-3.

[5] O. Caroe, *The Pathans, 500 BC-AD 1957* (Karachi: Oxford University Press, 1975), 202-3; J. Arlinghaus, 'The Transformation of Afghan Tribal Society:

Before the sixteenth century, there did not exist a separate tribe or sub-tribe which was actually called Bangash. The seventeenth-century *Makhzan-i Afghani* does not mention them at all. Most of the inhabitants of Bangash envisaged themselves to be the descendents of one Karlani, and were known under various sub-headings such as Orakzais, Turis, Malik-Miris, Baizis and Kaghzais. The latter three had entered the lower valleys of Kohat somewhere between the late fifteenth and early seventeenth century from the surrounding hills of Upper Bangash and, probably, were subsequently called Bangash to distinguish them from the other Karlanis residing in the Kohat area, mainly Orakzais and Khataks.[6] According to the eighteenth- and nineteenth-century Indo-Afghan sources, all the tribal groups in Bangash could be commonly referred to as Bangash (*qaum-i Bangash*) in accordance with their place of origin or residence.[7] This was not unlike the development of the Indian eponym Rohilla, i.e. the people from Roh, although the name Bangash referred to a more restricted territory and as such applied to a smaller audience of potential immigrants.[8] Although the category Bangash was more restrictively used in the Afghan context of Bangash proper, in India it could include all sorts of Afghan tribes who one way or another claimed their origin from this homonymous area, and this in its most imprecise and wide-ranging dimensions. Thus, we may conclude that the tribal eponym of Bangash came only into existence in sixteenth- and seventeenth-century Kohat and Bangash, but found its widest definition in eighteenth-century Hindustan. Unlike the Indo-Afghan label Rohilla, Bangash referred to an existing ethnic category in Afghanistan although its definition became wider and more diffused in India.

Muhammad Khan Bangash himself belonged to the Kaghzai line

Tribal Expansion, Mughal Imperialism and the Roshaniyya Insurrection, 1450-1600' (unpublished PhD thesis, Durham, N.C., Duke University).

[6] H.G. Raverty, *Notes on Afghanistan and Part of Baluchistan* (London: Eyre and Spottiswoode, 1888), 389-90.

[7] Hafiz Rahmat Khan, *Khulasat al-Ansab*, OIOC, Egerton 1104, fol. 84b; Muhammad Wali al-Lah, *Tarikh*, fols. 6a-b.

[8] For the Rohilla case, see Gommans, *The Rise of the Indo-Afghan Empire*, 163-71.

of the Karlani tribe of Bangash. He could trace his descent for more than four generations to one Daulat Khan, alias Haji Bahadur, of the Shamilzai of the Harya Khail.[9] Our most trustworthy source in this respect, the mid-nineteenth-century *Tarikh-i Farrukhabad* by Muhammad Wali al-Lah, presents us with a fairly careful description of the *nasab* of these Kaghzai Karlanis. According to most of the earlier Afghan traditions the Karlanis could not claim an Afghan descent since they were regarded, and regarded themselves as direct descendents from the prophet Muhammad or his clan, as Saiyids or Quraishis, who only through adoption and gradual accommodation had turned into Afghans.[10] Muhammad Wali al-Lah, however, mentioning the Saiyid claim of some Karlanis in passing, stresses their undisputed Afghan identity. He points out that although Karlani was originally a Sayyid, he had been adopted by one Amir al-Din, the youngest son of Sarban and eldest son of the well-known great Afghan ancestor Qais. Therefore, he claims, the Karlanis should be recognized as full Sarbani Afghans.[11] Muhammad Wali al-Lah is equally keen on delineating the origin of the name Kaghzai. The name must have given cause for confusion because there were also other Kaghzais who had no relationship whatsoever to the Karlanis but claimed to be the descendants of Sarwani. It was made clear that both should not be mixed up because the Karlanis acquired the name only because they had settled near the Sarwani Kaghzai in Upper Bangash. So, in fact, there were two unrelated variants of Kaghzais: Karlani Kaghzais and Sarwani Kaghzais.[12]

Muhammad Wali al-Lah's rationalizations may again underline the importance of open and achieved identities. The Nawabs were Afghans by adoption; they were Bangash because they had moved to India from the Bangash area; and they were Kaghzai because they had lived near the Sarwani Kaghzai. Only their Karlani identity was taken for granted and not explained properly. All of this made perfect sense to

[9] Muhammad Wali al-Lah, *Tarikh*, fol. 10b; Hafiz stresses the requirements for a proper *nasab* to be at least four generations (Hafiz Rahmat Khan, *Khulasat*, fol. 15a).

[10] Raverty, *Notes*, 380-8.

[11] Muhammad Wali al-Lah, *Tarikh*, fols. 9a-b.

[12] Ibid., fols. 10a-b.

the contemporaries since it alluded to existing customs and traditions and to a recognizable frame of reference. Within the wider category of Bangash Pathans, the Nawabs could claim a clear and recognizable *nasab* for themselves. The Bangash label served as an external marking against other Hindustani or Afghan groups like Rajputs, Rohillas or Afridis, and, at the same time, could appeal to all sorts of ethnical sub-groups from within the Bangash area. Within this fairly open category, the Nawabs could lay claim to a more exclusive and distinguished Karlani Kaghzai descent that underpinned their authority and status amongst the other local groups of Bangash Pathans, such as the Sarwani and Ustarzai Bangash.

The Nawab's attitude towards the other Bangash was always somewhat ambivalent. In order to raise their solidarity and group feeling he had to associate them with his government and with the inner circle of his sons and slaves (*khanazadas*). This, however, made them powerful and always potentially dangerous. Consequently, he also needed to dissociate himself from them and keep them on a safe distance. The ethnical nomenclature reflected this continuous need for association and dissociation: the Bangash label could serve the tendency of inclusion while, at the same time, the Kaghzai identity could stress the exclusiveness of the Nawab's *nasab*.

The Farrukhabad Household

The core of the Bangash powerbase in Farrukhabad was the domestic household: the Nawab's wives, sons and *khanazadas*. This latter group consisted primarily of his sons and his personal slaves or *chelas*.[13] Those Bangash who were not part of the core household were attached to it by being married to the Nawab's daughters. In general we may say that the Nawab made his sons, slaves and sons-in-law his most

---

[13]Khanazada literally means 'son of the house' or 'household born one' and indicates clearly the slave's inclusion in the household. For *khanazadagi* ethos under Mughals, see J.F. Richards, 'Norms of Comportment among Imperial Mughal Officers', in B.D. Metcalf (ed.), *Moral Conduct and Authority: The Place of Adab in South Asian Islam* (Berkeley: University of California Press, 1984), 262-7; and D.E. Streusand, *The Formation of the Mughal Empire* (Delhi: Oxford Universit Press, 1989), 146-8.

trustworthy instruments of power by bestowing land on them in jagir or *inam* and by entrusting them with the most crucial positions in the administration and the army.

The marriage rules of the Bangash household were relatively strict in order to maintain their exclusive status. Hence, beyond the household, the Bangash sub-tribe delineated the more or less endogamous marriage group. The Bangash Pathans declined, for example, to intermarry with Rohillas and even marriages with the esteemed Awadhi ruling family were considered problematic.[14] Although the Nawab's daughters were exclusively married to his Bangash brethren, rules for the male members of the family were less rigid. They sometimes married women from another distinguished Pathan tribe or accepted women from all kinds of people but as ladies and concubines of the harem these had to content themselves with inferior positions. For example, females could be presented as *dolas*, i.e. daughters given in marriage to a superior by way of a tribute.[15] Gifts of daughters could also confirm important political relations of the Nawab. This was the case with Raja Jaswant Singh, *zamindar* at Bhadoi, who had presented one of his daughters to Muhammad Khan. This indicated the Nawabs direct interests as the Raja was appointed to the charge of the *rahdari* from Benares to Allahabad and as such was responsible for keeping open this important supply line for Farrukhabad.[16] In this respect, the, size of the Nawab's harem was a direct reflection of the Nawab's attempts to build up a reliable following and alliances of his own.

Muhammad Khan had been married to a daughter of Kasim Khan, another Bangash *sardar*, who became his only legal wife and became styled as Bibi Sahiba. Besides, it was said, he held some 2,600 concubines in his private apartments, of which 900 lived in nine separate establishments according to their ethnic origin of 100 each.[17] According to Irvine, his son Nawab Qaim Khan (1743-8) had four

[14] *OIOC*, Orme Mss. o.v.173, fols. 171-2; Irvine, 'The Bangash Nawabs', part 1, 350.

[15] Shivdas Lakhnawi, *Shahnama Munnawar Kalam*, trans. S.H. Askari (Patna: Janaki Prakashan, 1980), 5.

[16] Irvine, 'The Bangash Nawabs', part 1, 330.

[17] Ibid., 339.

wives, of whom only the first was a Bangash.[18] Also Nawab Ahmad Khan (1750-71) married four wives whose descent is not altogether clear. His fourth wife, however, was an adopted daughter of Yaqut Khan, one of his slaves. Some of the Nawab's courtiers claimed that she was a natural daughter of the famous Khan Jahan Khan Lodi, the principal Afghan noble during the rule of Shah Jahan. As Irvine relates: 'The Nawab hearing this story fell in love with her without seeing her.'[19] Eventually, she became the mother of Ahmad Khan's successor, Dalir Himmat Khan Muzaffar Jang (1771-96) but, as he was known as the son of a slave girl, the legitimacy of his succession was strongly contested.

The female part of the Nawab's household could become very influential in state affairs. The Nawab's begams, his wives and daughters, were able to accumulate lots of treasure and land deposited with them as pensions by their male relatives. Besides, from the harem they could manipulate court politics by setting up for themselves a private network of *chelas*, some of whom could walk freely in and out of the *zanana*. Especially, at critical moments such as the Nawabi succession the principal widow-begam's influence could be decisive.[20] This was certainly true in the case of the Bibi Sahiba. After her last son Qaim Khan was killed in battle, she acted as the main representative of Bangash *nasab* and Bangash legitimacy. She could not only support the preferred successor by presenting him with sufficient funds to build his own following; above all, her approval was absolutely needed to sanction the Nawabi succession. At the death of Qaim Khan in 1748, the Bibi Sahiba sent for all her husband's sons and she directed the second eldest son, Ahmad Khan, to succeed his half-brother to the *masnad*. He refused the honour, however, because he knew that

[18] Irvine, 'The Bangash Nawabs', part 1, 373.

[19] Irvine, 'The Bangash Nawabs', part 2, *Journal of the Asiatic Society of Bengal* 48, 1 (1879), 123-4, 159.

[20] Cf. Begams at Awadh court: R. Barnett, *North India between Empires: Awadh, the Mughals and the British 1720-1801* (Berkeley: University of California Press, 1980), 100-2. For a similar case at Bhopal: S.N. Gordon, 'Legitimacy and Loyalty in some Successor States of the Eighteenth Century', in J.F. Richards, *Kingship and Authority in South Asia* (Madison: University of Wisconsin, 1978), 287-300.

she actually favoured Imam Khan, and therefore would not support him in the longer term. Consequently, Imam Khan became the new Nawab but it was actually the Bibi Sahiba herself, with the help of her late husband's *chelas*, who ruled in his place.

Two years later she was, however, confronted with grave difficulties because of her continuing conflicts with the Awadh court and, as a consequence, her general lack of funds. In these circumstances Ahmad Khan had been able to bring the Mau Afghans on his side. With their moral and material support he could now claim the Nawabship for himself. Still, he could not dispense with the approval of his step-mother. With Imam Khan imprisoned in Allahabad, the Bibi Sahiba decided to ask Ahmad Khan to come to Mau. After he showed his obeisance by presenting her with a tribute (*nazr*), she, in her turn, invested him as reigning Nawab with a robe of honour (*khilat*), after which all the other Afghans could present their offerings, again by way of *nazr*. Only now, Ahmad Khan seemed to be fully installed as the legitimate Nawab.[21] However, it should be kept in mind that *de jure* and *de facto* Nawabi rule not only hinged on the internal backing of the begams, the Bangash and other Pathan *sardars* or the *chelas*, but also, on the external mechanism of imperial power and authority.

Chelas

The Bangash Nawabs systematically recruited personal slaves who were entrusted with the most vital posts in the administration, the revenue collection and the army. These slaves played a significant role as a kind of artificial family in-group which was entirely attached to the person of their patron. The Farrukhabad slaves were basically identical to the elite military slaves of the Turko-Iranian world, called *mamluk* in Arabic, *ghulam* in Persian and *kul* in Turkish. Though technically meaning 'slave', these words carried a connotation not of enslavement or the servility of the Uncle Tom's kind, but, on the contrary, of power and dominance. The model for this kind of elite slavery was highly developed in the Middle East and Iran ever since the decline of the caliphate. Until the nineteenth century the *mamluk*

---

[21] Irvine, 'The Bangash Nawabs', part 2, 59-60.

system worked exceedingly well and succeeded in perpetuating and stabilizing the existing structures of the Islamic state.

In the Indian setting, the elite slaves of the Bangash Pathans and other Afghans were commonly referred to as *chelas*, i.e. disciples. This was consistent with the existing Indo-Mughal paradigm. Nonetheless, Muhammad Khan Bangash, who introduced the *chela* system from the start of his rule in Farrukhabad, preferred to call his *chelas*: *atfal-i sarkar* or children of the state. Other appellations include *ghalib bacha* or *ghazanfar bacha* which both refer to Muhammad Khan's personal progeny.[22] These names seem to indicate an Afghan approach to the slave as being regarded as the (adopted) son (*walad*) of his master-patriarch (*walid*). As such the master and his slave developed relations very similar to those of a family. Hence, the natural sons of the Nawab, were to be regarded as the brothers of the *chela*, like the *chelas* felt brothers of each other and the death of a *chela* had also to be revenged by his Bangash brothers. *Chelas* were as close to the person and body of the Nawab as were his sons. Like his own children, at a young age *chelas* were displayed in public on the lap of the Nawab. Similarly, the Nawab's principal wife, the Bibi Sahiba, observed no *purda* to the closest *chelas* of the Nawab. The *chelas* were also allowed to sit in *darbar* immediately next to the Nawab and his sons. Riding on his elephant, two *chelas* were always closely seated behind him. Not surprisingly, their preferential treatment could give rise to feelings of envy and annoyance among the Nawab's sons as the following story given by Irvine shows:

One day Bhure Khan [*chela* and *naib* of Allahabad] coming into darbar late, could find no place to sit; kicking away the pillow separating Muhammad Khan and Qaim Khan, he sat down between the Nawab and his son. Qaim Khan turned angrily to his father and said: 'You never respect me.' Muhammad Khan replied that he loved them as he did his sons. Qaim Khan got up in rage, and went off to his home in Amethi. Muhammad Khan then scolded Bhure Khan saying that he had lost confidence in him, for if while he was alive they did not respect his sons, who knew what they would do when he was dead. Bhure Khan putting up his hands, said: 'May God Almighty grant that I never see the day when you no longer live'.[23]

---

[22] Irvine, 'The Bangash Nawabs', part 1, 340-1.
[23] Ibid., 345.

Although *chelas* could never lay claim to the leadership of the Bangash clan, in the wake of their master, some of them were able to acquire great wealth and two of them were allowed to bear the title of Nawab themselves. Nawab Ahmad Khan Bangash himself was styled *Bara* (senior) Nawab, his *chela* Zulfiqar Khan, *nazim* of Shamshabad, was called *Majhla* (middle) Nawab and his *chela* Daim Khan, in charge of Shahpur-Akbarpur, was called *Chhota* (junior) Nawab.[24] Besides, like the Nawab's sons, many of the *chelas* were allowed to ride in *palkis* or on elephants.[25] All this may illustrate how the *chelas* were incorporated into the household of the Bangash Afghans. As adopted sons they became part of the patrimonial family of the Bangash ruler and as such were co-sharers of his realm. This was, however, only one side of the coin. What about the indigenous feedback to this Afghan model?

In a way, the life of a military *chela* fitted in well with the martial outlook of the Hindustani peasantry. Dirk Kolff has already stressed the importance and extensiveness of the armed peasantry in Hindustan, where, as late as the nineteenth century, military sports were very much a part of the daily life of the village.[26] Moreover, Rajput values stimulated the complete devotion in service to a rightful patron. As analysed by Norman Ziegler, the Rajput *chakari* bond to their Mughal patrons was a 'product of identification and obligation generated through the establishment of personal bonds and affiliations sanctioned by local custom, and the fulfilment of cultural aspirations and ideals, defined in local myth and symbol'. In this symbolism the kingdom or *thakurat* was a marriage between the ruler or *thakur* and the conquered land, which came to his care and protection. All his subjects were regarded as his children and in particular his servants and clients were his sons. In this sense, the Rajput ideal is the complete devotion to the *thakur*.[27]

[24] Irvine, 'The Bangash Nawabs', part 2, 160-1.

[25] Ibid., 69.

[26] D.H.A. Kolff, *Naukar, Rajput and Sepoy: The Ethnohistory of the Military Labour Market in Hindustan, 1450-1850* (Cambridge: Cambridge University Press, 1990), 17-31.

[27] N.P. Ziegler, 'Some Notes on Rajput Loyalties during the Mughal Period', in J.F. Richards (ed.), *Kingship and Authority in South Asia* (Madison: University of Wisconsin, 1978), 215-51.

As this Rajput outlook underscored the bond between Mughal patron and his Rajput client, it applied no less to the even more intimate relationship between the Afghan master and his Rajput *chela*. Of course, unlike the *chela*, the *chakar* could preserve his freedom, caste and religion. Nevertheless, the *chela*, as the adopted son of his master, came even closer to the ideal of utter and exclusive devotion. Indeed, *chelas* could turn into venerated personalities, local heroes, or otherwise become the centre of Rajput ritual. This was the case with one of Muhammad Khan's principal *chelas*, Dalir Khan. He was originally a Bundela Rajput but as *amil* in Bundelkhand he became an important *chela* of Muhammad Khan. After his death in a battle against his own kinsmen the following local custom grew up in Bundelkhand:

Every son of a Bundela, on reaching the age of twelve years, is taken by his father and mother to Maudan, where they place his sword and shield on Daler Khan's tomb. They make an offering, and the boy then girds on the sword and takes up the shield, while the parents pray that he may be brave as Daler Khan.[28]

In general, we may say that, on the one hand, the *chela* institution of Farrukhabad was originally a predominantly Turko-Persian phenomenon which, through Turkish and Afghan channels, was imported into Hindustan. On the other hand, its kinship idiom positively appealed to indigenous Rajput values which stressed their manliness and their utter devotion to a father-*thakur*. Let us now take a closer look at the main functions and characteristics of this particular Indo-Afghan variant of the *mamluk* system.

From the very beginning of their first settlement in Mau, the Bangash Afghans felt themselves in a very precarious position vis-à-vis other groups in the area, not the least with regard to their fellow Afghans. As mentioned already, in Mau the Bangash Pathans were in a distinct minority of only 100 to 600 men.[29] Muhammad Khan Bangash was only one among equals, and because other sub-tribes were far more numerous than his, he was even less equal than most

[28] Irvine, 'The Bangash Nawabs', part 1, 286.

[29] Ibid., 363; Nur al-Din Husain Khan Fakhri, [Tawarikh-i Najib al-Daula], trans. J. Sarkar, *Islamic Culture* 7, 4 (1933), 617.

of the other tribal leaders. After his foundation of Farrukhabad, Muhammad Khan was able to relieve the situation somewhat but it was very clear that without sufficient support in numbers there was no room for ambitious chiefs like him. In order to create a reliable and large following which shared his *nasab*, he could try to invite other Bangash Pathans from Afghanistan to come and join him in Hindustan. This could, however, not solve the problem entirely. First of all, the sources for recruitment were limited since even in Afghanistan proper, Bangash Afghans were not very numerous. Besides, these Bangash allies and sons-in-law were not always as reliable as one might hope for. They frequently became unruly elements or even, since they shared his *nasab* and his power, they could become rivals to the ruling Nawab himself. Thus like their western neighbours, the Afghan chiefs of Rohilkhand, the Bangash Nawabs tended to stress the importance of *nasab*, but unlike them, this was not sufficient to incorporate a large and reliable Afghan following. In these circumstances, the systematic and large-scale recruitment of personal slaves could offer a viable alternative for which they could have recourse to that old-established Turko-Persian formula of the *mamluk*.

Under Muhammad Khan Bangash the recruitment of personal *chelas* was dramatically intensified. During his rule some 4,000 of them were raised. The *chelas* were recruited at a very young age, of about 7 to 13 years old, from primarily Rajputs and also sometimes Brahmins.[30] Although according to their religious prescriptions, Muslims were not allowed to enslave fellow Muslims, this did not raise much problems in an overwhelmingly Hindu society. Hence, expensive and long-distance importation of slaves could be dispensed with and most of the *chelas* were locally taken in military campaigns against revenue defaulters or other enemies of the Nawab. Whenever one or another of his 'amils had a fight with a troublesome village, he was instructed to seize all the Hindu boys he could get and to forward them to the Nawab.

People were sometimes quite willing to let their children be adopted by a person in higher places. As a consequence of famine or other distress they could also be forced to do so because they lost all means

[30] Irvine, 'The Bangash Nawabs', part 1, 340.

to subsist their children any longer.[31] Hence, natural disasters and failing crops could strongly stimulate the slave trade as a result of a dramatic increase of the supply of slaves at public markets.[32] During the eighteenth century, especially the slave markets along the fringes of the Himalayas were famous for their large supply of slaves which were carried in large numbers down the hills for sale. Although the *chelas* were recruited from local sources, there were no serious problems of slaves remaining loyal to their own original family or kin. Once a *chela* always a *chela*: it was probably neither possible, nor attractive, to return to one's former home. Under Muhammad Khan the *chelas* preserved their caste-identity, and were married to slave girls from the same caste. Later, however, in Ahmad Khan's time, there was left just one corporate *chela* category. *Chelas* were never reported to have deserted the Nawab's ranks in order to join their former kinsmen. They were too much involved in their own careers which were entirely linked to the fate of their patron.

After the *chelas* were adopted, they converted to Islam, received a new name and were submitted to a regime of religious, literary and military training which was focused on the transformation of the recruit's identity, the development of his military and administrative skills and also on the drumming of his loyalty to the Nawab. When a boy could read and write he was taken before the Nawab who bestowed upon him, by way of *khilat*, one hundred rupees, a shield and a sword. From among the *chelas* of 18 to 20 years of age the Nawab selected 500 for his private bodyguard and trained them as a picked regiment.[33] Most of the others were entrusted with leading positions in the army such as recruitment officers (*bakhshi*), cavalry officers (*sawar-wala*), heads (*darogha*) of the stables and the elephant establishment or commanders of forts (*qiladar*). Some of them were

---

[31]In 1770 Shuja al-Daula send one of his eunuchs to Patna in order to 'purchase a number of those boys and girls who are starving at Patna and who he intends to bring up for the service of his household' (National Archives of India, New Delhi (NAI), Foreign and Political Dept., SC70, no. 17, 'Harper to Alexander, 15 April 1770.'

[32]I. Habib, *The Agrarian System of Mughal India* (Bombay: Asia Publishing House, 1963), 102, 110, 322-4.

[33]Irvine, 'The Bangash Nawabs', part 1, 341.

in the possession of large *jagirs*. They played a crucial role in the revenue collection as ministers (*diwans*), collectors (*amils*) and farmers (*ijaradars*), particularly at places where the collection was difficult as a result of widespread Afghan landholding interests in areas such as in Mau and Shamshabad. Some of them also figure as the Nawab's deputy (*naib*) in the provincial government of newly acquired territories. Others feature as architects of public buildings. Niknam Khan Chela, for example, laid out the fort of Farrukhabad and built many of its edifices. Some 17 *ganjs* were founded by *chelas*. We also come across them as founders of bazaars, caravanserais, bridges, step wells (*baolis*) and as planters of gardens. In all these capacities *chelas* played a pivotal role in stimulating cultivation and trade.

Apart from these critical military and administrative posts a score of other *chelas* were employed in the household and the harem. These were functions that derived their importance from their closeness to the body-person of the Nawab. Some *chelas* were allowed to approach the Bibi Sahiba unveiled such as attendants of the bath-rooms, keepers of the rosaries, attendants to help in the ablutions for prayers, for driving away flies, and numerous other personal servants who encircled the Nawab.[34] This *chela* encirclement of the Nawab's person in his private departments was even literally reproduced in the design of Farrukhabad city. Not only around the central fort the houses of his *chelas* were built, but also around the outer part of the city, residences were allotted to the Nawab's sons and *chelas*.

Despite of his intimate and close relations with his *chelas*, there was also a feeling of ambivalence and wariness towards them on the part of the Nawab. Being entrusted with the most crucial positions of the state, the *chelas* were extremely liable to abuse their power against the Nawab. Although they were deracinated and freed from their former ties to society, this did not deprive them from looking at their own personal interests or from being seduced to courtly intrigue; even when this was against the interests of the Nawab himself. With this in mind, the first Nawab frequently reallocated his *chelas* from one province to another as every *chela amil* was not allowed to

[34] Irvine, 'The Bangash Nawabs', part 2, 165.

stay more than two or three months in one province.[35] In this way
the Nawab hoped to screen his *chelas* from new vested and local
interests. Nevertheless, things could easily run out of control whenever
a *chela* became too successful in collecting the revenue. One example
of this is the case of Islam Khan, *chela* of Ahmad Khan, who was
appointed *faujdar* of Kasganj. At his arrival there, he started to procure
money from the local bankers and landholders in return for a bond
on the incoming revenue. Then he suddenly collected a following of
some 5,000 men and started to plunder some of the neighbouring
villages. On hearing of this threatening development, the Nawab
ordered Islam Khan to stop these dreadful activities immediately.
Meanwhile the *chela's* troops had risen to 10,000. Amongst them were
even numerous Pathans from Mau, Qaimganj and Shamshabad. No
wonder, Islam Khan declined the order by answering that he only
wanted to seat his patron on the throne of Delhi. The Nawab, however,
concluded that the *chela* had rebelled and he asked the Rajas of
Hathras and Bharatpur to crush the rebellion. After they had defeated
Islam Khan's army, the *chela* returned to Farrukhabad and presented
himself submissively to his master. He was not punished heavily and
after some time he was even restored to his former post of *bakhshi*.[36]

Similar apprehensions made the Nawab forbid *chelas* to construct
masonry structures. Nothing was to be built but with sun-dried bricks
and mud-mortar. Each *chela* was only allowed to build one single
brick room as a reception hall. In general, the Nawab appears to have
had a monopoly on building imposing brick buildings.[37] Exceptionally
trustworthy, however, were the eunuch *chelas*, many of whom were
entrusted with various responsibilities in the *zanana*. Since they had
no issue of their own and had lost much of their manliness, they were

[35] Irvine, 'The Bangash Nawabs', part 1, 346.

[36] Irvine, 'The Bangash Nawabs', part 2, 162-3.

[37] Perhaps, this explains why European travelers were not impressed at all
with the outward appearance of Farrukhabad, otherwise so famous for its riches.
But as Tieffenthaler observed around 1750, this was only a facade because from
the inside the houses were more luxurious and brick was used freely (J.
Tieffenthaler, 'La géographie de l'Hindoustan', in J. Bernouilli (ed.), *Description
historique en géographique de l'Inde,* vol. 1 (Berlin: C.S. Spener, 1786-91), 196.

more reliable than other *chelas*, as is indicated in the case of eunuch Yaqut Khan Chela, who was allowed to found seven *ganjs* and to erect numerous brick buildings for himself.[38]

The controlling measures of the Nawab were not sufficient to keep all ambitions of his *chelas* at bay. Especially after Ahmad Khan's government, they increasingly began to claim independent positions of their own. The most important example of this was Fakhr al-Daula. He started his career as a *chela* and *bakhshi* of Muhammad Khan but rose to real prominence under Ahmad Khan. He played a decisive role in Muzaffar Jang's successful succession in 1771 and was for three years the practical ruler of Farrukhabad. After Farrukhabad had become tributary to Awadh in 1775, the *chelas* remained active in court politics and in 1796 again played an important role in the Nawabi succession. Meanwhile they became more and more inclined to lead a luxurious and easy life. The British residents in Farrukhabad complained frequently about them as decadent and worthless people. With the decline of Nawabi power, some of them were able to keep land under their own control but others changed into nominal and absentee managers and left the real business to their local agents (*karindas*) who frequently embezzled the proceeds.[39]

What might in retrospect be said about the efficiency and success of the *chela* system? During the rule of the first three Nawabs it worked extremely well as a foundation of Nawabi power and even as an instrument of economic growth. The army and administration were successfully professionalised as they were staffed, so to speak, by enslaved 'young urban professionals' who were entirely dedicated to their careers in the service of the Nawab. Without a powerbase of their own they were less inclined to get themselves enmeshed in the

[38] Irvine, 'The Bangash Nawabs', part 1, 342. For eunuchs in India, see G. Hambly, 'A Note on the Trade in Eunuchs in Mughal Bengal', *Journal of the American Oriental Society* 94 (1974), 125-30. In 1764 Ahmad Khan adviced Shuja al-Daula to put more trust on his eunuchs (*khwaja-saras*). During the eighteenth century they remained extremely popular at the Awadh court (A.H. Sharar, *Lucknow: The Last Phase of an Oriental Culture* (Delhi: Oxford University Press, 1989), 30-2).

[39] Irvine, 'The Bangash Nawabs', part 1, 161.

usual Indian procedures of constantly shifting loyalties and alliances. In the long run, however, as their power and wealth increased, *chelas* also proved human after all. Under Ahmad Khan already the *chelas* had started to carve out positions of their own and some of them began to assume aristocratic lifestyles in which they indulged in singing and dancing performances and hunting. They had lands of their own, money of their own, houses and entire *ganjs* of their own, children and even *chelas* of their own. Most of them had been the personal slaves of Muhammad Khan. After his death they became automatically the slaves of his successor but felt naturally less allegiance to him. From the time of Muzaffar Jang, the Nawab had become entirely dependent on the *chelas* since they could secure his accession to the throne. Even during the successful years of Muhammad Khan and Ahmad Khan the *chelas* were only one instrument of power. At times, they had to be balanced by other groups, other *chelas*, other Bangash Pathans, Afridis or some other *naukari* retainers and mercenaries for sale at the military labour market. Probably aware of the growing dangers of a system which was so closely attached to the Nawab's person and hence could get easily out of hand, Ahmad Khan decided to reduce the *chela* establishment, recruiting only 400 new slaves during his entire rule (1750-71). As indicated by Seema Alavi, Ahmad Khan's rising star in the mid-eighteenth century politics of the Mughal Empire, was accompanied by a considerable change in the ideology which had welded together the Farrukhabad state. Bestowed with high Mughal honours, Ahmad Khan started to emulate Mughal processions, celebrated his birthday with the pomp and circumstance of the Mughal emperor and used these political rituals and celebrations to construct a new form of kingship to replace the Afghan and *mamluk* traditions of the earlier period.[40] While Ahmad Khan's *naukari* businesses escalated, his Afghan brethren and *chelas* faded into the background. But let us have a closer, chronological look at the Bangash Nawabs' rising fortunes at the military labour market and start again with Muhammad Khan.

---

[40] S. Alavi, *The Sepoy and the Company: Tradition and Transition in Northern India* (Delhi: Oxford University Press, 1995), 201-2.

NAUKARI

RISE

As one of the principal personages among the Mau Afghans, Muhammad Khan had led many plundering and freebooting expeditions especially into the Deccan and Bundelkhand. Most of the time he put himself at the head of 500 to 1,000 horsemen and became engaged by one or other local Raja who had to deal with rebellious landlords or peasants in his territory. Frequently, a kind of contract was agreed to in which part, usually one fourth, of the plunder, was reserved for the chief, the *jamadar*, and his troops, half of which was to be forwarded in advance.[41] The usual routine of such an expedition was to realise the contract and to maximize its gains by discrete mutual arrangement with the so-called rebels.[42] A show of force and, sometimes even actually fighting could be useful as this could raise the reputation of the gallant *jamadar* and, so, could strengthen his bargaining position vis-à-vis potential employers or rivals. Whenever violence did occur it was to be presented, rather theatrically, as a demonstration of the *jamadar's* military ability to handle his horses, guns and swords. Usually, however, violence was as much as possible avoided since it could harm the major assets of the expedition itself: horses, men and, to a lesser extent, elephants. Hence, *naukari* was a business of rational cost-benefit analysis in which violence contributed to the *naukar's* warlike profile but in practice had to be as much as possible avoided.

In 1712, Muhammad Khan could muster around 12,000 troops and had become a force to be reckoned with in Hindustan. During the struggle for the Mughal throne between Jahandar Shah and his nephew Farrukhsiyar, both claimants invited all major nobles and chiefs to join them. After he had sent one of his agents to find out which side was most likely to succeed, Muhammad Khan joined Saiyid Abdulla Khan, main supporter of Farrukhsiyar. At the ensuing

[41] Irvine, 'The Bangash Nawabs', part 1, 270. Note the similarity with the Maratha 'protection rent' of *cauth* (A. Wink, *Land and Sovereignty in India* (Cambridge: Cambridge University Press, 1986), 44-7).

[42] Irvine, 'The Bangash Nawabs', part 1, 271.

battle, the Bangash appears to have served bravely as a commander of the elephant corps (*jamadar-i fil-sawar*).[43] At the end of the day his political intuition had proven itself quite reliable. After Farrukhsiyar had gained the throne, Muhammad Khan was incorporated in the new Mughal body politic by being invested with a *khilat* and by being presented with an elephant, a horse, a sword, besides eight *jagirs* in Bundelkhand. At the same time he was raised to the rank of a *mansabdar* of 4000 and from that time he became styled as *Nawab*.[44] Suddenly, Muhammad Khan had changed from a petty *jamadar* into an important *mansabdar* and pillar of the imperial throne. Obviously, his expertise and contacts in Bundelkhand and Malwa were most welcome to the new emperor who did not have any effective control over this area because of increased Maratha incursions. In order to regain imperial rule in the area imperial policy tried to direct Muhammad Khan's attention especially to these turbulent areas. For this reason, he was given the *jagirs* in Bundelkhand and similarly, in 1730, he was appointed *subadar* or governor of Malwa.

These were not positions meant to be enjoyed without opposition but only paper appointments, intended and hoped to be realized in future. It was equally clear, however, that if he was to succeed in the Deccan, he needed a sure and more permanent base on which he could build his power on. For this reason, in 1714, Farrukhsiyar bestowed 52 Bamtela villages, not far from his home in Mau, in *al-tamgha* (on a permanent basis) on his wife, the Bibi Sahiba, as a kind of blood price for her father who had been killed by one of its Bamtela Thakurs. This was an excellent opportunity for Muhammad Khan to leave the parochial atmosphere of Mau once and for all. In order to underline his newly achieved imperial status he needed to emancipate himself from his fellow Afghans. In 1713 he had already founded Qaimganj and Muhamdabad, two new *qasbas* in the direct vicinity of Mau. One year later, however, he commissioned one of his *chelas* to build an extensive new city called Farrukhabad, along the Ganges 34 km east of Mau, in the new Bamtela territories he had gained through his wife. From this new basis the Bangash Nawab started to

---

[43] Muhammad Wali al-Lah, *Tarikh*, fol. 19a.
[44] Ibid.; Irvine, 'The Bangash Nawabs', part 1, 274.

carve out a territory of his own (*watan*), free from either Afghan or Mughal interference. The adjacent territories were relatively easily annexed, mostly sanctioned by bribing the local *qanungo* (accountant of revenue records) or by new grants or revenue-farm (*ijara*) leases from absentee *jagirdars* in Delhi to the Nawab or one of his *chelas*. This was done especially in those areas where revenue defaulters made the collection extremely difficult. After some time most of these lands fell completely under Nawabi control and were annexed to the Nawabi *watan*.[45] From the start, the local power of the Bangash Nawab was traded off against the legitimacy of the imperial nobles in Delhi. During this process the Nawab's position grew more and more powerful and, at the same time, more and more legitimate. At the end, he had come in a position to ward off any involvement of the old *jagirdars* from the imperial centre who could only acquiesce in this fate or invite other greedy outside powers to restore their lost incomes.

Meanwhile, Farrukhabad had not only become the central repository of the Nawab's wealth but soon developed as the regional hub of an extensive commercial network that linked the flourishing north-western trading network to the south and east. For example, horses, coarse cotton cloths, grain and indigo from Rohilkhand found their way to Awadh and Benares via Farrukhabad, where they were exchanged for more luxurious textiles from the eastern provinces. Via its extensive credit network, and facilitated by its mint, Farrukhabad was also able to attract new emigrants from Afghanistan. The Nawab made over large sums of cash by bills of exchange (*hundis*), via Lahore to Kabul in order to invite his countrymen to settle in Farrukhabad.[46] Also many Sufi and other intellectuals were enticed to come and settle in Farrukhabad. Some of these men attracted a large following of both Muslims and Hindus who visited their hospices or became their disciples. Many *khanaqas* were built to accommodate them and lands

---

[45] OIOC, Bengal Proceedings, Secret and Political Consultations, P/Ben/Sec/2, 'Willes to GG&C, 29 January 1787', fols. 148-92.

[46] Irvine, 'The Bangash Nawabs', part 1, 323-4. Cf. Farrukhabad coins found in Afghanistan (J. Rodgers, 'The Coins od Ahmad Shah Abdali', *Journal of the Asiatic Society of Bengal* 54 (1885), 72.

and large sums of cash were distributed amongst them.[47] According to Wendel, Farrukhabad had such a reputation as a home for holy men that it became commonly known as *Faqirabad*.[48] At that time, increased migration to Farrukhabad was part of the general drain of people and resources away from the old Mughal centres of Lahore, Delhi and Agra to the new capitals of the Mughal successor states. Obviously, all these new immigrants strengthened the local power base of their new patron as they gave him additional status or as they even fought actually with him in his numerous campaigns.[49]

Many of the conquered and newly acquired territories were given in *jagir* to the Afghan nobility and *chelas*. Some of them laid out large fruit and pleasure gardens but the overall revenue collection was mainly farmed out to those who had sufficient power and credit to ensure regular payment. As a result, most of the Afghans were absentee landlords who were not actively engaged in the land management themselves, as they were fully engaged as administrators and mercenaries.[50] The existing local establishment of *zamindars* and peasants could frequently be preserved although on top of it a new layer of landholders and their agents (*amils*) or farmers (*ijaradars*) was placed. This was possible since agriculture was sufficiently sure to sustain a new ruling elite without having to throw out the existing cultivators. In due course, the old *zamindars* and *chaudhuris* could even become great *ijaradars* as they took over responsibility and paid the arrears of unwilling neighbouring villages.[51] Other *zamindars*,

[47] For some examples, see S.A.A. Rizvi, *Shah Wali-Allah and his Times* (Canberra: Ma'rifat Publishing House, 1980), 182-3; and, Irvine, 'The Bangash Nawabs', part 2, 137, 157.

[48] F.X. Wendel, *Les Mémoires de Wendel sur les Jats, les Pathan et les Sikhs*, ed. J. Deloche (Paris: Ecole Française d'Extrême-Orient, 1979), 140.

[49] For the economic position of Farrukhabad, see Gommans, *The Rise of the Indo-Afghan Empire*, 42, 96, 128-33.

[50] OIOC, V/23/136, Selections from the Revenue Records North Western Provinces, vol. 3 (Allahabad: North-Western Provinces' Govt. Press, 1873), 311; E.T Atkinson (ed.), *Statistical, Descriptive and Historical Account of the North Western Provinces of India*, vol. 7 (Allahabad: North-Western Provinces' Govt. Press, 1874-84), 75, 105, 349.

[51] OIOC, V/23/136, Selections from the Revenue Records North Western Provinces, vol. 3, 412-13; Cf. E. Stokes, *The Peasant and the Raj: Studies in*

however, had to renounce all rights to their estates and had to contend themselves with only a small part of the produce as an allowance for their assistance in the revenue collection (*nankar*). Sometimes, especially when resistance was strong or when room was needed for large Afghan or *chela* colonies, such as in Kampil and Shamshabad, the existing landholders were completely ejected.[52] This was the exception, however, and normally there was no actual displacement of the old, mostly Rajput, landholders. Usually, they were merely degraded to local agents and overseers (*mir-dah*) in direct service of the new *jagirdar, inamdar* or *ijaradar* and were paid a fixed allowance or a percentage of the realized revenue (*hasil*) in cash or in land.[53] In general we might conclude that, except from laying out watered fruit and pleasure gardens, the conquering elite of Afghans kept themselves aloof from direct engagements in the soil and from their subject peasant population. The actual management of their *jagirs* was left to wealthy and powerful *amils* or *ijaradars* who were only willing to be involved because they felt it to be profitable. At the same time, however, the Afghans could concentrate their attention on the business of military *naukari*. So let us return to Muhammad Khan's military adventures.

APOGEE

In 1720 the new Emperor Muhammad Shah confirmed the Nawab in his control of the existing and acquired territories around Farrukhabad and increased his *mansabdari* rank to 7000. He was also

---

Agrarian Society and Peasant Rebellion in Colonial India (Cambridge: Cambridge University Press, 1980), 75; and I. Husain, 'Agrarian Change in Farrukhabad District, Late Eighteenth and First Half of Nineteenth Century: A Study of Local Collection of Documents' (Mimeo. Aligargh: Aligarh Muslim University, 1979).

[52] H.F. Evans, *Final Report of the Settlement of the Farrukhabad District* (Allahabad: North-Western Provinces' Govt. Press, 1875), 14-15; Atkinson, *Statistical, Descriptive and Historical Account*, 282.

[53] OIOC, Bengal Revenue Consultations Ceded and Conquered Provinces, P/90/39, 20-9-1803, 'Agent Farrukhabad to Secret. Brd. of Revenue, 20 September 1803'; P/90/40, 'Agent Farrukhabad to Secret. Brd. of Revenue, 30 December 1803.'

appointed *subadar* of the extremely rich and relatively easily-held province of Allahabad.[54] Both this province and his *watan* of Farrukhabad, as also Rohilkhand, were the main *naukari* recruitment centres for the intensive Bangash Nawab's campaigns in Bundelkhand and Malwa. Several times the Bangash Nawab collected a large following of freebooters in the north and tried to secure a firm footing in the area by everywhere establishing his military posts (*thanas*) under the command of his *chelas*. He also sought a Muslim alliance with Nizam al-Mulk who also tried to gain a hold on his provinces in the Deccan against continued 'infidel' Maratha incursions.[55] In 1736, however, due to continued Maratha opposition, in league with local *zamindars* and several imperial *jagirdars* who all had their own local interests, the Bangash involvement in Bundelkhand and Malwa came to a complete halt. As a consequence, the Allahabad province was never again entrusted to the Bangash Nawabs, who now began to redirect their interests more to the north-west.

Muhammad Khan died in 1743 and was succeeded by his eldest son Qaim Khan without any opposition as he was the only living son of Muhammad Khan and the Bibi Sahiba. Qaim Khan started by modifying the alliances made by his father. Internally he distanced himself from his father's intimates. The *chelas* became less important as he invited in the Afridi Afghans and made them the main base of his power. He also replaced the court from Farrukhabad to his personal fort at Amethi. At the Delhi court, Qaim Khan switched sides from the so-called Turani party, to the Irani, newly appointed *wazir*, the governor of Awadh, Safdar Jang. The latter produced for him an imperial *farman* which made over to him the entire territory of the Rohillas. In order to enforce this imperial order the Bangash chief marched into Rohilkhand. In a very short time he collected a force of 50,000 horse and foot, besides the contingents of his Bangash relations, all provided with elephants, and another 20,000 volunteers under neighbouring *zamindars* and Maratha freebooters. In spite of this massive force, the Bangash troops were defeated by the superior

[54] Irvine, 'The Bangash Nawabs', part 1, 305.
[55] Ibid., 309; Z. Malik, 'Muhammad Khan Bangash's Letters to Nizam-ul-Mulk', *Proceedings of the Indian History Conference* (1967), 176-85.

mobility of the Rohilla cavalry and the immense firing power of their infantry. Qaim Khan himself and many of the other Bangash and Afridi leaders were killed, sitting as open targets in their *haudas* on their elephants.[56]

As a result of this defeat, the Bangash Nawabs lost all the provinces north of the Ganges to the Rohillas. Moreover, without a strong leadership, Farrukhabad now risked being annexed to Awadh since Safdar Jang had only awaited an opportune moment. As Farrukhabad became more and more hemmed in between the troops and intrigues of its enemies, the Bangash Pathans rallied around their new leader Ahmad Khan, who started to collect funds by putting pressure on all the nearby merchants and bankers. Also the still powerful and wealthy Afridis supported him with cash on the condition that he would grant their leader, Rustam Khan, half of the territories that would be recovered. The result of this general subscription was gathered and publicly displayed in one of the Nawab's tents. Everybody willing to be enrolled in the Bangash service was allowed to take from it: one and a quarter anna for a footman, three annas for a horseman. As usual, the free military labour market again proved its smooth functioning: in no time an army was recruited of 12,000 horse and 12,000 foot. At the ensuing confrontation at Khudaganj (1750) Ahmad Khan had learned from his brother's experience as he took a seat in his *palki*, protected by the shield of his fellow Pathans. On the other side, Safdar Jang's *naib* and commander of his forces, Nawal Rai, was soon killed in his *hauda*, and perceiving what had happened, the whole army dispersed immediately. Because of their sudden retreat the plunder from the enemy's camp was enormous. Here, several wealthy merchants, still playing cards in their tents, were overwhelmed and subsequently blackmailed. As everybody wanted to be part of the victory, the morning after the 'battle' the Bangash army had swollen to even 60,000.

At the next showdown, Safdar Jang's army was again defeated, this time mainly because a large parts of its vanguard, under the command of Kamgar Khan Baluch, had been secretly in league with the Bangash and suddenly defected from the Awadh ranks. Since his main rival

[56] Irvine, 'The Bangash Nawabs', part 1, 380-1.

Rustam Khan Afridi was killed in one of the few encounters, Ahmad Khan had gained a double triumph as he was now the sole Afghan leader without openly having to share his territory. By winning these two battles, Ahmad Khan had clearly regained the initiative. In order to cash in these successes he now sent his agents to Awadh, Allahabad and Benares. After the bankers of the latter had paid a sufficient *nazr* of 2 crores Benares was left untouched and its Raja, Balwant Singh, was bestowed with the Nawab's *khilat*. Nevertheless, Awadh and Allahabad he had to take by force.[57]

In 1751, Ahmad Khan was clearly at the height of his power. He now could boast a following of nearly 100,000 men and access to a territory, equal to nearly 4 or 5 *subas*. Obviously, Safdar Jang, but also the emperor and the other imperial nobles began to feel a bit worried as they felt that Ahmad Khan could not possibly be refrained from proclaiming himself independent king. Consequently, a mammoth alliance of the Emperor, Safdar Jang, the Jats and the Marathas, secretly endorsed by the Rohilla leader Hafiz Rahmat Khan, was created against him. Learning about what had happened, Ahmad Khan immediately retreated his army into his own territories and entrenched himself at his fort of Fatehgarh. On his way back, he had lost most of his following as many of his mercenaries saw no further profit in taking up a strong resistance with him. Meanwhile, he had asked for help from the Rohillas and Saadalla Khan, obviously without the backing of Hafiz Rahmat Khan, marched in the direction of Fatehgarh to support the Afghan cause. What followed was a protracted siege during which endless negotiations were held. After a while, Saadalla Khan suddenly retreated from the scene and returned with his troops to Rohilkhand. Ahmad Khan left alone facing fearful odds, immediately decided to follow the example of his former ally. The following year after the monsoon, the Maratha and Awadh forces entered Rohilkhand but the Afghans had withdrawn already into the Tarai jungle at the foot of the Himalayas. During this campaign, news arrived from the west that Ahmad Shah Durrani had entered the Punjab in order to assist his endangered tribesmen. Feeling a

---

[57] For a detailed account of these events, see Irvine, 'The Bangash Nawabs', part 2, 60-77.

growing uneasiness about the whole situation, especially the Marathas became very keen on arriving at a rapid peace settlement. Both parties agreed that Ahmad Khan would be restored in his former position although he had to pass half of his territory to the Marathas. The latter were allowed to extract Rs. 30 lakh from these lands but subsequently had to hand this sum over again to the Bangash.[58]

From 1752 to 1770, as most leaders lived in a continuous awe of potential Durrani invasions, large parts of Hindustan experienced a period of relative calm. Durrani and Maratha incursions were redirected to the Delhi-Agra region south of the Ganges which left the major northern successor states of Rohilkhand, Farrukhabad, and Awadh undisturbed. This was even so for large tracks in the Doab, mainly under variable Afghan, Jat and Maratha rule. As we have observed before, the Durrani campaigns engendered new prospects for military service but also of long distance trade as new markets were opened up for Hindustani goods in Afghanistan, Persia and Central Asia. Besides, the successor states in northern India could strengthen their position thanks to a renewed and sure supply of warhorses and manpower from the Durrani territories.

What might in retrospect be concluded about the Bangash Nawab's relations with the imperial court and the status he derived from this? From the time of his rise to the *mansabdari* status in 1712, Muhammad Khan had been constantly engaged in the court politics of Delhi. Although the Bangash Nawab, like fellow Afghan Ali Muhammad Khan Rohilla, has been frequently related to the Turani faction at court, this point should not be overstated. Muhammad Khan started his career by being attached to the parvenu Saiyid brothers under Farrukhsiyar. At the right time, following the declining star of the Saiyids, the Bangash Nawab switched his loyalty from them to the Turani noble Muhammad Amin Khan, Itimad al-Daula. During his campaigns in the Deccan he, obviously, tried with fair words to placate Nizam al-Mulk to his side by stressing his traditional bond with the Turanis and the Muslim mission of *jihad*. In practice, however, this seems to be only a matter of articulation and form. In 1747, Muhammad Khan's successor Qaim Khan thought it opportune to

[58] Irvine, 'The Bangash Nawabs', part 2, 77-123.

change sides again because, this time, the Irani noble Safdar Jang had become the imperial *wazir* and an alliance with him held out many new promises. And indeed, as related before, Safdar Jang now procured for his new Bangash ally an imperial *farman* in which Rohilkhand was entirely made over to him. In general we may say that loyalties at the royal court remained always extremely conditional and uncertain. This was as true for the so-called Irani and Turani factions as for the Afghan solidarity feelings: defection was the rule, patriotism the exception.[59]

How did the involvement in Mughal politics affect the Nawab's prestige at home and amongst his fellow Afghans? Muhammad Khan Bangash had risen to a first-grade *mansabdar* and *subadar* over important provinces like Allahabad and Malwa. However, to his contemporaries he continued to make a rather plain and soldier-like impression. He always wore clothes of the coarsest stuff and in his audience hall and his house the only carpets were rows of common mats. On these mats the Nawab sat together with his fellow Pathans and *chelas* and all other persons and guests high or low, all enjoying the same simple meal of *pilao*. At these occasions the Nawab always excused himself as being merely a soldier. The contemporaries were amazed to observe so much discrepancy between his great wealth and power and the simplicity of his personal habits. This roughness and general lack of *adab* sometimes could become rather embarrassing, in particular during imperial audiences at court, most notably in 1739 when he had to present himself before Nadir Shah, and had to live up to the expectations of one of the principal nobles at the imperial court. Muhammad Khan's clumsiness in matters of courtly etiquette was articulated by the fact that he pretended not to understand a single word of Persian, for which he had to be accompanied by one of his sons.[60] Clearly, first and foremost, Muhammad Khan considered himself a soldier among soldiers, a Pathan among Pathans.

The second generation of the Bangash family was much more

[59] Cf. for example: J. Sarkar, *Fall of the Mughal Empire*, vol. 1: *1739-1754* (London: Sangam, 1988), 250; Z. Malik, *A Mughal Statesman of the Eighteenth Century, Khan-i Dauran, Mir Bakhshi of Muhammad Shah 1719-1739* (Aligargh: Aligarh Muslim University, 1973), 56; R. Joshi, *The Afghan Nobility and the Mughals, 1526-1707* (Delhi: Vikas Publishing House, 1985), 195.

[60] Irvine, 'The Bangash Nawabs', part 1, 332, 338.

cultivated and fully acclimatized to the Nawabi lifestyle and the etiquette and ceremony of the Indo-Persian court. Muhammad Khan's successor Qaim Khan on occasions used to adorn his personal fort of Amethi sumptuously with canopies of precious broadcloth and gold curtains. No one's horse, *palki* or elephant was allowed to enter into the fort and all visitors had to dismount at the gate.[61] Ahmad Khan Bangash, even lived for much of his time in Delhi where he enjoyed a luxurious life and started to collect precious books and pictures from the imperial stores.[62] His increased engagement at the Delhi court was also reflected in his appointments as *mir-bakhshi* and first noble of the reign: *amir al-umara*.

Ahmad Khan began to consider himself as the leader of the Indo-Afghan nobility. The Rohilla leaders more or less acknowledged, albeit grudgingly, his first-rate status as being a principal Mughal *mansabdar*.[63] The Rohillas were not allowed to marry his daughters, were offered his *khilats* and had to accept his precedence in protocol. The Rohillas were generally looked down upon as parvenu slaves and horse merchants, whereas the Bangash, mostly seated on their elephants, regarded themselves to be real Pathans entrusted with high *mansabs* earned through their gallant service to the imperial throne. This became particularly and visibly clear at general meetings and visits. In 1752, for example, Ahmad Khan sent his son Mahmud Khan together with Hafiz Rahmat Khan for peace negotiations to the *wazir* Shuja al-Daula. Everything had to be neatly arranged in order to meet the requirements and sensitivities of everybody's rank and status. On this occasion in front of the *wazir's* tent an enclosure was erected which consisted of three courts. On his way to meet the *wazir*, the Nawab's son was allowed to pass the first two courts on his elephant but at the third court he had to change for a *palki*. Nevertheless, the other chiefs had to dismount their elephants already at the first, and to leave their *palki* at the second court. After they had entered the

[61] Irvine, 'The Bangash Nwabs', part 1, 373.

[62] OIOC, Bengal Proceedings, Secret and Political Consultations, P/Ben/Sec/2, 'Willes to GG&C, 29 January 1787', fols. 148-92.

[63] Nur al-Din Husain Khan Fakhri, [Tawarikh], *Islamic Culture* 7, 3 (1933), 433; Ghulam Hasan Samin Bilgrami, *Indian Antiquary* 36 (1907), 15. For Rohilla aversion, see Irvine, 'The Bangash Nawabs', part 1, 376; part 2, 128.

third court on foot the young Nawab followed and was helped by them out of his *palki*.[64] Under Ahmad Khan even the Awadhi Nawab was considered to be of inferior rank.[65] Obviously, this was strongly contested. After Shuja al-Daula's defeats against the EIC in 1764 and 1765, he had to escape to Farrukhabad. Being a humble visitor of Ahmad Khan made him, however, extremely circumspect. In one of their meetings the Bangash chief took a pearl necklace, once worn by his brother Qaim Khan, and put it round Shuja's neck. If he had accepted the gift it would have been interpreted as a clear sign of subservience, as a *khilat*, and, naturally, Shuja was strongly irritated about it as he quickly took it off and left the place.[66]

The Durrani emperor confirmed the foremost position of the Bangash Nawabs. Although Ahmad Khan had attended slowly and with only a very small force to support the Durrani forces at Panipat, Ahmad Shah bestowed large plots of land in the Doab on him and at court he gave him the right of first entry, preceding all of the other *amirs*.[67] Later, however, the overall status of the Bangash chiefs declined again, since, on the one hand, the Durranis started to raise the power and pretensions of the Rohilla generalissimo Najib al-Daula and, on the other hand, the Mughals became more and more reliant on their eastern vassals: Shuja al-Daula and the East India Company.

All in all, in spite of their proper *nasab* and strong regional powerbase, the legitimate authority of the Bangash Nawabs, both inside and outside their own territories, still required Mughal or, to a lesser extent, Durrani sanction. The Bangash had gained their supremacy and their territory thanks to their *naukari* services to the Mughals and, with that, had acquired rank and status within the Mughal hierararchy. This had amply increased their standing amongst their fellow Afghans, not only because of imperial recognition but also because through their patrons they could perhaps get access to court politics and to imperial *farmans* and *sanads*. Obviously, the

[64] Irvine, 'The Bangash Nawabs', part 2, 115.

[65] OIOC, Bengal Proceedings, Secret and Political Consultations, P/Ben/Sec/2, 'Letter mother Muzaffar Jang, 14 March 1787', fol. 518.

[66] Irvine, 'The Bangash Nawabs', part 2, 144-5.

[67] Ibid., part 2, 128.

Bangash Nawabs being completely embedded into the existing Mughal hierarchy had much to loose and only hesitantly answered to the Durrani claims of imperial sovereignty.

DECLINE

Towards the end of his lifetime, Ahmad Khan became blind and more than before relied entirely on his principal *chela* and *bakhshi*, Fakhr al-Daula. When he died in 1771 the Nawabi succession became hotly contested. Ahmad Khan's oldest son, Mahmud Khan, born from the Nawab's principal begam, was considered the only legitimate son. He had died, however, some years before and his son, Himmat Khan, was still a minor. Ahmad Khan's second son was Dalir Himmat Khan *alias* Muzaffar Jang. He was only fifteen years old and his mother, although she claimed Lodi descent, was regarded as a mere slave girl from the *zanana*. However, Muzaffar Jang's claims to the *masnad* were propagated and endorsed by Fakhr al-Daula against the wishes of the late Nawab's brothers. The ensuing conflict was decided by Muzaffar Jang's mother who handed over 70,000 gold mohurs from her treasury to the *chela*.[68] He also melted down all the silver of the *haudas* and other furniture.[69] With this money he was able to hire a sufficient following to checkmate the other Pathans and to pay a proper *nazr* to the Mughal Emperor, who subsequently acknowledged the legitimacy of Muzaffar Jang's succession.

Initially, the new Nawab was under the complete control of Fakhr al-Daula. The *chela* tried to improve the financial position of the state by reducing the army and by raising the revenue on the rich. This policy brought him in conflict with the other great *chelas* of the court and influential Pathans, who, in a coalition with the young Nawab, succeeded in killing him. This temporary coalition again brought the old Pathan nobility to the fore, as Ahmad Khan's brother Khudabanda Khan now managed to dominate state affairs.[70]

[68] OIOC, Bengal Proceedings, Secret and Political Consultations, P/Ben/Sec/2, 'Willes to GG&C, 29 January 1787', fols. 148-92.

[69] Irvine, 'The Bangash Nawabs', part 2, 154.

[70] OIOC, Bengal Proceedings, Secret and Political Consultations, P/Ben/Sec/2, 'Willes to GG&C, 29 January 1787', fols. 148-92.

Meanwhile, Farrukhabad had become increasingly involved in the politics and intrigues of the Awadhi court. During the years 1771-4, Muzaffar Jang had taken part in Shuja al-Daula's campaigns against the Marathas and the Rohillas, which recovered much of the lost territories in the Doab. Shuja, however, played upon the Nawab's authority problems at home and managed to annex not only all the lands in the Doab but large parts of Farrukhabad state as well. In return, he had been willing to support the claims of Muzaffar Jang against his Pathan rivals. In 1774, with the help of Awadhi troops, Muzaffar Jang ousted his uncle Khudabanda Khan and numerous other Pathans from Farrukhabad and Mau and posted his own officials in their place. Although Nawabi rule appeared to have been fully restored, Muzaffar Jang had become totally indebted to the Awadhi court. This new relationship also found its reflection during one of the Nawab's visits to Awadh, where he joined in the Muharram festivities and openly converted to Shiism.[71]

In 1775, the new Awadh ruler, Asaf al-Daula, decided to make no further bones about it and wanted to seize the entire Farrukhabad territory. Due to the mediation of the British agent in Lucknow this was avoided but instead it was agreed that Farrukhabad had to pay an annual tribute (*nazrana*) of 4½ lakhs, the payment of which was to be supervised by a Lucknow agent (*sazawal*) present at Farrukhabad. In addition, a deputy to the Nawab was to be appointed to control the Nawab's affairs. In fact, the *sazawal* started to control the revenue affairs of the Nawab and in due course became the *de facto* ruler in Farrukhabad. The *niyabat* was a major fund raiser as it became an office which could be directly purchased in Lucknow and, as a result, personal changes occurred frequently.[72] All in all, in 1775 Farrukhabad had lost almost three quarters of its former territory, and its revenue had fallen from about 60 to 40 lakh in 1761 to about 10 lakh in 1775.[73] Farrukhabad had changed from a major Sunni Indo-Afghan power into a minor Shi'a puppet state of Awadh.

[71] Atkinson, *Statistical, Descriptive and Historical Account*, 174.

[72] For a detailed account of these affairs, see OIOC, Bengal Proceedings, Secret and Political Consultations, P/Ben/Sec/2, 'Willes to GG&C, 29 January 1787', fols. 148-92.

[73] For revenue figures, see Irvine, 'The Bangash Nawabs', part 2, 157; OIOC,

In the wake of the Awadh intervention, the East India Company also entered Farrukhabad politics and as a tributary state of Awadh it was incorporated into the British subsidiary political system. In 1777 British troops were stationed in Fatehgarh and three years later a British agent was send to Farrukhabad to take charge of the annual tribute, which was now earmarked for the Company's treasury. As such, the British resident replaced the *sazawal* as manager of the Nawab's finances. In the 1780s the state of the latter had been worsened again. The court and its household had become isolated from the rest of the state and the incoming revenue had further decreased. According to the British resident, John Willes, this was due to the fact that: 'the Pathans seized and enclosed whatever land they chose, the aumils absconded, the zamindars resisted and the revenue of course fell down'. Only with the help of British troops, the Nawab was able to collect the revenue which was, however, increasingly siphoned off by the local *zamindars* and revenue farmers who entrenched themselves in their mud forts. Because of the annexation of large territories by Awadh, many *zamindars* controled lands which overlapped the newly created borders. At the time of collection, most of them retreated with their assets across the border to Awadh.

The borderlands, both in- and outside Farrukhabad state, were almost exclusively under the revenue farm of one of the principal eunuchs of the Lucknow court, Almas Ali Khan, who, obviously, protected his *zamindars* and instigated them to escape the revenue authorities from Farrukhabad. Almas Ali Khan had even for some years been appointed as *sazawal* himself during which time he had been able to strengthen his hold on the area and its *zamindars*.[74] From the perspective of Willes:

Add. 60337, Shee Papers, 'Shee to Hastings, 15 July 1780', no fol.; OIOC, Home Miscellaneous, 219, 'Minute GG, 22 May 1780', fol. 541; OIOC, Bengal Proceedings, Secret and Political Consultations, P/B/13, 'Willes to GG&C, 30 July 1786', fol. 550.

[74] OIOC, L/Parl/2/20, House of Commons, 'Articles of Charge of High Crimes and Misdemeanors against Warren Hastings, 4 April 1786', 5; Bengal Proceedings, Secret and Political Consultations, P/Ben/Sec/2, 'Willes to GG&C, 10 February 1787', fol. 117.

I have every where found the ryotts and lesser zemeedars poor and oppressed whilst the wealthy possessed of strong holds have opposed government by open force and only paid the most moderate rents [and] the whole country will be divided between the neighbouring powerful aumils, the refracting zamindars and banditti of robbers; and the Patans who might be made useful subjects will fly from the scene of anarchy.[75]

However, to the British resident, this state of so-called utter anarchy and confusion was entirely the fault of the Nawab and his personal ambiance of fraudulent and dissimulative *chelas*. The Nawab, he opined:

is inclined to keep the generality of his relations in the most abject poverty and to squander whatever he can get on a set of worthless and abandoned cheelahs. To preserve to the Nabob from their rapacious hands any considerable sum has been a difficult task. They have claimed with tumult assignments on the country and the Nabob, whose disposition is a compound of imbecility, cunning and malevolence, has been happy in distressing his diwan by the grant of their demands. It is and has been long the misfortune of this country, that whoever in office endeavours to render the Nabob respectable, or restore his country to order is sure of experiencing every opposition.[76]

Willes was particularly exasperated about the decadence of the Nawab's lifestyle. After three generations there appeared to be not much left of the former soldatesque image of the first Bangash Nawab. Again, Willes was of the opinion that all the intrigue at court and the abstention of the landholders originated from the 'imbecility and folly' of the Nawab, who:

ridiculous as it may appear, from the time that he rises, which is not till three in the evening, he is solely occupied in blowing thro' a tube for the purpose of encreasing the heat of a blowing furnace which he has been persuaded convert copper into gold.[77]

---

[75] OIOC, Bengal Proceedings, Secret and Political Consultations, P/B/13, 'Willes to GG&C, 30 July 1786', fol. 551; OIOC, Bengal Proceedings, Secret and Military Consultations, P/A/8, 'Willes to GG&C, 24 April 1785', fols. 340-56.

[76] OIOC, Bengal Proceedings, Secret and Political Consultations, P/B/13, 'Willes to GG&C, 30 July 1786', fol. 551.

[77] OIOC, Bengal Proceedings, Secret and Political Consultations, P/B/14, 'Willes to GG&C, 3 October 1786', fol. 253.

Obviously, the Nawab had a different view about what had gone wrong. According to his own analysis, the main trouble was Awadhi and British interference. The *sazawals* and British residents had deprived him from all his means to support his relatives and following. Besides, the brutality of Willes had been an affront and an open challenge to his status and authority. He had not only sold the Nawab's guns and elephants but he had also insulted him by installing his brother as *naib*. As the Nawab pointed out:

It is an established principle among the chiefs in Hindostan, not to intrust the management, or control as naib of their affairs, to their own brother or son, on account of several apprehensions.[78]

General unrest was also caused by the unemployment of his fellow Pathan chiefs. Due to the expansion of the Pax Britannica the Nawab could not any longer direct their attention to raids outside his own territories. For the Pathan mercenaries there was a general loss of employment. Hence, dissatisfaction among the Pathans increased and could only turn inwardly against the ruling Nawab. In order to appease them he begged the Company authorities to be allowed to raise a larger regiment on a permanent footing.[79] In short, however, his solution was fairly simple: in order to end the general distress and to increase his income he had to be relieved from all further British and Awadh involvement.

Although the Nawab's point of view made much more sense than the unbalanced indignations of the British residents, the picture which they presented shows that it was already far too late for a real amelioration. The revenue of the state had shrunk to an utter minimum of 9 lakh. Almost all what had remained of the Farrukhabad territory was beyond Nawabi control. In general, there was no sharp decline in agricultural produce but the benefits remained in the pockets of the local *zamindars* and *ijaradars*. The foremost revenue farmer, Almas

[78] OIOC, Bengal Proceedings, Secret and Political Consultations, P/Ben/ Sec/2, 'Muzaffar Jang to GG, 12 February 1787', fol. 106.

[79] OIOC, Home Miscellaneous, 219, 'Minute GG, 22 May 1780', fol. 565. Similar pleas were raised by the Nawab of Rampur. For general unemployment among the Afghans a few decades later, see R. Heber, *Narrative of a Journey through the Upper Provinces of India: From Calcutta to Bombay* (London: J. Murray, 1828), 138.

Ali Khan, who acquired large territories in Farrukhabad, linking them to those in the Doab and Rohilkhand, started to carve out a principality of his own which held great attraction to the surrounding *zamindars* and peasants. In addition, the general trade pattern of northern India had also changed radically with the expansion of British power and the demise of the Rohilla state. Long-distance overland trade relations with the west were interrupted and the important Indo-Afghan horse trade was redirected to the south. Besides, trade was lured away from Farrukhabad towards the new local *ganjs* of Almas Ali Khan and various British army commanders.[80] Also the slump in the *naukari* trade of Rohilkhand and Farrukhabad affected the cash transports to the north-west and brought about unemployment and dissatisfaction among the Afghans. During the 1770s and 1780s there was an overall lack of cash in the area and coin was strongly devaluated.[81] Besides, the the collapse of the Rohilla courts and trading network caused a general slump in the demand for luxury products and other consumer goods from the eastern and southern provinces. One of the main losers was the city of Farrukhabad which had been the central entrepôt linking the Indo-Afghan trading network to the east and the south. Governor-General Warren Hastings in 1780 expressed it as follows:

the capital which but a very short time ago was distinguished as one of the most populous and oppulent commercial cities in Hindostan at present exhibit nothing but scenes of the most wretched poverty, desolation and misery.[82]

Only later during the century many of the losses were compensated by increased trade in indigo, grain and cotton from the Doab and Rohilkhand, but now the general trade pattern had been completely turned upside down.[83] Farrukhabad had become more and more

[80] NAI, Foreign and Political Dept., sr.no. 3, 'Nabob Vizier to Rajah Gobind Ram, 15 February 1781.'

[81] OIOC, Add. 60337, Shee Papers, n.d, 'Shee to Wheler', no fol.; NAI, Foreign and Political Department, sr.no. 1, 'Bristow to GG&C, 29 July 1776.'

[82] OIOC, Home Miscellaneous, 219, 'Minute GG, 22 May 1780', fols. 543-4.

[83] For a general survey of Farrukhabad trade in the late eighteenth- and early nineteenth century, see A Siddiqi, *Agrarian Change in a Northern Indian State*

incorporated into the eastern economies of Awadh, Bihar and Bengal
as it became an important channel of communication with the now
more peripheral areas of the Doab and Rohilkhand. Again, economic
redirection had coincided with political and even religious realignments.
As trade became reoriented to the east, the Indo-Afghan courts turned
from their Sunni Afghan allegiances with the Durranis and Rohillas
in the west, towards full Shia Irani affiliations to Awadh in the east.

The tributary relation with Awadh was ended in 1801 when the
annual *nazrana* was directly transferred to the Company. One year
later the Nawab ceded the entire sovereignty of Farrukhabad to the
British government who settled upon him and his heirs an annual
stipend of Rs. 108,000.

CONCLUSION

SLAVES IN INDIA

The history of the Bangash Nawabs is an interesting case study of
Indian military slavery. In fact, it represents a last flickering of a six-
century old phenomenon. The first Muslim raids into India in the
twelfth century and the eventual conquest and occupation of
Hindustan at the end of the thirteenth century was for a great deal
accomplished by Turkish *mamluks* or *bandagan* in the service of the
Ghaznavid and Ghurid dynasties in Afghanistan and Khorasan. As a
result, the Turko-Persian blend of the *mamluk* system was introduced
in India through Ghaznavid and Ghurid channels and became the
mainstay of the Delhi Sultanates until it underwent a perceptible
decline from the fourteenth century onwards.[84] Indigenous elements
became increasingly important, first as elite slaves themselves but later
as free Muslim or Hindu allies. Therefore in India, the *mamluk* system
was only for a relatively short time able to succeed and as such it

---

(Oxford: Clarendon Press, 1973); and C.A. Bayly, *Rulers, Townmen and Bazaars:
North Indian Society in the Age of British Expansion 1770-1870* (Cambridge:
Cambridge University Press, 1983).

[84] A. Wink, *Al-Hind. The Making of the Indo-Islamic World*, vol. 1: *Early
Medieval India and the Expansion of Islam 7th-11th Centuries* (Leiden: E.J. Brill,
1990), 23.

always remained a predominantly Middle Eastern phenomenon, only incidentally imported, but never completely at home in the Indian setting.

Under the Mughals, slaves played a minor part in the central administration and the army. Akbar employed, however, a small amount of household slaves which he started to call *chelas* because he disliked the term *bandagan* (*banda* meaning slave) under which heading *mamluks* were called during the Delhi Sultanate. According to Abul Fazl, he believed that mastership belonged to no one but God and therefore he borrowed the name *chela* from the Hindu *bhakti* and the Indianises Sufi tradition, signifying the complete attachment of a faithful disciple or pupil (*chela* or *murid*) to a holy man (*guru* or *pir*). Here the relationship is not one of ownership but one of unqualified and unconditional love. In due course, though, *chela* became the current Indian name for a *mamluk*-like or *ghulam*-like slave. This departure from a thoroughly Turko-Persian, Islamic expression reflected not only the decreased importance of slaves under the Mughals but also Akbar's reorientation to a thoroughly Indianized, Indo-Islamic culture.[85]

It is beyond the scope of this essay to explain India's short-lived and incidental experience of the *mamluk* system into full detail, but it seems that at the root of it lies the prevalence of a very extensive, free military labour market on the subcontinent as driven by *naukari*. In principle, this notion of *naukari*, being the free, negotiable service to one's patron-employer, appears to stand in juxtaposition to the idea of slavery, being the enforced, unconditional service to one's master-owner. At the same time, though, both notions of service stood at right angles with the emotional loyalty based on kinship ties and religious conviction. In other words, the loyalties and alliances based on *naukari* and military slavery had a simple cold logical outlook based on calculated selfish best interests and devoid of tribal and religious biases. This was also the prevailing attitude at India's military labour market in which the loyalty of allies was not so much directed at their patron's tribe or person, but more at his purse. Of course, the idiom of tribe, religion or caste played its role in the formation and

[85] Abul Fazl Allami, *The A-in-i Akbari*, vol. 1, trans. H. Blochmann, H.S. Jarrett (Delhi: New Taj Office, 1977-8), 263-4.

rationalisation of alliances but what strikes most is the repeated tendency to cut across rigid party lines. This extreme pragmatic, no-nonsense attitude of Indian politics was underpinned by the sheer endless availability of both cheap military labour and ready cash to pay for it. In these circumstances, *naukari* made slavery almost super-fluous and rather expensive in comparison with self-trained free retainers. Moreover, the imprinted loyalty of the *chela* itself continuously tended to dissolve into the conditional *naukari* alliance of the free mercenary.

Nonetheless, during the eighteenth century we still find *chelas*, some of them eunuchs, entrusted with important positions at the local Indian courts, most notably in Awadh and the Afghan principalities of Rohilkhand and Bhopal.[86] Among the Afghans, the number of *chelas* which one was able to sustain, to a large extent contributed to their overall status.[87] They were, however, not employed on a large scale and, as a rule, were recruited locally. Outside India, however, the *mamluk* institution was still flourishing and at some places even found renewed vigour as under the Durranis in Afghanistan, the Mamluk Pashaliq of Baghdad,[88] the Bu Said Sultanate of Muscat,[89] and the Uzbek Ming Khanate of Khoqand.[90] In eighteenth-century India, there are only two examples of the large scale recruitment of *chelas*. The case of the Nawabs of Mysore, who not only recruited from local sources but also bought military slaves from Karim Khan Zand's Persia, is interesting but still awaits further examination.[91] The other exception is the *chela* system of Farrukhabad.

[86] For the *chelas* at the court of Awadh, see M.N. Fisher, *A Clash of Cultures: Awadh, the British and the Mughals* (Delhi: Manohar, 1987), 53-6; at Bhopal, see J. Malcolm, *A Memoir of Central India* (London: Farbury and Allen, 1823), 121.

[87] Wendel, *Les Mémoires*, 118.

[88] T. Nieuwenhuis, *Politics and Society in Early Modern Iraq* (Den Haag: M. Nijhoff Publishers, 1982), 13, 25 (mainly Georgians and Circassians).

[89] Abdul Sheriff, *Slaves, Spices and Ivory in Zanzibar* (London: Currey, 1987), 37 (mainly Africans and Baluchis).

[90] T. Saguchi, 'The Eastern Trade of the Khoqand Khanate', *Memoirs of Research Department of the Toyo Bunko* 24 (1965), 64 (more than 20,000, mainly Kirghiz).

[91] M. Wilks, *Historical Sketches of the South of India*, vols 1 and 2 (Madras: Higginbotham, 1869), vol. 1: 406-7, 527, vol. 2: 392.

Slaves in Farrukhabad

The case of Farrukhabad epitomises India's experience with military slavery. Being introduced from a Turko-Persian context, military slaves were highly instrumental in conquering and establishing a new homeland for their Afghan masters. Initially, slaves served well to counterbalance the power of the Pathan chiefs. At the same time, as implied by the label *chela*, slavery was reformulated in indigenous Indian terms focussing on Rajput heroism and spiritual devotion. Meanwhile, on the imperial level, the Bangash Nawabs were most successfully beginning to exploit the endless opportunities offered by *naukari*. This gained them enormous riches and honours which further strengthened their position both in Farrukhabad and among the Mughal nobility. In this situation, slavery lost a great deal of its appeal as even the *chelas* themselves could hardly be expected to withstand the immense gravitational force of the military labour market.

The end the eighteenth century, though, witnesses a sudden slump in the Nawab's *naukari* trade. At the same time, we are faced with a remarkable revival of the Farrukhabad *chelas*. Being deprived of its extensive external military service networks, the court now increasingly turned in on itself, giving free reign to the 'indoor' *chelas* and begams in control of the domestic treasury besides to all kinds of new businessmen in charge of collecting the local land revenue. The intervention by Awadh and the British further undermined the external reach and the internal legitimacy of the Nawab. So within half a century, the rustic image of the courageous warrior Muhammad Khan, loyally served by his hero-*chela* Dalir Khan, changed into the decadent picture of the effeminate dandy Muzaffar Jang, treacherously manipulated by his courtier-*chela* Fakhr al-Daula. Far from being a colonial construction only,[92] this 'orientalised' picture – indeed created by the British residents at Farrukhabad – was also a factual indication that the days of the free military labour market of northern India were numbered.

---

[92] For a discussion of this aspect, see R. O'Hanlon, 'Issues of Masculinity in North Indian History: The Bangash Nawabs of Farrukhabad', *Indian Journal of Gender Studies* 4, 1 (1997), 1-19.

CHAPTER 9

The Warband in the Making of Eurasian Empires*

Ogni città riceve la sua forma dal deserto a cui si oppone.

ITALO CALVINO[1]

Any royal authority must be built upon two foundations. The first is might and *group feeling*, which finds its expression in *soldiers*. The second is *money*, which supports the soldiers …

IBN KHALDUN[2]

INTRODUCTION

THE COHESION OF Eurasian empires before the age of European hegemony is often conceived as being the result of a particular form

*This Chapter profited enormously from the critical comments of my fellow travellers in the so-called Horizon project on Eurasian Empires funded by the Netherlands Organisation for Scientific Research (NWO), more in particular Maaike van Berkel, Jeroen Duindam, Richard van Leeuwen, Peter Rietbergen and Willem Flinterman. I also cherish the ongoing discussions on the topic with my Leiden colleagues Gabrielle van den Berg, Remco Breuker, Maurits van den Boogert and Henk Kern. David Robinson's and Tom Allsen's insights proved extremely helpful at the very beginning and the very end of the writing process. Finally, I am also grateful to the organizers and participants of two conferences, one at the Centre for Global History at Oxford (January 2014) and one at Pembroke College at Cambridge (December 2014), which allowed me to share some of my still immature thoughts with a critical audience, among them in particular Ali Anooshahr.

[1] 'Each city receives its form from the desert it opposes' (Italo Calvino, *Le città invisibili* (Torino: Einaudi, 1972), 8.

[2] Ibn Khaldun, *The Muqaddima. An Introduction to History*, transl. F. Rosenthal (London: Routledge and Kegan Paul, 1978), 246. Italics are mine.

of imperialism, driven by the demands of nationalism and the industrial revolution, and realized thanks to new gunpowder technology.
The effective combination of these three elements produced a paradigm
of what contemporary historians would consider to be a successful
empire or the 'optimal imperial outcome'. According to Charles Maier,
the latter occurs 'when subject nations and their leaders voluntarily
emulate the metropole's values and tastes'.[3] This is exactly what we
experience today, as both the last remaining empires and the more
recent nation states all attempt, to a greater or lesser extent, to follow
the European paradigm that encompasses national cohesion, economic
growth and military strength. This raises the critical question, though,
of whether imperial success has always been built on these three pillars.

In this essay I will address this problem by trying to detect a
similarly universal but different pre-modern paradigm; a paradigm
that is, however, closely related to the ideas and ideals of pre-modern
historians. This more emic approach forces me to take seriously those
elements that contemporary observers considered important for
success. With only a few exceptions, all of them seem to agree that
strong empires start with a strong group of warriors. For much of the
second millennium of the Common Era there were two powerful
imperial paradigms that dominated large parts of the globe's eastern
hemisphere. The first was the practical example provided by the
astonishing success of the Mongolian 'world-conqueror' Chinggis
Khan (c. 1162-1227); for centuries after, his was the story that other
conquerors strove to emulate as it promised endless wealth and a
glory that was truly imperial. It was clear to all that this Chinggisid
model entailed, at the bare minimum: nomadic mobility based on a
large number of well-bred warhorses mounted by well-trained horse-
archers.

Yet even more crucial than sheer horsepower was the cohesion of
the conquering band of warriors and their loyalty to its leader. Here
we come to the second paradigm for this study, the well-known
cyclical theory provided by the North-African historian Ibn Khaldun
(1332-1406). Inspired by some imperial best practice both in West

---

[3] C. Maier, *Among Empires: American Ascendancy and its Predecessors*
(Cambridge Mass.: Harvard University Press, 2006), 66.

and Central Asia, Ibn Khaldun elaborated on what he considered the secret of imperial success: the cohesion (*asabiya*) of the conquering warband. At the very beginning, when the latter was still roaming in the desert, *asabiya* was at its strongest, and actually enabled the easy conquest of sedentary societies that lacked such cohesion. In due course, though, due to the debilitating conditions of settled life, the conquering elites lost their cohesion and could only wait for their unavoidable defeat against new nomadic conquerors from the desert, who had a much stronger *asabiya*. Obviously, Ibn Khaldun's cyclical ideas are far from unique in global history; for example, they may remind one of Polybius' theory of predictable constitutional cycles. Elaborating on the latter, as if preparing the way for Ibn Khaldun, Tacitus made a juxtaposition of the virtuous hardy barbarian against the decadent city-dweller. Similarly, Confucianist historians were deeply aware that dynasties rose and fell like man himself, obeying a cycle of life and death that governed all animate beings. Hence, it was assumed that an imperial regime like that of the Chinese Ming would follow a general pattern: after the political and military vigour of its youth (fourteenth century), a mature middle age of peace and stability would ensue (the fifteenth century), to be succeeded by feebleness and, eventually, fatal decline (the sixteenth century). As such, various late-imperial rulers did not passively accept their perceived fate but anxiously attempted to freeze the process with various imperial rescue missions and restorations, even though they knew very well that their dynasty was bound to end sooner or later.[4] Being a man of his own time and place, Ibn Khaldun gave the theory a nomadic dynamic and brilliantly systemized and theorized the pre-existing wisdom of the influential Islamo-Persian historical tradition to organize imperial time into three stages: (1) conquest based on tribal cohesion; (2) highpoint based on justice; and (3) decline based on moral regression. Most rulers were very much aware of this scheme and anxiously tried to situate themselves in either the first or the

[4] Cited from F. Wakeman Jr., *The Fall of Imperial China* (New York: The Free Press, 1975), 55-71. See also M.C. Wright, *The Last Stand of Chinese Conservatism: The T'ing-Cheh Restoration, 1862-1874* (Stanford: Stanford University Press, 1957), 43-68.

second stage. In such a paradigm, the mere suggestion of moral decline was to be avoided at all costs as this would automatically lead to their fall.

At the heart of the practical Chinggisid model and the theoretical model of Ibn Khaldun stands the nomadic warband: a group of loyal nomadic warriors that follow their leader in the construction of an empire. My two central aims in this Chapter are (a) to study the development of the nomadic warband's cohesion before and, in particular, immediately after a conquest, and (b) to study the role of the nomadic warband in creating imperial cohesion beyond the warband, particularly its position in the imperial organization and, to a lesser extent, its remuneration and cohesion. This immediately raises the important question of the phenomenon's relevance across time and space. How relevant are the two models for Eurasia as a whole? Is not the idea of the warband too general, and do we really need the Chinggisid model and that of Ibn Khaldun to explain it? My main argument holds that both models provide some very important keys for understanding processes of Eurasian conquest and state-formation by nomadic warbands. Hence, in contrast to what seems to be the much more universal phenomenon of the *comitatus* as analyzed by Christopher Beckwith and many others, this contribution focuses more in particular on the specific features of the *nomadic* as well as the *post-nomadic* warband within the very specific spatial limits of the Central Asian Arid Zone.[5]

SPACE: WARZONES AND FRONTIERS

It is my contention that the applicability of both models is determined by the *longue durée* geopolitical conditions of the Eurasian macro-region. Considering the importance of (semi-)nomadic groups and the central role of the warhorse in our period, I will differentiate between four *military* zones, each with a different balance between

---

[5] For the *comitatus*, see C.I. Beckwith, *Empires of the Silk Road: A History of Central Eurasia from the Bronze Age to the Present* (Princeton: Princeton University Press, 2009), 17-19. Cf. P.B. Golden, 'Some Notes on the *Comitatus* in Medieval Eurasia with Special Reference to the Khazars', *Russian History/Histoire Russe* 28 (2001), 153-70.

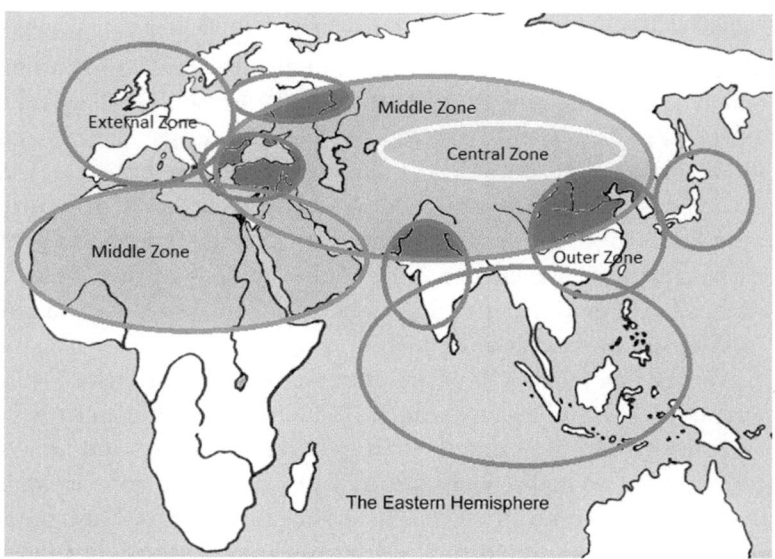

Map 3: Eurasian Military Zones

Central Zone : Nomadic: Central Eurasian Steppes
Middle Zone   : Semi-Nomadic: Middle East – Iran – Turkistan – Northeast Asia
Outer Zone   : Post-Nomadic: China – India – Eastern Europe – Anatolia
External Zone : Sedentary: Western and Central Europe – Southeast Asia – Japan

nomadic and sedentary ways of life, each with a different logic and relevance regarding our two models.

These four zones will be employed as a spatial framework in which the making and cohesion of imperial warbands will be analysed. Although the main temporal framework will be Ibn Khaldun's concept of cyclical time, I will attempt to detect some more general developments in linear time. At the base of this study, however, is my assertion that the warband became particularly powerful and effective in areas with a nomadic frontier and, in particular, after the end of the first millennium CE. This was the time when nomadic warriors, in particular the Mongols, Turks, Afghans and Jurchens, gained the advantage over the armies of the settled societies that surrounded them. As they were able to tap into the increasing resources of these settled societies through trade, plunder and tax, horse-based nomadic warbands gained unprecedented power, particularly when they managed to take the best of both worlds by carving out their empires at the very transition of the desert and agricultural zones. Although this was, at its earliest and most relevant point, focussed particularly on the Central, Middle and Outer Zones, this development of increasing horse-based militarization even affected, albeit indirectly, the External Zone, and as such it is often discussed under the label of a 'medieval' or 'feudal' mutation. One of the main objectives of this chapter is to detect both spatial and temporal patterns in the development of the nomadic warband and how its changing organization played a role in the making and unmaking of empires.

The starting hypothesis of this study is the idea that the nomadic warband is a critical tool to create and sustain those Eurasian empires that surrounded the Central Asian Arid Zone. During and after a conquest the nomadic warband both extends and reproduces itself in order to encompass allied and subjugated groups. Under the conditions usually found in a sedentary empire, the original open warband runs the risk of being overstretched which, in due course, may reduce it to an isolated, much more closed, and purely military institution, one increasingly dominated by the imperial bureaucracy. Hence the warband should not be studied as a given, static phenomenon but as an institution of transition. It emerges and expands under conditions of nomadic or semi-nomadic raiding, called *qazaqliq* by contemporaries, before markedly changing under the settled conditions of a sedentary economy. Such a transition can be pinned down on the map. It often occurs when moving from the Central Zone of full nomadism into the mixed economy of the Middle Zone or, as in the rare case of Chinggis Khan, straight into the Outer Zone. More usually, though, the most crucial transition occurs on the very frontier of Middle and Outer Zones, and most clearly at those places where both were still within striking distance of the Central Zone; in other words, at those areas where the Middle Zone most sharply demarcates the interface between the desert and agricultural environment. Hence, this chapter will highlight the genesis of the Chinese, Indian, Russian and Middle Eastern empires at the four main crossroads of the Central, Middle and Outer Zones, in, respectively, Manchuria, Khorasan, Ukraine and Anatolia. As I will briefly discuss below, lacking such an interstitial zone with Central Eurasia, North Africa in the southern Middle Zone deserves a different analysis.

Although all regions shared a highly dynamic nomadic frontier, the specific geopolitical conditions varied significantly, and this had different effects on the way a nomadic warband amalgamated with an empire. Comparing the various frontiers with each other, the late Owen Lattimore made the insightful observation that the Great Wall of China demarcates a relatively sharp transition from sedentary agriculture to pastoral nomadism.[6] Indeed, immediately north of the Wall and the Gobi Desert, the steppes of Central Mongolia provided the most favourable assembly-ground for the great nomadic hordes that repeatedly changed the course of world history. As such, the Chinese frontier can be characterized as a relatively fixed *outer frontier* between the settled Han Chinese and the nomadic Mongols. It was only in the last two centuries that agriculture managed to expand, in fits and starts, into the northern steppes. Further east, though, in what is now Manchuria, there emerges a more interstitial region of river valleys and plains in the south, and forests and mountains in the north and east, which extend all the way to Siberia and Korea. Albeit in different ways, this mixed economy is quite common across the Middle Zone as a whole. For example, to the south-west, the Middle Zone continues into the oasis and desert landscapes of Turkistan between the Hindu Kush and Elburz Mountains, extending quite naturally through Khorasan and Iran towards the Middle East. Further south, the north-west frontier of India is not bordered by a steppe plateau but is instead encircled by mountains, while its irrigated and unirrigated lands are not set off from each other in large blocks but interconnect with each other. Moreover, the adjacent 'pastoral' economies of Afghanistan, Baluchistan, Sind, Punjab and Rajasthan are more often semi-pastoral, and as such are distinct from the

---

[6]The following section builds on Owen Lattimore's insightful comments albeit, considerably adjusted for the purpose of my own research questions. See his *Inner Asian Frontiers of China* (Hong Kong: Oxford University Press, 1988), but more particularly his *Studies in Frontier History: Collected Papers 1928-1958* (Paris: Mouton and Co., 1957), 148-54.

increasingly pervasive nomadism that can be observed when moving from the savannahs of the western Middle Zone towards the open steppes of the Central Zone. Although agreeing with Lattimore, I would add that in India the gradual nomadic transition created various *inner frontiers* which did not support a purely nomadic society but still critically facilitated very powerful post-nomadic empires with a highly dynamic military labour market based on *qazaqliq*. As in the case of Manchuria's interaction with the extremely prosperous China, this process of empire building was energized by an extremely rich sedentary economy along India's fertile river valleys and coastal regions. Although slightly less so than with China's outer frontier, India's inner frontiers proved relatively stable until the nineteenth century, which makes both of them stand out from the constantly *retreating nomadic frontier* that we discover when examining Russia.

Compared to China and India, the Russian frontier finds itself somewhere midway between these two. Like China, it has a relatively sharp distinction between the northern forest zone, dominated by settled agriculture, and the vast Pontic-Caspian steppes and savannahs to the south, dominated by nomadic pastoralism. In between, though, there is a narrow but long latitudinal stretch of mixed landscape with oak woodlands, meadows and grasslands. At some points, the grassland of this area extends deep into the north, providing the natural pathways used for nomadic incursions. The mixture of woodland, steppe and savannah created a semi-nomadic livestock economy alongside rainfall agriculture. Although the nomadic way of life retreated more and more towards the southern savannahs, the society that it replaced remained extremely mobile; people may have lived in houses, but they were nevertheless ready to migrate from one place to another. Not unlike India's semi-arid frontier zones, this produced the warlike society of the Ukraine based on *qazaqliq*, in another, but related, Russian idiom, the vagabondage of the Cossacks.

What differentiates the Russian frontier from the Chinese and Indian is its shifting character, which is described by the Russian writer Bogdanoff, as cited by Lattimore:

Russian extensive rainfall agriculture, capable of being combined both with the grazing of livestock and with the exploitation of the forest, rapidly, though superficially, conquered enormous territories – the Russian could carry on his general farming wherever he pleased.[7]

Indeed, throughout our period, the Russian nomadic frontier pushed ever further towards the south and south-east. In addition to the southern flow of Russia's main river systems, this was stimulated by the fact that the temperature, soil and moisture of the southern forest-steppe zone were more conducive for agriculture than those of the northern forests.[8] Overall, though, Russian agriculture was very poor compared to that of China and India, a basic fact of life that was considerably aggravated by the devastating effects of ongoing plagues and other epidemics. For nomadic raiders, the trouble with Russia's sedentary economy was that it never produced the kind of economic surpluses that are so distinctive for its Chinese and Indian counterparts. As a result, Russians were relatively poor peasants, and so their Tatar neighbours could only be equally poor nomads. For them the north, lacking both grazing opportunities and material resources, offered no incentives whatsoever for a permanent conquest. Hence, from the nomadic point of view, the vast Russian forest zone lacked natural anchorage points. Whereas in the east, conquering Beijing or Delhi announced the beginning of an empire, in the west, the taking of Moscow never really paid off. The Mongols and Tatars knew perfectly well what later European conquerors would later learn to their utter dismay: the capture of Moscow leads nowhere!

Moving to the western parts of the Middle Zone, this region also had a different frontier dynamic. As already noted, the ecological conditions in the Middle East are not all that different from those of Turkistan and Khorasan. In this entire region, the contrast between

---

[7] Cited by Lattimore, *Studies in Frontier History*, 154.

[8] For a survey of the ecological conditions in Russia and its southern frontier, see R. Pipes, *Russia under the Old Regime* (London: Weidenfeld and Nicolson, 1974), 1-27; D.J. Shaw, 'Southern Frontiers of Muscovy, 1550-1700', in J.H. Bater and R.A. French (eds), *Studies in Russian Historical Geography*, vol. 1 (London: Academic Press, 1983), 117-42; J. Ledonne, 'The Frontier in Modern Russian History', *Russian History* 19 (1992), 143-54; D. Moon, 'Peasant Migration and the Settlement of Russia's Frontiers, 1550-1897', *The Historical Journal* 40 (1997), 859-93.

steppe and sown can be as sharp as in China, but the irrigated agriculture is centred on oases, or strung alongside rivers that have steppe or desert on both sides. Moving into Anatolia, agriculture becomes more widespread and expands over time, but, in our period at least, it also continued to support a semi-pastoral economy that linked the towns and villages to each other and to the more open grazing lands of the Caucasus and Iran. This Anatolian frontier is thus reminiscent of the inner frontiers of India and Russia; more Indian than Russian, though, since it shares the dispersed and recurrent characteristic of the first against the more contiguous, retreating features of the latter. Overall, the sedentary economy of Anatolia was not as rich as that of India, but also not as poor as that of Russia. Hence, sustaining an extensive empire in Anatolia was only possible if one could also exploit the agricultural and commercial resources of the eastern Mediterranean.

Coming to North Africa, despite the pervasive presence of a nomadic frontier and *pace* Ibn Khaldun, there is one important feature that makes this region as a whole very different from the Turko-Mongolian part of the Middle Zone: its scale. The Middle East itself has never been able to support the sheer mass of horsepower that was produced in Central Eurasia. Of course, Arab and Bedouin nomads had a huge impact on Middle Eastern societies, in a manner very much in line with Ibn Khaldun's logic, but this primarily represents an internal dynamic, and so was not directly linked to the Chinggisid explosion which reverberated so deeply in the heartlands of China, India, Russia and the northern Middle Zone. As in the case of Europe, Central Eurasian warbands could indeed threaten but never really conquer North Africa, partly because the latter was protected by Central Eurasian slave armies, but, more importantly, because it simply lacked the space and the resources to attract and accommodate them. Hence the North-African warband comes closer to being one of the smaller and more isolated household troops of the External Zone than the huge and more open Chinggisid warband. As I will try to demonstrate, although the latter is an extremely potent category for analysing the creation of Eurasian empires in China, India, Russia and the Turkish Middle East, we should make a clear-cut distinction between its various Turko-Mongolian avatars on the one hand, and the various functionally related but genealogically different imperial guards on the other.

Time: The Warhorse Millennium and the Rise of the Centre

What was the secret behind the Chinggisid success story? For most onlookers, the most obvious explanation was the quality and quantity of the Eurasian warhorse. Indeed, the Arid Zones of Eurasia were the natural breeding grounds for the world's best warhorses.[9] In the same way as the ascent of infantry warfare is linked to the wider story of the Rise of the West, the age of the horse warrior is embedded in the story of what we may call the Rise of the Centre. But this also raises the question to what extent this 'Rise' was really unique for the Chinggisid era.

Domesticated about 6,000 years ago, the warhorse started its huge impact on world history about 2,000 years later, drawing the war-chariots of the (mostly) Indo-European conquerors who swept across the great sedentary civilizations of the Middle East, India and China. A few centuries later, the chariots were replaced by the first riding nomads who, from their natural habitats in the Eurasian deserts and steppes, started to breed rather small but sturdy warhorses. At about the beginning of the Common Era, horse breeding penetrated the new Eurasian empires that started to stretch across the various frontier zones between the nomadic and sedentary worlds: from the Roman and Parthian empires in the west, through the Kushana and Shaka empires in the south, to the Xiongnu and Han empires in the east. Kept and fed more and more in stalls, horses gradually became larger

---

[9] The idea that the Arid Zone is an important historical category and that the organization of post-nomadic empires depended on their capacity to tap the horsepower that was produced there stems from my earlier work; see, in particular: 'The Silent Frontier of South Asia, *c.* 1100-1800 AD', *Journal of World History* 9 (1998), 1-25 and 'War-horse and Post-Nomadic Empire in Asia, *c.* 1000-1800', *Journal of Global History* 2 (2007), 1-21 and more recently 'Continuity and Change in the Indian Ocean Basin', in J. Bentley et al. (eds), *The Cambridge World History*, vol. 4: *The Construction of the Global World, 1400-1800 CE*, part 1: *Foundations* (Cambridge: Cambridge University Press, 2015), 182-210. See also Chapters 2 and 4 in this volume.

and stronger, specifically geared for the more heavily armored cavalry which increasingly accompanied the lighter variety of mounted archers. Only in Western Europe and Japan, far beyond the Eurasian steppes, did the heavy knightly individual become the dominant military brand, heralding a 'feudal' age at the end of the first millennium CE.

Slightly earlier, from the sixth century onward, in regions in or immediately bordering on the Eurasian Arid Zone, the earlier development towards ever heavier cavalry was halted by the sudden emergence of two new incredibly powerful nomadic powers: the Arabs in the Middle East and the Turks in Central Eurasia. Due to the sheer number and quality of the Central Eurasian horse, it was in particular the Turks who were able to dominate military practice in a vast area that stretched from the Hungarian plain to the Great Wall of China in the northern steppes zone, and from Egypt to southern India in the southern desert zone. By far the most prominent ethnic group among them was the Qipchaq Turks who, in the thirteenth century, provided state-of-the-art horse warriors, and even rulers, to states as far apart as the Kingdom of Hungary, the Mamluk Sultanate in Egypt, and the Delhi Sultanate in India. Much later, the most powerful of the post-nomadic empires, the Ottomans and the Mughals, continued this already rich tradition of Turkish empire-building, and even the Iranian but heavily Turkified dynasty of the Safavids could only follow in the footsteps of their Seljuq (eleventh-twelfth centuries), Qara Quyunlu and Aq Quyunlu (fourteenth-fifteenth centuries) predecessors and build their power on the Turkish man- and horsepower found in their northern territories. In all these cases, Turkification automatically implied militarization, and a growing tendency towards Turkish or Turkified rulers across the whole Eurasian continent, who conspicuously demonstrated their outstanding martial qualities, be it in actual practice on the battlefield or imagined in sumptuous rituals, heroic poetry or austere architecture. In Northeast Asia, the machismo of the Turks finds an almost perfect parallel in the military prowess of the Mongolian Khitans and Tungunsic Jurchens who provided the horsepower and dynasties for, respectively, the Liao and the Qara Khitai, and the Jin and Qing empires.

However, the absolute highpoint of Central Eurasian power was not a Turkish but a Mongol achievement: the unprecedented thirteenth-century expansion of the nomadic empire under the brilliant leadership of Temüjin, better known as Chinggis Khan. But was there really that much difference between Mongol and Turkish expansion? Our use of the contraction 'Turko-Mongol' is partly informed by the contemporary Arabic and Persian sources which repeatedly convey the belief that Mongols and Turks belonged to one and the same race. For example, the historian Rashid al-Din (1247-1318), who was very close to the scene, stressed very much unity in diversity:

> Although the Turks and the Mongols and their branches are similar and their language is of the same origin, the Mongols being a kind of Turks, there is yet much difference and dissimilarity between them.... These Mongols were just one people amongst the Turkish peoples.

Rashid al-Din also suggests that it was simply the strongest in the group who determined that group's ethnic identity, as he adds that when the Mongols rose to such power and eminence, 'most of the Turkish peoples were called Mongols'.[10] Moving further westwards, we find a similar ethnic amalgamation taking place with the emergence of the term Tatars, which did *not* refer to the more specific Mongol 'tribe' that fought Chinggis Khan, but was a more generic term indicating a broad mixture of primarily Turkish, but also other ethnic groups, all of them Muslims, beyond the southern frontiers of the sedentary and Christian states of Muscovy and Poland-Lithuania. Indeed, whether Mongols, Turks or Tatars, all these groups built their power on the capacity to mobilise more horse warriors than ever before. Despite some advances in equipment, such as the stirrup, technology and tactics basically remained the same. The key weapon remained the composite bow, tactics continued to be based on a combination of heavy and light cavalry, the latter of which was extremely mobile, wheeling around the enemy while delivering continuous showers of deadly arrows against them. Hence, the reason this has been called a horse-warrior revolution derives not from any

---

[10] D. Ayalon, 'The Great Yasa of Chingiz Khan: A Reexamination (Part C1)', *Studia Islamica* 36 (1972), 126; D. Ayalon, 'The Great Yasa of Chingiz Khan. A Reexamination (Part C2)', *Studia Islamica* 38 (1973), 149-50.

quality but their sheer quantity. So if there was a revolution, it really was a revolution of size: the Mongols merely provided the most spectacular example of a much older and wider Turko-Mongolian development that started as early as the seventh century but, for more than a millennium, continued to have a tremendous impact on patterns of state-formation in the entire Eurasian continent.

How can the fact that Turko-Mongolian conquerors managed to operate the biggest cavalry armies that the world had ever seen be explained? It is my contention that it was primarily because of their organizational skill that they were able to tap into the rich agrarian and commercial resources of the sedentary worlds that they first plundered and subsequently conquered. As such, it rather follows the first part of 'the circle of justice' as described in Iranian advice literature: 'there is no kingship without an army, no army without revenues'.[11] Indeed, conquering an empire was one thing, ruling an empire quite another. The main challenge was how to link the revenues of the sedentary world to the nomadic armies, which was extremely important, as the nomadic conquerors could not risk giving up their trump card: their horsepower.

Here we may speculate whether the earlier 'frontier states' of Turks, Khitans and Jurchens, as well as other semi-nomadic, 'cooked' (*shou*) dynasties, paved the way for the Mongols, who could simply build on an already proven infrastructure to bridge the nomadic and sedentary constituents of their polities.[12] This is certainly suggested in the early Song accounts of the Mongols who are depicted as a 'new kind of northerners': true nomads (noble savages), not yet spoilt by the corrupt pseudo-nomadic officials of the Jurchen Jin.[13] Whatever the case may be, these Jurchen and Turkish frontier states clearly

[11] It continues as follows: 'there is no revenue without subjects, no subjects without justice, no justice without a king'.

[12] See the suggestive comments on the Qara Khitai Empire ('The Qara Khitai established an empire in Central Asia that for the first time joined the worlds of China, the Inner Asian nomads and Islam') by M. Biran, *The Empire of the Qara Khitai in Eurasian History: Between China and the Islamic World* (Cambridge: Cambridge University Press, 2005), 204-6.

[13] C.D. Garcia, 'A New Kind of Northerner: Initial Song Perceptions of the Mongols', *Journal of Song-Yuan Studies* 42 (2012), 326, 330.

showed the Mongols how to manage a complicated balancing act in which one had to keep one leg in the nomadic world in order to procure horsepower, while setting the other firmly in the sedentary to be able to collect sufficient revenues to pay for it. Obviously, this exercise required specialized administrators, sophisticated people of the pen, who were both outsiders and insiders, and who were able to read and write in the language of the conquerors *and* the conquered. Before turning to our four concrete cases of Turko-Mongolian empire-building in China, India, Russia and the Middle East, let me first elaborate further on our two models in order to make the latter more operational and testable.

THEORY AND BEST PRACTICE: IBN KHALDUN AND CHINGGIS KHAN

IBN KHALDUN'S MODEL: THE SWORD AND THE PEN

For nomadic conquerors, becoming administrators themselves was never a serious option as there was always the apprehension that they would lose their martial prowess and, with that, their very identity as true Turks or Mongols. Examples abound that show the nomads' fear of devitalization which they believed would inevitably accompany the sedentarization process. Take, for example, Bilge (r. 717-34), the ruler of the second Turkic khaganate, who advised his people to remain apart from the settled world, as recorded in the eight-century Orkhon inscription:

Because some ignorant people accepted this invitation and came near the plain in order to settle (in China), many of your people are dead. If you go into that country, O Turkish people, you will die. But if you dwell in the land of the Ötüken and send caravans and convoys, and if you stay in the forest of Ötüken, where there is neither wealth nor trouble, then you will continue to preserve an everlasting empire, O Turkish people, and you will always eat your fill.

In the unfortunate case that one could not avoid such an encounter with the sedentary world, it was better to stick to old customs, as was the approach of Xiéli or Illig Khagan (r. 620-30), the last ruler of the Eastern Turkic Khaganate, who had to surrender to the Tang but still managed to: 'Pitch his felt tent in the middle of the palace, (before he) fell into a state of profound sadness, and could endure his fate. Surrounded by the people of his household, he chanted plaintive airs and wept with them.'[14]

This is repeated frequently in later sources, all of which boast of the martial qualities of the Turkish nation, and among which perhaps the most graphic depiction is that of the eleventh-century Arab historian Ibn Hassul:

> The most amazing thing about them [the Turks] is that nobody has ever seen a pure Turk (*turkiyyan khalisan*) who had been afflicted by effeminateness (*takhnith*), and this in spite of the fact that this disgrace is general, and this affliction is common among all the peoples we saw, especially those of Gilan. If, however, one does find traces of effemination (*ta'nith*) in any Turk in his speech, in his hints, in his dress or in his jewellery, he is proved to be one of the Turks of a mixed breed, who thoroughly mingled with the race of their neighbours, the local inhabitants of those lands.[15]

In the Islamic world, the threat of being assimilated into a settled society was often expressed in the dualism between Mongol law (*yasa*) and Islamic law (sharia).[16] Indeed, the specific underlying principle of the so-called Chinggisid *yasa* was the maintenance of a nomadic military culture that was fundamentally opposed to the values of settled society.[17] Equally prominent, though, is the discourse that distinguishes between the civilized Tajik or Persian and the rustic Turk or Turani, as reflected most famously in Firdausi's *Shahnamah*, and as quintessentially formulated in a Turkish proverb provided by the eleventh-century Turkologist Mahmud al-Kashghari: 'Just as the effectiveness of a warrior is diminished when his sword begins to rust,

---

[14] J. Cuisenier, 'Parenté et organisation sociale dans le domaine turc', *Annales. Économies, Sociétés, Civilisations* 27 (1972), 932. Translation by E. Forster.

[15] D. Ayalon, 'The Mamluks of the Seljuks: Islam's Military Might at the Crossroads', *Journal of the Royal Asiatic Society* 6 (1996), 314.

[16] See for example the recent discussion in G. Burak, 'The Second Formation of Islamic Law: The Post-Mongol Context of the Ottoman Adoption of a School of Law', *Comparative Studies in Society and History* 55 (2013), 579-602.

[17] See also the comment by Ayalon, 'The Great Yasa of Chingiz Khan (Part C2)', 135.

so too does the flesh of a Turk begin to rot when he assumes the lifestyle of an Iranian.'[18]

Much later, Qara Usman (r. 1378-1435), the founder of the Aq Quyunlu Empire, could only agree, as he advised his sons: 'Do not become sedentary, for sovereignty resides in those who practice the nomadic Turkmen way of life.'[19]

As we have seen, the most sophisticated exponent of the idea of a process of effeminization was Ibn Khaldun. For him, the unavoidable decline of group feeling (*asabiya*) among the conquering tribe was linked to the declining role of the people of the sword and their replacement by the people of the pen:

It should be known that both 'the sword' and 'the pen' are instruments for the ruler to use in his affairs. However, at the beginning of the dynasty, so long as its people are occupied in establishing power, the need for 'the sword' is greater than that for 'the pen'. The same is the case at the end of the dynasty when its group feeling weakens and its people decrease in number under the influence of senility. The dynasty then needs the support of the military. The dynasty's need of the military for the purpose of protection and defence is as strong then as it had been at the beginning of (the dynasty) when its purpose was to become established. In these two situations 'the sword' thus has the advantage over 'the pen'. At that time, the military have the higher rank. They enjoy more benefits and more splendid fiefs.

In the mid-term of the dynasty, the ruler can to some degree dispense with 'the sword'. His power is firmly established. His only remaining desire is to obtain the fruits of royal authority, such as collecting taxes, holding (property), excelling other dynasties, and enforcing the law. 'The pen' is helpful for (all) that. The swords stay unused in their scabbards, unless something happens and they are called upon to repair a breach. The men of the pen have more authority. They occupy a higher rank. They enjoy more benefits and greater wealth and have closer and more frequent and intimate contact with the ruler.[20]

In this long quote, Ibn Khaldun links the fortunes of the peoples of sword and pen to the natural cycle of empires. The group feeling among the warriors

---

[18] M.E. Subtelny, *Timurids in Transition: Turko-Persian Politics and Acculturation in Medieval Iran* (Leiden: E.J. Brill, 2007), 29.

[19] J.E. Woods, *The Aqquyunlu: Clan, Confederation, Empire* (Salt Lake City: The University of Utah Press, 1999), 17.

[20] Ibn Khaldun, *Muqaddima*, 213.

of the desert enables them to conquer the settled societies around them. After a conquest, though, more and more of the people of the sword will be replaced by people of the pen, which may optimize the management of the empire, but will also soften the group feeling of the conquering elites and thus make them liable to renewed conquest from nomadic outsiders with a stronger group feeling.

What is perfectly clear from all this is that the elite groups living on the edge of the deserts and steppes of the Arid Zone were well aware that warriors tended to lose their freshness and vigour after conquest. As indicated by Ibn Khaldun, the problem was often perceived as a power struggle between the warlike people of the sword, who were often direct descendants of the conquerors, and the civilized people of the pen, often recruited from the conquered. For example, the sword-pen dichotomy was a major theme in the Persianate political wisdom literature (*akhlaq*) that flourished immediately after the Mongol conquest. The most influential of these traditions is that of the thirteenth-century writer Nasir al-Din Tusi who stresses that the king should always keep an equable mixture of the four classes of mankind – very much like the four humours of the human constitution. As well as the men of the pen and the men of the sword, these consisted of the men of negotiation and men of husbandry.[21] As a comment on Nasirean ethics, later mirrors for princes confirmed the symbiotic relationship between pen and sword, although they increasingly tended to prefer the pen over the sword, as, for example, was the case with the scholar-bureaucrat Kashifi (d. 1504-5), who actually warns against the latter: 'Men of the pen never aspire to take over a kingdom, whereas men of the sword often do; moreover, men of the sword empty a sultan's treasury, while men of the pen fill it.'[22]

The sword-pen relationship is hardly exclusive to the Islamic world. In the case of China, a similar tension is expressed in the dichotomy between *wu* (the martial domain) and *wen* (the civil domain), which became particularly acute during the conquests of the Khitans (Liao), Jurchens (Jin), and Mongols (Yuan), and later under the Manchus

[21] Nasir al-Din Tusi, *The Nasirean Ethics by Naṣir ad-Din Ṭusi*, trans. G.M. Wickens (London: George Allen and Unwin, 1964), 230, 327.

[22] M.E. Subtelny, 'A Late Medieval Persian Summa on Ethics: Kashifi's Akhlaq-i Muḥsini', *Iranian Studies* 36 (2003), 605.

(Qing). It is interesting to read the words of the Yuan historiographer who almost repeats Ibn Khaldun's observation when he states:

> Jin established the dynasty by use of *wu*. In this it did not differ from Liao. But it was able to establish institutions drawing on both Tang and Sung, achieving in some things what Liao had not attained. This was accomplished with *wen*, not with *wu*. [Confucius in the *Tso*] *Commentary* said: If one says it without *wen*, he will not be able to practice it extensively.[23]

The Yuan historiographers saw *wen* as a process by which the initial *wu* of the conquerors was civilized. So *wen* involved learning specific skills: the creation of a chain of command, the adoption of civil speech, the use of administrators, the extension of central authority, the creation of a bureaucracy, the accumulation of books, the welcoming of the learned, the recognition of cultural tradition, the spread of education, the rise of the literati to *de facto* leadership, and, finally, the creation of a cultural legacy. All this was seen as the achievement of the civil order, of benefit both to the state and people, and as such it was sharply contrasted with military rule. Like Ibn Khaldun, the Jurchens themselves were very much aware of both the historical sequence and the natural complementarity of *wu* and *wen*. This is illustrated by one of the questions in the first Jurchen-language examination that they introduced: 'Our dynasty settled all under heaven with spirited *wu*; His Majesty is bringing comfort to all within the seas with *wen* virtue. *Wen* and *wu* are both employed.'[24]

At about the same time, another version of *wu* was influential in the creation of empires on the north-western side of the Eurasian steppes, and particularly along the fringes of the Russian forest belt. In thirteenth- and fourteenth-century Muscovy, it gave rise to the introduction of a dual administrative system between, on the one hand, military tasks, which were in the hands of the so-called *basqaq*, and, on the other civilian tasks that were in the hands of the *daruga*.[25] If Russia's dual administration was built on Mongol antecedents, we

---

[23] P.K. Bol, 'Seeking Common Ground: Han Literati under Jurchen Rule', *Harvard Journal of Asiatic Studies* 47 (1987), 487. Based on the *Jin shih*.

[24] Idem, 488.

[25] D. Ostrowski, *Muscovy and the Mongols: Cross-cultural Influences on the Steppe Frontier, 1304-1589* (Cambridge: Cambridge University Press, 1998), 36-63.

may wonder about the nomadic nature of this duality. Was it really that different from the experience of the European Middle Ages where a similar distinction between pen and sword emerged between, on the one hand, the knights who held land by right of blood and, on the other, the clerics of profession and ordination?[26] What made the European situation different is shown by the situation at its very fringes, where the kings of Hungary, Georgia and, indeed, Muscovy attempted to fight the growing power of the landed nobility by inviting nomadic Qipchaqs (Cumans) into their territories and making marriage alliances with them.[27] This suggests that it was much easier for the mounted warriors in Europe to become firmly rooted in their landholdings; as nomadic horse warriors were close to non-existent, there was no reason for them to be concerned about losing their nomadic purity. Consequently, this meant the dichotomy between sword and pen was much less tense and, as such, there is no discourse that comes anywhere near the dichotomy between nomadic Turk or Mongol *versus* sedentary Tajik or Chinese. Indeed, as we will see below, the foremost purpose of the nomadic warband after a conquest was to forestall the process of gentrification in which the imperial aristocracy settles permanently and takes root in landed properties.

THE CHINGGISID MODEL: SCALE AND ETHNIC ENGINEERING

How did nomadic conquerors cope with this unavoidable dichotomy between the military and administrative organization in their newly-won empires? The vehicle that had gained them their empires was

[26] R.I. Moore, 'The Transformation of Europe as a Eurasian Phenomenon', *Medieval Encounters* 10 (2004), 89-90. See also the discussion in *Past and Present* following T.N. Bisson's 'The "Feudal Revolution"', *Past and Present* 142 (1994), 6-42, which demonstrates how European historiography is still captivated by national perspectives and traditions.

[27] N. Berend, 'Cuman Integration in Hungary', in A. Khazanov and A. Wink (eds), *Nomads in the Sedentary World* (Richmond, Surrey: Curzon, 2001), 110-11. King László IV (1272-90) was half Cuman and during his reign Cumans gained unprecedented importance, as *neugerii* or members of his military bodyguard. See also N. Berend, *At the Gate of Christendom: Jews, Muslims and "Pagans" in Medieval Hungary, c. 1000-c. 1300* (Cambridge: Cambridge University Press, 2001), 145, 183.

the *keshig*. It was a conscript-based elite made up of the personal followers and guardsmen of the war-leader who trained and paid them, often by distributing booty among them. These non-tribal companions (*nökörs*) constituted the latter's personal household, and were recruited not on the basis of family or a hereditary position but purely on their loyalty and talent. All this gives the impression of something rather modern, and sounds more like the latest recruitment policy of a present-day multinational. The same modern spirit can be found in a well-known classical-Chinese verdict on the barbarians: 'these people despise the old and take joy in the strong'.[28]

As stressed already, the most important ingredient of 'the strong' was the warhorse. Its importance is indicated in some of the earliest titles given within the *keshig* of Chinggis Khan, which included grooms, herders of horses (*agtacin*) and, more specifically, herders of geldings (*adugucin*).[29] Interestingly, and perhaps significantly, it was in the course of retrieving horses stolen from his family that Chinggis Khan started his *keshig* in the early 1180s by recruiting Bo'orchu, the very first of his many *nökörs*.[30] But more important than its equine nature is the fact that all titles within the *keshig* express a physical closeness to the leader. This is shown in titles such as cooks, chamberlains and other so-called *ichki*'s (insiders), which were most clearly manifested in the rotation of personal guard duties, which created another set of honourable titles: night-guards and day-guards.

[28] The information on the *keshig* is entirely based on the pioneering work of: T.T. Allsen, 'Guard and Government in the Reign of The Grand Qan Möngke, 1251-59', *Harvard Journal of Asiatic Studies* 46 (1986), 495-521; S.M. Grupper, 'The Barulas Family Narrative in the *Yuan-shih*: Some Neglected Prosopographical and Institutional Sources on Timurid Origins', *Archivum Eurasiae Medii Aevi* 8 (1992-4), 11-99; Woods, *The Aqquyunlu*; P.B. Golden, '"I will give the people unto thee": The Činggisid Conquests and their Aftermath in the Turkic World', *Journal of the Royal Asiatic Society* 10 (2000), 21-41; C. Melville, 'The *Keshig* in Iran: The Survival of the Royal Mongol Household', in L. Komaroff (ed.), *Beyond the Legacy of Genghis Khan* (Leiden: E.J. Brill, 2006), 135-64; and Subtelny, *Timurids in Transition*. Much of this goes back to much older literature inspired by B. Vladimirtsov, *Le régime social des Mogols; Le féodalisme nomade* (Paris: Librairie d'Amérique et d'Orient, 1948).

[29] Grupper, 'A Barulas Family Narrative', 39.

[30] Allsen, 'Guard and Government', 513.

Historians have interpreted the *keshig* as a typical Indo-European *Männerbund* or *comitatus*. Thus, the phenomenon seems not to be particularly typical for Turko-Mongolian nomads.[31] Indeed, there is an overall scholarly consensus that the *keshig* has a much older history, and even could be quite universal in nature, as it can be traced far beyond Central Eurasia, and even beyond the Indo-European heartlands.

Staying aloof from the controversies surrounding its origins, it is my contention that the *keshig* as an open, meritocratic institution fares best in the socio-political context of (semi-)nomadic vagabondage, in Turkish expressed as *qazaqliq*.[32] Another term related to *qazaqliq* is the Arabic word *ghaza* – not to be confused with its later theological rationalization of *jihad* as in the case of early Ottoman history. Both refer to the adventurous life of a charismatic hero in the wilderness who, through his success, is able to attract a following of retainers called *qazaq*, cossack, *ghazi*, or any other label that indicates free association. The *keshig* loses its strength, though, the moment it gains a foothold in sedentary society. This also explains the well-known Tatar strategy of *kazak cikmak*: withdrawing to the steppes in order to regain your strength, a phenomenon that will immediately be familiar to those with a knowledge of Indian epics, where princes conquer kingdoms after the return from exile in the wilderness.[33] Although *qazaqliq* can be seen as universal, it is at its most effective in the specific geopolitical conditions of the Arid Zone, hence its ongoing association with its Indo-European or Central Eurasian past and the repeated fear of losing one's strength by being assimilated into the settled world. Obviously, this fear makes the necessary linking of sword and pen all the more challenging for nomadic conquerors who not only want to conquer a sedentary empire, but also want to rule one without losing the original strength and cohesion of their *keshig*. This was quite a challenge since the *keshig* is so closely associated

---

[31] Most forcefully in Beckwith, *Empires of the Silk Road*.

[32] Subtelny, *Timurids in Transition*, 28-32.

[33] H. Inalcik, 'The Khan and the Tribal Aristocracy: The Crimean Khanate under Sahib Giray I', *Harvard Ukrainian Studies* 3 (1979), 452.

with *qazaqliq*, a way of life that could not be tolerated under the new sedentary order.

Hence, it is my contention that there was something unique about the empires that were carved out by nomadic or semi-nomadic conquerors. In cases where the imperial household was not able to repeatedly recruit fresh nomadic horse-power, it had to accept that its military power would be increasingly based on a hereditary and rooted military aristocracy. On the other hand, imperial households that were built on significant numbers of highly mobile mounted warriors had a much tougher hold on the society and could more easily disregard hereditary rights and privileges. Thus, for these so-called *post-nomadic* imperial households it was crucial to retain the 'spirit of the steppes', which involved (a) the meritocratic recruitment of mobile horse warriors, and (b) their (re)organization into artificially constructed groups that were as closely attached to the *keshig* of the ruler as possible. For the purpose of the present argument, both the *keshig* and those freshly fabricated, *subsidiary* military groups are perceived as nomadic warbands. Even after conquering or settling in sedentary surroundings, some of these 'conscriptive' units, such as the Manchu Banners or the Russian Cossacks, retained their highly meritocratic and personalized (i.e. non-tribal) spirit, and as such can be seen as post-nomadic avatars of the nomadic warband.

So the important question remains: how to govern after a nomadic conquest? The obvious answer would be to convert the *keshig* into an imperial army and administration while, at the same time, retaining its core ingredients that would ensure ongoing personal loyalty and counteract the assimilative pull of settled civilization that threatened to draw the nomadic elite into its snare. In line with the pioneering work of Peter Golden, we suggest that it was indeed Chinggis Khan who perfected the institution of the *keshig*, and did so on an unprecedented scale,[34] by a thorough reshuffling of the 'tribal' components of his following.[35] The result of this was the creation of

---

[34] Golden, "'I will give the People'", 21-41.

[35] The term 'tribe' is highly problematic. The common term used in the early Mongol sources is *irgen*, which indicates 'a community of common "shape, form, vocabulary, dialect, customs and manners"' (L. Munkh-Erdene, 'Where did the

decimal military units as artificial tribal formations, and which consisted of a well-engineered mixture of original tribes and other, defeated, groups.[36] The leaders of these subsidiary warbands of 10,000, 1,000 and 100 – here for mere convenience called *tümens* after its biggest contingent – had to send their brothers, sons, and the best of their own companions and mounts as, respectively, hostages and security to the chief Khan or Khaghan. The hostages were incorporated into the imperial *keshig*, sharing a tent and table with the Kaghan. At the same time, their sisters and daughters often married into the dynastic line, some of them serving as wet-nurses, meaning their sons shared their mother's milk with the sons of the Khaqan, thus being turned into foster-brothers of the royal princes.[37] Interestingly, this policy in which numerous women of *nökörs* married into the dynastic line is still visible in the Y-chromosomal lineage that goes back to Chinggis Khan and can be traced in 8 per cent of the male population of the Central Eurasian region. The existence of this Chinggisid genetic lineage is as telling for Chinggisid patrilineal descent rules as the non-existence of other such lineages is for the Chinggisid policy of ethnic engineering.[38]

In other words, Chinggis Khan's innovation consisted firstly of the expansion of his *keshig* from a mere 80 night-guards and 70 day-guards in 1189 to a permanent administrative unit of 10,000 men in 1206, which amounted to about one-tenth of all active military forces. At the same time, the crucial meritocratic principle of the *keshig* was maintained or, as the *Secret History of the Mongols* has it, 'those shall be enlisted who are clever and attractive of appearance'. As big as the *keshig* grew, it remained a personal army, subordinate and loyal to Chinggis Khan alone. As big as the empire grew, the old practice of

---

Mongol Empire come from? Medieval Mongol Ideas of People, State and Empire', *Inner Asia* 13 (2011), 211-37 – partly cited from Thackston).

[36] Of course, the decimal system as such was not a Chinggisid invention as it was used already by the Xiongnu.

[37] Subtelny, *Timurids in Transition*, 35.

[38] T. Zerjal, Y. Xue, G. Bertorelle et al., 'The Genetic Legacy of the Mongols', *American Journal of Human Genetics* 72 (2003), 717-21; see also E. Heyer, P. Balaresque and M.A. Jobling, 'Genetic Diversity and the Emergence of Ethnic Groups in Central Asia', *BMC Genetics* 10 (2009), 49.

organizing repeated physical contact was maintained; this was described by the thirteenth-century Walloon Globetrotter William of Rubruck, who reported that all nobles stationed anywhere within a two-month journey of the capital Qara Qorum were obliged to assemble each summer at the court in order that Möngke, one of Chinggis Khan's successors, might drink and sup with them, bestow garments and presents upon them, and display his great glory.[39] Meanwhile, members of the extended *keshig* of 10,000 served as a pool of loyal managers, both for the military and for the civil administration. As Thomas Allsen so aptly observes, the *keshig* remained 'the training ground, proving ground and recruitment ground for central government personnel'.[40] As such, it served as a top-layer of various kinds of special, personal envoys (*wakils*) and controllers (*darughachis*) who supervised the existing hierarchy of both military and civil administration. Chinggis Khan was known to have expressed it as follows: 'My guardsmen are higher than the external commanders of thousands. The escorts of my guardsmen are higher than the external commanders of hundreds and of tens.'[41] Thus, a crucial part of the system was a structural doubling as well as the overlapping of functions; shadow officers from the *keshig*, who served as a check on the regular administrator, also had to keep an eye on their closest colleagues with similar functions. As is often observed by later analysts, Mongolian officials were generalists rather than specialists.

Chinggis Khan's second major innovation was his widely implemented policy of ethnic engineering: the breaking up of old patterns of ethnic organization and reconstituting them into new subsidiary warbands such as the *tümens*. This is what Golden calls the 'nökörization of the tribal fighting force'; the imposition of the non-tribal, personal principle of the warband onto the existing tribal organization.[42] This is forcefully argued by Lhamsuren Munkh-Erdene, who demonstrates the existence of a category of people called felt-

[39] Allsen, 'Guard and Government', 518.
[40] Ibid., 517.
[41] Grupper, 'The Barulas Family Narrative', 44-5, 57.
[42] Golden, '"I will give the People"', 23.

tent, or *ulus* in Mongolic, who only became Mongol *ulus* after being administratively organized into 95 units of a thousand. As a result, the term *ulus* itself has not an ethnic but a political meaning, that of a 'community of the realm'; that is, a political community formed by the state.[43]

As will be discussed below, there is growing doubt as to whether the pre-existing structure of society was really based on 'tribes', and so it must be wondered to what extent Chinggis Khan was really that innovative in this regard. However, the difference between the ideas of the original core warband of the ruler (*keshig*) and these new subsidiary warbands (*tümen*) was that the latter was not freely but forcibly associated with the Khaghan, a process engineered by the latter, who was assisted by the men and women of his own *tümen*, the *keshig*. In principle, though, the *tümen* signifies the successful repetition and trickling down of the *keshig* at the level of the ruler's followers, and as such will be studied here as an important and integral ingredient in the legacy of the Chinggisid, and the Turko-Mongolian, warband.

TRANSITION

Central to the present argument is the view that many of the Eurasian empires of the period 1200-1800 were not only inspired by long-standing Roman, Islamo-Persian or Chinese traditions – a fact all too often stressed by our present-day area studies – but also by the nomadic legacy of the *keshig* as created by Chinggis Khan. Although the *keshig* was an ancient institution that was to flourish particularly in the socio-political context of Central Eurasian *qazaqliq*, Chinggis Khan expanded the institution to an imperial level and imposed it ruthlessly on both the old aristocracies and the new imperial administration. It should not be forgotten, however, that even Chinggis Khan soon made compromises with his own model. Although in principle the members of the *keshig* derived their honour and position from their talent and personal loyalty to Chinggis Khan, the latter made sure that his new world order was perpetuated:

[43] Munkh-Erdene, 'Where did the Mongol Empire Come From?', 211-37.

As for my 10,000 personal guardsmen, who have come selected to become personal servants in my presence from the ninety-five thousands, the sons who will have sat on my throne henceforth – to the descendants of my descendants – shall consider these guardsmen as a legacy and give them no cause for dissatisfaction and take excellent care of them.[44]

Although this arrangement undermined the meritocratic principle of the nomadic warband, it also explains the endurance of the Chinggisid tradition; for example, S.M. Grupper has shown how Timur's legitimacy was based on ancestors who had served as *nökörs* in the Chinggisid *keshig* for five generations. But by far the most crucial substance of that Chinggisid tradition was the model of the warband which was, often in hidden, submerged ways, implemented by most of the post-Chinggisid conquerors who carved out their own domains at the interface of the nomadic and sedentary worlds. Some of this enduring institutional legacy has been studied quite extensively – for the Timurids by Grupper and Maria Subtelny, for Iran by John Woods and Charles Melville.[45] The current work will gratefully build on these pioneering endeavours and make further comparisons and connections between the *keshig* phenomenon within the various post-nomadic empires at the sedentary fringe of the Eurasian Arid Zone, including Russia, Islamicate India, and the Middle East.

Taking the *keshig* as a starting point, we will particularly focus on the way its martial spirit was extended from the inner circle of the warband towards the military organization and military culture of the post-nomadic empires as a whole. This will necessarily involve a discussion of the transition of the warband into the imperial army and its relationship with the imperial bureaucracy. This bring us back to Ibn Khaldun and what he observed as the three main ingredients of group feeling: pedigree (*ansab*), religious devotion (*din*), and military slavery (*mawali*; *mamalik*).[46] In all three cases, these elements

---

[44] Grupper, 'The Barulas Family Narrative', 45-6.

[45] See also the pertinent comments by M. Biran, 'The Mongol Transformation: From the Steppe to Eurasian Empire', *Medieval Encounters* 10 (2004), 339-61.

[46] Based on a less well-known part (i.e. beyond the *Muqaddima*) of his *Kitab al-'Ibar* as analysed and translated by D. Ayalon [D. Ayalon, 'Mamlukiyyat', *Jerusalem Studies in Arabic and Islam* 2 (1980), 321-49.]

should be understood in their widest connotations, whether it produces real or imagined feelings of belonging, whether it takes an orthodox (e.g. *jihad*) or a heterodox (e.g. *ghulluw*) form, and whether it relates to the army or to the martial culture in the society as a whole. In addition to Ibn Khaldun's three components, a fourth ingredient, that of material rewards, from loot to cash salaries, should be added. As indicated in the quote opening this chapter, Ibn Khaldun was more than eager to recognize the importance of the 'money and taxation' that paid for the warband. Indeed, it once again demonstrates the fact that the pre-modern Chinese, Indian, Russian and Ottoman empires surrounding Central Eurasia did not originate from either the nomadic or the sedentary world but were very much the result of the unescapable interaction and transition between the two.

Last but not least, we should take into account the possibility that Ibn Khaldun was altogether wrong about *asabiya*. This is suggested by the South Asian experience so eloquently expressed by the late Indologist Jan Heesterman who stresses that it was *not* the desert but sedentary agriculture that enabled the construction of extensive networks of kin and marriage, often using genealogical formulas to express the repartitioning of rights in the soil and its produce. By contrast, the cohesion of nomadic groups depends on immediate, albeit momentary, success and easily breaks down in case of failure. Hence, the nomadic warband lacks, as a basic principle, primordial ties which persist over time. It is equally understandable that the nomadic warband is likely to look for patronage and leadership outside its own ranks, with a settled magnate, who has a better chance of leading them to success and supporting them in case of failure – in other words, to provide the continuity (read *asabiya*) – the warband lacks on its own.[47]

Similar is suggested by the recent work of Ali Anooshahr, who has demonstrated that Ibn Khaldun was just one – and the most systematic – of a series of historians – Baihaqi and Nizam al-Mulk to name two others – who endowed the history of Islamic dynasties with a universal, tripartite, cyclical quality, each embodied by an

---

[47]J.C. Heesterman, 'Warrior, Peasant and Brahmin', *Modern Asian Studies* 29 (1995), 644.

idealized monarch.[48] In the first stage, we find the simple and austere ghazi leading a life of vagabondage with his brothers in arms. The second stage involves the climax of the dynasty, represented by a righteous king who rules over a prosperous and orderly realm. In the third stage, decline sets in because of a debauched and ineffectual ruler and who, through mismanagement, loses his throne to a different warband of fierce ghazis who are endowed with a high level of group cohesion. All this suggest that the work of Ibn Khaldun and his colleagues should not only be read as historical analysis, but also, and perhaps more importantly, as a literary model of a programmatic nature. Their chronicles are less about what happened in the past than prescriptive texts that provide a programme for the present and the future. In other words, they don't tell us how rulers actually behaved but how they *should* behave in order to avoid, as long as possible, the inevitable final stage, that of moral decline and foreign conquest. In this scenario, it should not be a surprise that every late-dynastic ruler was keen to demonstrate that he had not yet entered the third stage or that, by presenting himself as a *ghazi*, he had actually opened a new dynastic cycle.

From this perspective, Ibn Khaldun's notion of *asabiya* should be perceived as a rather romantic post-conquest rationalization of the early period of empire. In fact, looking at various important examples, it appears that group cohesion on the basis of pedigree, devotion or slavery was rather weak and never unchallenged. It was only *after* a conquest that sedentary administrators made *asabiya* into a sophisticated tool to forge new bonds of loyalty between the ruler and his former warband. To turn the latter into an effective imperial army, the ranks of the warband had to be opened up to outsiders on the basis of a new ideological model. For someone like the sixteenth-century Ottoman historian Mustafa Ali it was no more and no less than divine

---

[48] A. Anooshahr, *The Ghazi Sultans and the Frontiers of Islam: A Comprehensive Study of the Late Medieval and Early Modern Periods* (London: Routledge, 2009), 13, 44; A. Anooshahr, 'Mughals, Mongols, and Mongrels: The Challenge of Aristocracy and the Rise of the Mughal State in the *Tarikh-i Rashidi*', *Journal of Early Modern History* 18 (2014), 559-77; A. Anooshahr, 'The Rise of the Safavids according to their Old Veterans: Amini Haravi's *Futuhat-e Shahi*', *Iranian Studies* (2014), 1-19.

favour that made for a successful ruler or *sahib-i zuhur*, i.e. 'the manifest one'. Indeed, unconsciously echoing Thomas Aquinas's famous words 'if justice is taken away, what are kingdoms but robberies', he considered the main difference between a bandit and a legitimate ruler purely a matter of chronology: once a *sahib-i zuhur* had established a legitimate dynasty, his successors ruled by right of heredity.[49] One can almost hear Chinggis Khan's affirmative 'yeah'.

With Mustafa Ali, also modern historians increasingly start to doubt the traditional narrative of the first imperial stage. As we have seen already, Chinggis Khan started to stress the kinship ingredient of his extended *keshig* only *after* his first conquests. In a recent monograph David Sneath convincingly dismisses the idea of tribe and kinship as the most important ingredient of cohesion among Central Asian groups, even before Chinggis Khan.[50] Munkh-Erdene elaborates on this and cites Thomas Allsen, who had previously stressed that the use of kinship rhetoric was 'designed to enhance political unity, not authentic descriptions of biological relationship.'[51] Before this, Colin Imber strongly argued against the so-called ghazi-thesis to explain the early rise of the Ottomans. Others, like Lindner, had already questioned the ideological content of the ghazi narrative and instead stressed a much more pragmatic meaning, that of a phenomenon that was basically geared towards the making and redistribution of spoils.[52] After dealing with the Ottomans and Mughals, Anooshahr is now attempting to show that the earliest Safavid warband was forged less by the religious devotion of its members to an extraordinarily

[49] C.H. Fleischer, *Bureaucrat and Intellectual in the Ottoman Empire: The Historian Mustafa Ali (1541-1600)* (Princeton: Princeton University Press, 1986), 281, 290.

[50] D. Sneath, *The Headless State: Aristocratic Orders, Kinship Society, and Misrepresentations of Nomadic Inner Asia* (New York: Columbia University Press, 2007).

[51] Munkh-Erdene, 'Where did the Mongol Empire Come From?', 221.

[52] C. Imber, 'The Ottoman Dynastic Myth', *Turcica* 19 (1987), 7-28; R.P. Lindner, *Nomads and Ottomans in Medieval Anatolia* (Bloomington IN: Research Institute for Inner Asian Studies, Indiana University, 1983). For the Ottoman discussion on the ghazi-thesis, see C. Kafadar, *Between Two Worlds: The Construction of the Ottoman State* (Berkeley: University of California Press, 1994).

charismatic leader than the fact that it simply made the best of specific historical circumstances and good fortune.[53]

Where, then, does all this leave Ibn Khaldun's *asabiya* as an analytical framework? What does this imply for the idea of nomadic cycles? What if chronicles do not so much *describe* but *make* history? My provisional answer would be that nomadic cycles do still exist but that these cannot be explained by *asabiya*, as this was an ex-post-facto rationalization of more random historical circumstances. Hence, rather than the various elements of group cohesion, it is horse-based military superiority that accounts for the recurrent nomadic conquests. At the same time, historians should read chronicles as instruments of fashioning and disciplining both the ruler himself and his following. As chronicles tend to represent the administrators' points of view, their contents are less about the emergence of empire as about its continuation. In other words, most chronicles reflect Ibn Khaldun's second imperial stage after the divergence between the people of the sword and the people of pen, and thus they increasingly represent the latter's highly polished views on the former. As a consequence, bonds of kinship, devotion and military slavery should be studied as post-conquest, sedentary *alternatives* to the nomadic warband. Although each of these would deserve at least a separate monograph, in this chapter they will only feature in the background of what should be considered its main topic: the specific case of the *nomadic* warband. Keeping these considerations in mind, let us now discuss our four case studies of nomadic imperial state-formation, beginning with the most successful and powerful by far.

CHINA: FROM JURCHENS TO QING

JURCHENS: KESHIG AND MUKUN

Travellers who journeyed eastwards along the Silk Road usually, at some point, turned southward and headed for the wonders and riches of China. If, however, that temptation was ignored and the eastern trail continued to be followed, one arrived in an area that is now

---

[53] Anooshahr, 'The Rise of the Safavids'.

called Manchuria, and which extends into Korea. The people of this area occupied themselves either with hunting and fishing in the forests or with raising cattle and agricultural work in the plains. Although oxen dominated their own mixed economy, they exported huge numbers of horses to China. Sitting on the edge of the exchange between the desert and agricultural lands, the Jurchen inhabitants of Manchuria proved ideal commercial and political intermediaries between the Mongols and Chinese.[54] In fact, this intermediate position gave them more political agency in Chinese history than any of the purely nomadic people of the steppes, including Chinggis Khan and his Mongols. In the last millennium, Manchuria gave birth to three imperial dynasties that ruled either northern China, as in the case of the Khitan-Liao (907-1125) and the Jurchen-Jin (1126-1234), or China as a whole, in the case of the Manchu-Qing (1644-1911).[55]

Returning to Chinggis Khan's Mongols, their expertise was in conquest. For sustaining their conquests, however, they turned to the example of the Jin dynasty, which gave them the institutional tools to establish a dynasty that ruled the whole of China under the name of Yuan (1271-1368). As we have stressed in the introduction, the key concept by which Chinggis Khan was able to bridge the gap between his Mongol past and Chinese future was his personal warband

[54] Here one could add Korea, which was heavily influenced by Mongol institutions; see D.M. Robinson, *Empire's Twilight: Northeast Asia under the Mongols* (Cambridge Mass.: Harvard University Press, 2009); R.E. Breuker, 'And now, Your Highness, we'll discuss the Location of your Hidden Rebel Base: Guerillas, Rebels and Mongols in Medieval Korea', *Journal of Asian History* 46 (2012), 59-95.

[55] Ch'i-ch'ing Hsiao, *The Military Establishment of the Yuan Dynasty* (Cambridge Mass.: Harvard University Press, 1978); H. Franke, 'The Chin Dynasty', in H. Franke and D. Twitchett (eds), *The Cambridge History of China*, vol. 6: *Alien Empires and Border States, 907-1368* (Cambridge: Cambridge University Press, 1994), 215-320; I. De Rachewiltz, 'Personnel and Personalities in North China in the Early Mongol Period', *Journal of the Economic and Social History of the Orient* 9 (1966), 88-144; T. Allsen, 'The Rise of the Mongolian Empire and Mongolian Rule in North China', in H. Franke and D. Twitchett (eds), *The Cambridge History of China*, vol. 6: *Alien Empires and Border States, 907-1368* (Cambridge: Cambridge University Press, 1994), 321-414.

or *keshig*. Although we should acknowledge that the *keshig* as such was nothing extraordinary at the time, the main challenge was to stretch its size in such a way that it could incorporate new groups without undermining its overall cohesion, which was based on personal loyalty and intimacy. Since not all new manpower could be accommodated in the imperial warband, other *subsidiary* warbands were engineered from a wide variety of ethnic groups, of which only the leaders were represented in the imperial warband, while their sons became hostages and their daughters became spouses. Although Chinggis Khan and his successors exploited the warband's potential to its maximum and without precedent, they only succeeded because they could stand on the shoulders of their immediate, more 'cooked' predecessors: the Jurchen-Jin.

To expand on this, we should first of all take account of the fact that Chinggis Khan's Mongol tribe had been tributary to the Jurchens. Chinggis himself had the Jin honorary title of 'chief of hundred.' Being very close to the imperial experience of the Jin, Chinggis Khan must have learned from the way the former had used the Jurchen institution of the *mukun*. Although dictionaries give definitions of this term that vary from clan, family, village, herd and tribe, its political meaning at that time comes very close to that of household, but in the context of conquest and rule, even closer to that of the warband, and in many respects it was a precursor of the Manchu Banner (*niru*) system. In fact, the Jin had used the pre-existing idea of the *mukun* to set up a socio-military organization called *meng-an mou-ke*. Although the word *mou-ke* derives from *mukun* it also had the politico-military meaning of 'the leader of one hundred men'. The word *meng-an* derives directly from Manchu *minggan* and means thousand. Hence we have something that seems very old and Central Asian and yet also very close to Chinggis Khan's famous decimal military system. Indeed, although the *meng-an mou-ke* system originally served as a comprehensive socio-political system which organized the entire Jurchen population under emperor Aguda (r. 1115-23), it soon was extended to become the most important military and political means of control over all subjugated peoples. Although the term suggests kinship, fixed hierarchy and numbers, the reality was much more flexible and in reality it is fundamentally similar to Chinggis Khan's conscripted warband in its most extended, post-conquest form. For example, we know from the *Secret History* that Chinggis Khan created 95 *mingghan* (Mongolian) from the greatly enlarged manpower pool available to him in 1205-6.[56]

Going into even greater detail, the *mukuns* consisted of a mixture of households, and included their slaves and cattle. The numbers indicated neither exact amounts nor a fixed hierarchy beyond the fact that several *pu-li-yan* (unit of 50) formed a *mou-ke*, several *mou-ke* formed a *meng-an*, and several *meng-an* formed a *wanhu*, literally a unit of 'ten thousand households'. It is crucial in order to understand the *meng-an mou-ke* system as a controlling institution to highlight that the Jin emperor had his own personal *mou-ke* recruited from the other units, which constituted his own imperial guard. This organizational system was retained under the Yuan dynasty and each was considered a self-sufficient socio-economic community that either managed or worked their own territories as a kind of appanage, and which only later developed into a much less flexible, land-based system of hereditary military households.

Coming to the Yuan, we have already seen that under Chinggis Khan the *keshig* not only served as an imperial guard, but had a wide range of functions including being an imperial domestic service bureau, a hostage camp, and an academy for young leaders; in many ways it was thus a rudimentary imperial administration.[57] In due course, Chinggis Khan's *nökörs* obtained hereditary commands under the generic name of *noyan* (plural: *noyad*). Their titles were conferred by direct investiture or patent (*yarligh*) and they received a tablet as a token of authority, as well as expensive gowns in order to create and strengthen a sense of corporate identity.[58] When a *nökör* who served in Chinggis Khan's *keshig* joined Chinggis with his household, he was actually bestowed with a household of his own. It was not uncommon for Chinggis to allow old tribal allies to keep their own ethnically homogenous group, with all their dependants, to confirm its chief as

[56] Allsen, 'The Rise of the Mongolian Empire', 345.

[57] Hsiao, *The Military Establishment*, 34; this sections builds thoroughly on idem, 9-50.

[58] Reference MARIE.

noyan after a summary census, and to turn it into a *mingghan*.[59] More generally, though, original households were regrouped and given a uniform, if rudimentary, military skeleton. As such, those qualified were either organized into new units with Mongols or northern Chinese as their officers, or were distributed among original Yuan contingents. At the very heart of the organization, *nökör* were appointed as so-called *darughas*, or seal-bearers, to oversee other parts of the nascent imperial organization.

All this does not mean that the traditional consanguineous principle of 'tribal' organization was totally disregarded, but ethnic labels were of a more nominal character and, as later in the case of the Manchu Banners, stimulated the start of a new process of ethno-genesis. In this way, the chiliarchy became the basic political unit and the cradle of officialdom under the Jin, Mongol and Yuan empires, with the *keshig* at its centre and as the most privileged group. In due course, though, when the empire expanded further south and had to incorporate more and more Chinese allies, the *keshig* was outnumbered by the imperial guards and other Chinese units originating from the south. Hence, several decades after the conquest, it became very apparent that the extended warband, so typical of and well-suited to the Middle Zone, would reach it natural limits in the Outer Zone of sedentary China. Hence, soon after a conquest, there were increasing pressures to transform the Jurchen and Mongol warband organization into a Chinese administrative system along the lines of the proven Confucian ideas of *wen*.[60] For the conquering elites, this shifting from the Mongolian custom or *yasa* could be easily translated into traditional Sinic notions of losing martial strength (*wu*) and dynastic decline. Marco Polo, who stayed in China between 1275 and 1292, observed something similar:

[59] Allsen, 'The Rise of the Mongolian Empire', 347.

[60] For the Jin, see Jing-shen Tao, 'The Influence of Jurchen Rule on Chinese Political Institutions', *The Journal of Asian Studies* 30 (1970), 121-30; for the Yuan, see J. W. Dardess, *Conquerors and Confucians: Aspects of Political Change in Late Yüan China* (New York and London: Columbia University Press, 1973), 7-31, and G. Mangold, 'Das Militärwesen in China unter der Mongolen-Herrschaft' (PhD thesis, Ludwig-Maximilians Universität, Munich, 1971).

Now they are much debased and have forsaken some of their customs, for those who frequent Catai keep themselves very greatly to the ways and to the manners and to the customs of idolaters of those regions and have very much left their law.[61]

Interestingly, though, the notion of a Chinggisid kind of *keshig* even continued under the non-nomadic Ming dynasty (1368-1644).[62] Like their near contemporary Safavid colleagues in Iran, the emergence of Zhu Yuanzhang was based on millenarian religious fervour that gave the conquering group its initial cohesion.[63] Despite his celestial name of Hongwu, i.e. 'vastly martial', Zhu's early warband never attained the nomadic strength of its Jin and Yuan predecessors. From the very beginning, the early Ming rulers focused much of their efforts on limiting and controlling the power of military leaders and institutions.[64] Much sooner and much more than in the case of their nomadic predecessors, the Ming military became increasingly managed by civilian officers. Although Mongols and Jurchens continued to provide manpower to the various imperial guard units, the Ming *keshig* became more aloof from the outside world and as such could not play its earlier, post-nomadic role of incorporating new groups. Hence, its social status decreased and its function became focused purely on the military, and particularly the internal security of the empire. Instead of the *nökörs*, the important linkages between court, military and bureaucracy, as well as between the centre and the provinces, including its military households, increasingly became the domain of a new elite corps of eunuchs who served both as military

---

[61] Hsiao, *The Military Establishment*, 31.

[62] D.M. Robinson, 'The Ming Court and the Legacy of the Yuan Mongols', in D.M. Robinson (ed.), *Culture, Courtiers, and Competition: The Ming Court (1368-1644)* (Cambridge Mass.: Harvard University Press, 2008), 393-6. I am very grateful to David Robinson for elaborating so insightfully on these pages by email (March 2015).

[63] J.W. Dardess, 'The Transformation of Messianic Revolt and the Founding of the Ming Dynasty', *The Journal of Asian Studies* 29 (1970), 539-58. I assume that the fact that both the Ming 'Red Turbans' and the Safavid Qizilbash wore red headgear, is mere coincidence.

[64] K. Filipiak, 'The Effects of Civil Officials handling Military Affairs in Ming Times', *Ming Studies* 66 (2002), 1-15.

commanders and administrative supervisors.[65] According to the official Ming chronicle, the *Ming-shih*:

The eunuchs in the Ming period were sent out in charge of military expeditions, to supervise the army and the garrisons. They spied on the officials and the people, and secretly controlled all great authority. All this begin in Yung-lo's reign [r. 1403-1424].[66]

By the end of the Ming dynasty, what existed of the central military arm was largely in the hands of eunuchs. The remarkable military renaissance under the Wanli emperor (r. 1573-1620) was an attempt to give the empire a new lease of life without recourse to Barbarian *wu*, seeking strength instead in the neo-Confucian wisdom of a new generation of military commanders like Wang Yangming and Qi Jiguang, who could combine the ideals of *wen* and *wu* by increasing the personal discipline of their soldiers.[67]

MANCHU BANNERS

After the Ming, both the later Jurchens (the Manchus) under the Later Jin (1616-36), and the Qing dynasties (1636-1911) reintroduced the Jurchen *mukun* as their prime organizational unit and the fundamental criterion of Manchu identity.[68] Meanwhile, the idea of an imperial *keshig* lived on in the imperial clan of the so-called Aisin Gioro which was, however – and much more than the earlier *mukun* – based on 'real' ancestry consisting primarily of the descendants of the *de facto* Qing founder Nurhaci (r. 1616-26) through the main and collateral patrilineal lines, and which were traced back to his grandfather Giocangga. The only exceptions were the so-called Princes of the Iron Cap, non-kinsmen who had rendered extraordinary service

[65] F. Mote, 'The Ch'eng-hua and Hung-chih Reigns, 1465-1505', in D. Twitchett and F. Mote (eds), *The Cambridge History of China*, vol. 7: *The Ming Dynasty, 1368-1644*, part 1 (Cambridge: Cambridge University Press, 1988), 370-7; R.B. Crawford, 'Eunuch Power in the Ming Dynasty', *T'oung Pao* 49 (1961), 115-48.

[66] Crawford, 'Eunuch Power', 126.

[67] Filipiak, 'The Effects', 1-15. See also the forthcoming PhD thesis of Barend Noordam, Leiden University.

[68] P.K. Crossley, *The Manchus* (Oxford: Blackwell Publishing, 1997), 28-9.

to Nurhaci and his son and successor Hong Taiji (r. 1626-43). There was also the Imperial Guard, which consisted primarily of Manchus taken from the Banners, but, as in the case of the Ming, its function was limited and did not extend beyond protecting the emperor.[69]

Although the situation of *keshig* and guard under the Manchus was closer to Ming than to Jurchen antecedents, the warband system was actually reinvigorated through the re-introduction of *mukuns* in the new form of Banners. Indeed, in principle, a Banner was not that different from the earlier *meng-an* under the Jin and Yuan. Indeed, like the *meng-an*, a Banner was much more than a military unit; it was a social formation and a political structure which encompassed peoples of many different backgrounds. Membership was based on patrilineal inheritance, marriage or adoption. According to the main authority on the topic, Mark Elliott, we should not regard the Banners solely as an army, but rather as a sub-order of society that was defined primarily, but not exclusively, by an inherited duty to furnish professional soldiers of unimpeachable devotion to the dynasty, which in exchange supported the entire population registered in Banners (about 2 million in 1644), both materially and morally, through money, food, and housing, as well as privileged access to power, for their entire lives. Interestingly, the Manchu banners survived much longer than the Jin and Yuan *meng-ans*, partly because the Qing themselves lasted longer, and partly because the Qing decided to segregate the Banners from their Chinese surroundings while, at the same time, upholding their connections with its recruiting grounds in the north.

Following the precedents of other Central Asian conquerors like Aguda and Chinggis Khan, the Banner-system was introduced by the new Manchu leaders Nurhaci and Hong Taiji at the very start of their conquest. They also imposed the new ethnonym of Manchu on their Jurchen followers as it became organized into the first Eight Banners. Although the Banner-system started as an exclusively Jurchen organization, it included Mongols as early as the rule of Nurhaci, who were at first still redistributed under the Manchu Banners. In

[69] M.C. Elliott, *The Manchu Way: The Eight Banners and Ethnic Identity in Late Imperial China* (Stanford: Stanford University Press, 2001), 79-81.

1635, when Mongol forces had grown to number around ten thousand, Hong Taiji decided to remove the Mongol companies from the Manchu Banners and establish eight separate Mongol Banners, containing a total of 80 companies. Some Mongol units, however, were still incorporated into the Manchu Banners. The new Mongol Banners remained subject to the Manchu colour-Banner chiefs of which they were a part, and in the years 1637-42 this procedure was repeated for the Chinese (Hanjun) troops, creating a triple system of Eight Banners:

Each banner is divided into three sections. The tribes that were originally Nurhaci's ... make up the Manchu [section]. The various bow-drawing peoples from the Northern Desert ... form the Mongol [section], while the descendants of the people of Liao[dong], former Ming commanders and emissaries, those from the other dynasty who defected with multitudes [of soldiers], and captives are separately attached to the Hanjun.[70]

The Banner Army derived its name from the individual Banners or units of which it was composed, each identified by its own distinctive coloured flag. In a graph it looks like this:

	Yellow	Red	Blue	White	Yellow – Red Border	Red – White Border	Blue – Red Border	White – Red Border
Manchu								
Mongol								
Chinese								

Despite the labels, ethnicity was never crucial in determining loyalty. As one popular saying had it: 'Never mind who is Manchu and who is Han, but ask who is a Bannerman and who is a civilian.'[71] Nevertheless, as mentioned earlier, after the Banners developed into hereditary institutions there was room for new processes of ethnogenesis based on exclusive Banner membership, and as such the gap

[70] M.C. Elliott, 'Ethnicity in the Qing Eight Banners', in P.K. Crossley, H.F. Siu and D.S. Sutton (eds), *Empire at the Margins: Culture, Ethnicity, and Frontier in Early Modern China* (Berkeley: University of California Press, 2006), 45.

[71] Elliott, 'Ethnicity in the Qing Eight Banners', 46.

between the Chinese within and without the Banners actually increased.

As explained by Edward McCord, one significant feature of the Qing military system was its careful elaboration of checks and balances, aimed at preventing the concentration of military power in a manner that might present a threat to dynastic rule. Firstly, in addition to the Eight Banners, there was the Green Standard Army, a predominantly Han Chinese force that was more than 200 per cent bigger, housed in small garrisons scattered throughout the country and largely modelled on the military organization of the preceding Ming, in which the purely military chain of command overlapped with the civil administration and the supervisory powers of provincial governors. In a slightly different way, a similar overlap occurred in the internal structure of the Eight Banners; while each Banner had its own separate command structure and bureaucratic administration, the much less scattered Banner garrisons in the main administrative centres were formed not by one Banner but by a combination of forces taken from a number of different Banners. In major military campaigns, special expeditionary forces were formed by combining a number of different units from both Eight Banner and Green Standard Armies. Even under the Qing, the commanders of such campaigns were often appointed not from the officers of any of its component forces, but from the ranks of the civil bureaucracy. This again diluted the military in such a way as to hinder the accumulation of military power in the hands of any one official or military power.[72]

This balancing act was part and parcel of each and every Eurasian dynasty. What makes the Qing Banner system unique is that it represents the continued existence of the warband under fully sedentary conditions. At the very start of its existence it served to incorporate and reshuffle important allied and conquered groups. At the same time, for almost three centuries the Banners helped to preserve the identity and virility of the conquering elites as a whole

---

[72] E. McCord, *The Power of the Gun: The Emergence of Modern Chinese Warlordism* (Berkeley: University of California Press, 1993), 20-2; based on Wu Wei-ping, *The Development and Decline of the Eight Banners* (PhD thesis, University of Pennsylvania, 1969).

and provided a segregated pool of loyal personal *nökörs* who could be employed in both military and civilian capacities. As before, under the Jin and Yuan dynasties, this gave rise to a dual administration in which the official hierarchy of military commanders and civilian officers was permanently supervised by Bannermen as the emperor's personal agents. As argued elsewhere, the dual system was a typical post-nomadic variety of *apartheid*, not instigated by some 'modern' urgency to distinguish between ever more professional armies and ever more rational bureaucracies, but by the need of post-nomadic rulers to keep the cavalry core of the army as loyal, fit and ready as possible.[73] Since the Manchu military was as much as possible kept apart from the existing Chinese bureaucracy, there was also no need to find complicated compromises that would actually undermine both sides of the system.

Nonetheless, even under the Qing rulers the system found its natural limit when administrative units based on different, often well-established, standards of recruitment began to take over while, at the same time, a centralized system of stipends on the basis of individual Banner service was introduced which consequently moved away from a situation where garrison lands were collective appanages. Even the Qing could not avoid the increasing bureaucratization. As under the Ming, the Qing *keshig* became a more isolated and more exclusively military institution that was essentially geared towards internal security, once again giving rise to a now-familiar anxiety of imperial decline which, according to the old elites, could only be stopped by reinvigorating the 'old ways' and cleaning up the ranks through purifying programmes of Manchuization, as was the case particularly during the long reign of the Qianlong emperor (r. 1735-96).[74] As we will see, the 'purifying' policies of this fourth official Qing emperor were quite similar in their bases to those of the fourth official Mughal emperor Aurangzeb (1658-1717); after one and a half

[73] J. Gommans, 'Warhorse and Post-Nomadic Empire in Asia, *c.* 1000-1800', *Journal of Global History* 2 (2007): 1-21.

[74] M.C. Elliott, 'Ethnicity in the Qing Eight Banners', 27-58; P.K. Crossley, 'The Conquest Elite of the Ch'ing Empire', in W.J. Peterson (ed), *The Cambridge History of China*, vol. 9.1: *The Ch'ing Empire to 1800* (Cambridge: Cambridge University Press, 2002), 313-60.

centuries and three generations of dynastic rule, both emperors attempted to save their empires by stressing moral rearmament and a return to so-called fundamental values, Manchu ones for the former, Islamic for the latter.[75] This remarkable dynastic parallel immediately evokes Ibn Khaldun and demands a closer look at what seems to be another intriguing case of his dynastic cycle.

INDIA: FROM MONGOLS TO MUGHALS

TIMURIDS: KESHIG AND TÜMEN

Just like the Jurchens in Manchuria, the Afghans and Turks in Khorasan and Turkistan were ideally situated to serve as political and military brokers between the desert and cultivated lands. As in the case of Manchuria, control over these interstitial areas was the key to an Indian empire, as is demonstrated by the sequence of Indo-Turkish and Indo-Afghan sultanates in both Hindustan and the Deccan. Although it would certainly be worth comparing these Turkish and Afghan patterns of state-formation with those of the Jurchens, I will limit myself here to an investigation of the various Timurid states that were carved out along the various waystations connecting Turkistan and Khorasan in our Middle Zone to India in our Outer Zone. We will start with a discussion of Timur (r. 1370-1405) himself, and then move to his Bayqara and Mughal descendants, in Khorasan and Hindustan respectively.

Looking somewhat closer at the administrative practices in Timur's empire, we can observe a great deal of continuity with the Chinggisids. As far as the *keshig* is concerned, it seems that the institution continues, although the term itself disappears from the annals. If we look in greater detail, though, it seems that the so-called *tümen* comes quite close to exhibiting the classic features of the *keshig*, and very close to the Jurchen idea of the *meng-an mou-ke* system. As we have already seen in the introduction, under Chinggis Khan this was part of the decimal ranking system and referred to a military unit of 10,000. In principle, the *tümen* was not a tribal but a specifically engineered group with a mixed ethnic background whose leader was a trusted

---

[75] See my 'Warhorse and Post-nomadic Empire', 18.

member of the Chinggisid *keshig*. In due course, many of these *tümen* leaders received appanages that were supposed to support the *tümen*.[76] From these, *tümen* leaders were often able to gain territorial rights and develop extensive households of their own. As a result, a *tümen* began to signify both a household and territory. For example, Timur himself started his career as the hereditary *amir-i tümen* of Kesh. When he started to create his own empire, the relatives and other followers of his orginal *tümen* (read *keshig*) supplied the imperial leadership. Very much in line with the Chinggisid tradition, Timur divided his realm into four sections, each governed by the households of one of his sons but commanded by members of his followers' families. Each of his sons had a guardian (*ateke*) appointed to them in order to keep an eye on them. In principle the sons and their households were treated in the same way as the other leaders and their *tümen*. Timur regularly interfered in the make-up of the *tümen* by removing troops from the control of the traditional leaders and placing them under the command of his own intimates and early comrades-in-arms. Other control mechanisms included the rotation of guard services at court, taking relatives of leaders as hostages, and replacing *tümen*-leaders with other members of the same family. Another factor which is a clear reminder of those of the original *keshig* is Timur's policy of building entirely new *tümen* that were intimately attached to his own person through the use of labels such as *khanabachagan-i khass* (personal children of the house) and *bandagan-i khass wa muqarraban-i dargah* (personal slaves and intimates of the court) which even suggests bonds of adopted kinship and slavery, notions that persisted into Akbar's period but which, as we will see, received a rather mixed reception in India.

Another important consequence of the success of Timur's *tümen* was that, one way or another, it had to accommodate the growing financial administration of the conquered territories. As in the case of earlier Mongol rule in China, this entailed the emergence of a new civilian branch of government focusing on administrative and financial

---

[76] To distribute large parts of conquered lands as appanages to large military households was a well-known practice under Mongol rule, although in principle all income from all districts had to be shared by all the participating groups.

matters. These responsibilities could not be managed by the classic *keshig*, as this consisted primarily of military and more intimate household posts like chamberlain (*ichik-aqa*), chief taster (*bokavalbegi*) or chief arms-bearer (*qorchibashi*), to name but a few. The challenge was to incorporate expert civilian elements to govern the freshly conquered territories without compromising the effectiveness of the warband. Hence, the new Persian-dominated financial department, or *diwan-i aʿla*, was kept detached from the administration of the primarily Turkish military department which kept its own scribal organization, the *diwan-i tovachi*. In general, though, both divisions were still part of the ruler's warband and there were no rigid functional divisions; indeed, it was a conscious policy to maintain a certain vagueness of ranks and confusion of duties. Hence, as in the Mongol tradition, all officials remained generalist: scribes serving as commanders, commanders as scribes.[77]

Overall then, the imperial apparatus under Timur and his immediate successors was not that different from that of the Chinggisids nor that of the early Jin and Yuan. In fact, the basic idea of the *keshig* as a loyal military core consisting of relatives and personal followers continued to be manifested in the institution of the *tümen*, while the ruler's personal *tümen* was expanded to accommodate a new financial department. The leaders of the financial and military branches were also members of the ruler's household, and the ruler could control and manipulate his followers by awarding positions and honours such as *amir*, *bahadur*, *tovachi* (troop inspector) and *darogha*. These ranks indicate above all personal service to the ruler, often with no clear hierarchy and having a great deal of functional overlap. Indeed, the result of conquest was Ibn Khaldun's 'expected' divergence between pen and sword. However, the split was not allowed to be complete and both branches continued to operate within the ruler's extended household (read *keshig* or *tümen*), which also incorporated the empire's other *tümen*-leaders. Some newly created *tümen* were able to achieve

---

[77] For Timur's imperial organization, see B.F. Manz, *The Rise and Rule of Tamerlane* (Cambridge: Cambridge University Press, 1989), and her *Power, Politics and Religion in Timurid Iran* (Cambridge: Cambridge University Press, 2007).

a more intimate, personal relationship with the ruler and may have provided him with an inner army of intimates (*khass*) within the extended household. The correct functioning of this system depended on whether the ruler could collect sufficient revenues and was able to keep his own people on the move. Both were only conceivable under the (semi-)nomadic conditions of the mixed sedentary-nomadic frontier zones of Iran, Afghanistan and Turkistan.

When Timur was succeeded by his son Shahrukh in 1409, the transition from a Turko-Mongolian nomadic empire of conquest to a Perso-Islamic sedentary empire based on agriculture continued. This was reflected in Shahrukh's decision to move the capital from Samarqand to Herat. He also officially proclaimed the abolition of Turko-Mongolian customs in favour of Islamic law. As in other Turko-Mongolian states in the region, highly educated Persians, who hailed from old provincial families, played a key role in developing the bureaucratic administration of the Timurids and in promoting agricultural values. The Bayqara Timurids of Herat tried to bridge the two worlds by patronizing a Timurid artistic renaissance which involved the promotion of the Chaghatai language and *belles lettres*.

It is important to stress, though, that as long as the Timurids remained in Khorasan, they were not able to set up a stable civil administration with a solid fiscal base. Although they tried to limit the random distribution of revenue-free lands (*soyurghal*) to the Timurid military elite, the latter continued to dominate the administration and precluded the rise of a more autonomous state bureaucracy. In the end, the Timurids had no choice but to invest in their capital and its immediate environment, trying to maximize agricultural production on the basis of intensified irrigation. For the Timurids, though, even urban splendour and agrarian success had its downside as it would inevitably undermine their martial capacities. At least, this was what the young Babur wrote when he saw his Herat-based uncles indulging so excessively in the pleasures of settled life:

They were good at conversation, arranging parties, and in social manners, but they were strangers to soldiering (*sipahiliq*), strategy, equipment, bold fight and military encounter.[78]

[78] Subtelny, *Timurids in Transition*, 39.

The same was true for most polities in the Middle Zone. Whether slightly earlier under the Ilkhanids (1256-1335), during the reign of Ghazan (1295-1304), or under their contemporaries the Qara Quyunlu (1375-1468) or the Aq Quyunlu (1378-1501), all attempts at administrative centralization and reform towards more bureaucratization suffered the same fate against a coalition of nomadic-military and religious elites. Of course, one way to make themselves less dependent on their nomadic following was to take recourse to military slaves as advocated by the Seljuq minister Nizam al-Mulk (1064-95) and practiced earlier not only by the Seljuqs, but also by the Samanids, Ghaznavids and Ghurids in Afghanistan and the Mamluks in Egypt, all of whom mainly recruited Turkish slaves from Central Asia.[79] It appears, though, that following ongoing Mongol and Turkmen immigration, the Khorasan-Iran region as a whole experienced a nomadic resurgence which may have made this a less viable option than before, or indeed afterwards, as demonstrated by the reintroduction of slaves (*ghulams*) under the Safavids (1501-1735).[80] As we will see later, the neighbouring Ottomans also turned to military slavery, but shifted the main recruiting grounds from Central Asia to Eastern Europe. Indeed, it seems that the increasing wealth of the sixteenth century created a new window of centralizing opportunities for Turko-Persian dynasties in general.[81]

Before turning to military slavery, the Safavids had successfully started to use religious devotion to instil loyalty in their primarily

---

[79] Woods, *The Aqquyunlu*, 14. On the hesitant implementation of Ilkhanate reforms, see R. Amitai, *The Mongols in the Islamic Lands: Studies in the History of the Ilkhanate* (London: Ashgate, Variorum Collected Studies, 2007).

[80] D.T. Potts, *Nomadism in Iran: From Antiquity to the Modern Era* (Oxford: Oxford University Press, 2014), 213. On this question, see also Ann K.S. Lambton, *Continuity and Change in Medieval Persia: Aspects of Administrative Economic and Social History, 11th-14th Century* (London: I.B. Tauris, 1988). Despite the current revisionism that stresses Mongol agency in what seems to be an Iranian literary renaissance, the economic picture of the Mongol impact is still rather bleak.

[81] See e.g. L. Geevers, 'Challengers of the Safavid Dynasty: Abbasi Dynastic Centralization and the Bahrami Collateral Line (1518-1712)', *Journal of the Economic and Social History of the Orient* 58 (2015), 293-326.

Turkish following, now restyled as *Qizilbashes* or 'Redheads', a term derived from their distinctive crimson headgear. In fact, it was this Sufism that had accompanied the process of Turkish migration, settlement and state-formation in Iran. Under Turkish and Iranian influence there was an increasing convergence in the terminology of Sufism and kingship. All of a sudden, Sufi dervishes began to preface their names with titles appropriate to rulers such as *mir* and *shah*, their shrines called royal courts (sing. *dargah*) and their headgear considered crowns (sing. *taj*).[82] Almost every fifteenth- and sixteenth-century polity in the Middle Zone between Anatolia and India became infused with Sufi ideas of love and devotion. Whether rulers looked for the spiritual support of dervishes or, as in the case of the Safavids, claimed spiritual leadership for themselves, Sufism offered a wonderful tool for transforming a nomadic warband into a religious one.

With this observation, we are back at the heart of Ibn Khaldun's theory and may, along with him, conclude that, at least throughout the Middle Zone, both military slavery and religious devotion were considered two very forceful sedentary alternatives to, or even antidotes against, the nomadic warband of the Chinggisid kind. As such, the Safavid model was eagerly embraced by the Mughal emperor Akbar in the idiosyncratic form of *din-i ilahi* and as a remedy against the autonomous and egalitarian tendencies of his original Timurid and Turkish followers.[83] This bring us back to the later Timurids, better known as the Mughals, and it will be interesting to see to what extent

---

[82] N. Green, *Sufism: A Global History* (Oxford: Wiley-Blackwell, 2012), 99; see also J.R. Perry, 'Ethno-linguistic Markers of the Turko-Mongol Military and Persian Bureaucratic Castes in Pre-modern Iran and India', in I. Schneider (ed.), *Militär und Staatlichkeit. Beiträge des Kolloquiums am 29. und 30.04.2002 Halle 2003 (Orientwissenschaftliche Hefte 12; Mitteilungen des SFB 'Differenz und Integration' 5)*, 119. On the development of Sufism in relationship with royalty, see also the works of A.T. Karamustafa, *God's Unruly Friends: Dervish Groups in the Later Middle Period 1200-1550* (Salt Lake City: University of Utah Press, 1994) and J. Paul, 'Scheiche und Herrscher im Khanat Čaġatay', *Der Islam* 67 (1990), 278-321.

[83] J. Gommans, *Mughal Warfare: Indian Frontiers and High Roads to Empire, 1500-1700* (London: Routledge, 2002), 70-1.

the warband of its founder Babur was built on that of Timur and how it was transformed under his successors in India.

BABUR'S WARBAND

Very much like Nurhaci would do around a century later in Manchuria, Babur started his Indian empire in 1504 in Afghanistan: areas very much at the transition between our Middle and Outer Zones. His career is the romantic story of a young prince who, despite many mishaps, rises from Central Asian rags to Indian riches. Looking at his early years as a wandering vagabond in Central Asia, one is immediately struck by the similarities with the early careers of his celebrated forebears. Well-versed in the history of his own ancestors, Babur himself must have been very much aware that his history was one that repeated many earlier histories of empire builders. The tripartite division of his autobiography into Mawarannahr (i.e. Transoxania), Kabul (i.e. Afghanistan) and Hind (i.e. India) neatly follows the literary transition from the upbeat cheerful spirit of vagabondage or *qazaqliq*, when still in the steppes of Central Asia, to the more detached, melancholic mood of exercising *istiqbal* (sovereignty) after conquering India. There can be no doubt that Babur himself knew exactly how to frame his story. Of course, there was also the old Mongol custom or *törä* which, far from offering a fixed set of norms indicating that charisma (*sölde*), could be bestowed on anyone who proved his mettle on the battlefield.[84] Indeed, people turned to *törä* in appreciation of military valour, often in compensation for kin or age that may have been lacking, as in the case of Babur, when it was said, that 'he is young in years, but he is great according to the *törä* because he has conquered Samarqand several times by dint

---

[84] For *sölde* and *törö* in the Central Asian context, see the important Russian contributions by T.D. Skrynnikova, partly translated in: 'Sülde – The Basic Idea of the Chinggis-Khan Cult', *Acta Orientalia Academiae Hungaricae* 46 (1992/3), 51-9; 'Power among Mongol Nomads of Chinggis Khan's Epoch', in N.N. Kradin, D. Bondarenko and T. Barfield (eds), *Nomadic Pathways in Social Evolution* (Moscow: Russian Academy of Sciences, 2003), 135-57; 'Die Bedeutung des Begriffes törö in der politischen Kultur der Mongolen im 17. Jahrhundert', *Asiatische Studien* 63 (2009), 435-76.

of sword'.[85] For Babur, it was exactly this kind of military capacity that was missing among his settled and degenerate uncles in Herat, who have been referred to previously. Although he admired the carefree sociability of their court, he also despised the excessive *fesq u fujur* (immorality and debauchery),[86] and he and his successors knew very well that if they were to survive as a political power, such dynastic decay had to be avoided at all costs.

Whatever one may think of its Chinggisid origins, Babur's early warband did not constitute the tightly controlled, highly disciplined inner circle or *keshig* along the lines of the Chinggisid model.[87] In principle, though, its structure was similar. Like many Chinggisid khans or Timurid mirzas, Babur had a warband of his own consisting of both relatives (*biradaran wa khweshan*; brothers and relations) and retainers (*mulazaman/naukaran*), who were linked to each other through marriage and companionship. The latter was sustained in recurrent social events ranging from hunts to drugs-and-drinking bouts. In the early days of Babur's wanderings, the warband came closest to its original meaning: a small band of freely recruited warriors without much functional differentiation within the group. *Pace* Ibn Khaldun's ideas about tribal *asabiya*, and despite its small size, the cohesion of this group was never particularly strong as it depended entirely on the success of its leader. Without success, there were no spoils to be shared, and hence no reasons to be loyal.

In due course, when moving to Kabul to carve out his own small kingdom, Babur's warband started to expand and we begin to detect a structure that comes fairly close to that of the early Timurids and also that of the Jurchen idea of the *mukun*. First of all, Babur's warband incorporated the chiefs (*beglar*; Persian: *umara'*) of allied warbands with their own following (*jam'iyat*) of relatives and liegemen. Some of these begs were long-standing tribal leaders and had recently joined Babur as so-called Guest (*mihman*) Begs. Another pre-existing group

[85] Gulbadan Begam, *Humayun-nama*, ed. and trans. W.M. Thackston (Costa Mesa: Mazda Publishers, 2009), fol. 5a.

[86] S.F. Dale, *The Garden of Eight Paradises: Babur and the Culture of Empire in Central Asia, Afghanistan and India (1483-1530)* (Leiden: Brill, 2004), 81, 167.

[87] Dale, *The Garden of Eight Paradises*, 90.

of some standing was the so-called Andijani Begs, who apparently originated from Babur's homeland. Others had climbed the inner ranks by starting as young trainees or *ichkilar* (Persian: *jawanan*) before Babur was willing to turn them into full begs, called *ichki-beglar*. Like the so-called Baburi begs, these were Babur's own creatures and as such were probably part of his *naukari* (from *nökör*) following. In addition, Babur created a new inner army or *khasah tabin* (Persian: *tabin-i khasah*) consisting of so-called *yiqitlar*: individual warriors without pedigree who had no following of their own.[88] This guard corps comes very close to Timur's personal *tümen*, mentioned above. Moving from Samarqand to Kabul and Kabul to Delhi, the proportion of these trained servants would only increase. Not surprisingly, Babur's early Mongol allies, many of them of real or imagined Chinggisid descent, detested the growing importance of *tarbiyat* or 'training' against *nasab* or pedigree. Of course, they knew perfectly well that *tarbiyat* had always been a crucial ingredient of the Mongol warband, i.e. an important bond between a Mongol warlord and his retainers. What they deplored, though, was its further extension to actually replace Chinggisid lineage as the base of empire.[89] Overall, the structure of Babur's extended warband is quite similar to that of Timur. What seems missing, though, is a separate financial department. At this early stage of state-formation financial positions are still hidden behind the military façade of the warband.

MUGHAL MANSAB

Turning to Babur's son, due to his ongoing campaigns Humayun (r. 1530-56) was, like his father, hardly in a position to expand the warband into an imperial structure. What is new, however, is that he was very able to design his God-given empire and construct its organization on the basis of some highly esoteric cosmological and alchemical reasoning as revealed in his dreams.[90] In the *Qanun-i*

---

[88] A.R. Khan, 'Gradation of Nobility under Babur', *Islamic Culture* 60 (1986), 79-88.

[89] A. Anooshahr, 'Mughals, Mongols, and Mongrels'.

[90] A. Afzar Moin, *The Millennial Sovereign: Sacred Kingship and Sainthood in Islam* (New York: Columbia University Press, 2012).

Humayuni by Khwandamir, we find a whole range of original functional and hierarchical arrangements applied to the pre-existing warband organization of his father. For example, in introducing a new alchemical hierarchy of twelve golden arrows (*tir-i mutalla'*), there are echoes of the warband's ranks, including the mentioning of the term *ichkiyan*. The term 'beg' is not used; instead, the word *mansab* (position, rank), a rather loose term that would become much more prominent and more rigidly codified under his son Akbar, was applied.

Indeed, despite the conquests of Timur, Babur and Humayun, it was only the latter's son Akbar who managed to establish a settled empire on the Indian subcontinent, one which lasted for more than two and a half centuries. In contrast to their Qing counterpart in China, the Mughal warband seems to have evaporated almost immediately upon its arrival in the plains of Hindustan. Of course, we can see some kind of continuity in Babur's personal unit of the *yikitlar* in Akbar's standing army of a few thousand *ahadis*, who were equipped with several horses and had a reputation for being excellent archers. These gentlemen-at-arms, as the late William Moreland calls them, were single men (from *ahad*, one) who had no following of mounted retainers themselves. Apart from the *ahadis*, other household troops of the Mughals were composed of large number of *ahsham*: all sorts of rag-tag foot-retainers (*piyadagan*) comprising clerks, runners, gate keepers, palace guards, couriers, swordsmen, wrestlers, slaves, palanquin bearers, etc. These included infantrymen consisting of a few thousand musketeers (*banduqchis*) commanded by 'captains of ten' (*mir-dahs*). In addition, there was the important elephant unit, as well as the imperial stables for war-horses and pack-animals. The artillery was also part of the imperial household, as was also the case in the Ottoman and Safavid armies. Last but not least, probably inspired by the Safavid example, Akbar decided to start a kind of imperial Sufi order of his own. This *din-i ilahi* was meant for those amirs whose devotional service to Akbar required them to give up their life (*jan*), property (*mal*), religion (*din*), and honour (*namus*) for which in return they received Akbar's seal and portrait. Although only a few amirs became actually disciples of Akbar, the general way of expressing one's loyalty and service to the emperor became very much influenced by the Sufi idiom of love and devotion, even in the emperor's absence seeking his proximity in their dreams. Although slavery already evoked the notion of religious devotion for Akbar this was not enough:

His Majesty, from religious motives, dislikes the name *banda*, or slave; for he believes that mastership belongs to no one but God. He therefore calls this class of men *chelas*, which Hindi term signifies a faithful disciple.[91]

Despite these *chelas*, what is strikingly missing in the household segment of the Mughal army is a substantial number of military slaves. Military slavery had been an important element during the early Delhi Sultanate but began to disappear again during the fourteenth century. Hence, the Mughal *chelas* should not be mixed up with the mamluk phenomenon of Ottoman janissaries or Safavid *ghulams*. The *chelas* were only few in numbers and held no regular military functions.[92]

In numbers and weight, Akbar's inner army was dwarfed by the new *mansabdari* establishment which was ideally suited to exploit the huge military labour market that was India. Akbar had introduced this new scheme, where hierarchy was established by means of ranks or *mansabs*, and it had become necessary because rank and remuneration had to be linked to a pre-existing revenue system based on the so-called *iqta* (later, under the Mughals, called *jagir*), in which every holder of an *iqta* had the right to collect the revenue of an assigned piece of land in exchange for certain military or administrative services. In principle, the idea was not connected to the equal distribution of large tracts of conquests, as had been the case under the *tümen*, but involved, in theory at least, conditional and temporary rights to accurately-assessed revenue proceeds (*jama*). Thus, by individualizing the exploitation and redistribution of the agrarian resources as controlled by local imperial officials, the *iqta*-system provided the

[91] Abul Fazl Allami, *The A-in-i Akbari*, vol. 1, trans. H. Blochmann and H.S. Jarrett, 2nd revised edition by D.C. Phillott and J. Sarkar (Delhi, rpt. 1989; 1st published 1927-49), 263-4; Persian text, vol. 1, ed. H. Blochmann (Calcutta: Bibliotheca Indica 58, 1872-7), 427-9. For an insightful recent study, see W.R. Pinch, 'The Slave Guru: Masters. Commanders, and Disciples in Early Modern South Asia', in J. Copeman and A. Ikegame (eds), *The Guru in South Asia: New Interdisciplinary Perspectives* (London and New York: Routledge, 2012), 64-80.

[92] Gommans, *Mughal Warfare*, 49, 61, 83.

ruler with a sophisticated system of top-down and bottom-up control. To make it work though, the army had to be accordingly organized to match the system. Hence, each military unit with a direct relationship to the ruler, from the great warlord to the individual horseman, received temporary rights to collect the revenues in a defined territory.

The *mansabdari* establishment was managed by a separate financial department which also administered the distribution of *jagirs*. Like the departments for the army and the revenue, the imperial household kept its own administration, which also managed the crown-domain (*khalisa*). Although some functional overlap continued, a division of three separate elements had emerged: *manzil* (house), *sipah* (army) and *mulk* (land).

Tracing the location of our warband, it may still be seen in the ongoing role of the household to bring officers close to the emperor and to transform many of them into loyal sons of the house, or *khanazadas*. All major imperial officers were bestowed with a *mansab* by the emperor himself. *Mansab* had a dual capacity and represented both real military strength (i.e. the number of troops) and the emperor's personal favour. So on paper, one could still imagine *mansabdars* as 'begs' of their own subsidiary warbands. At court *mansabdars* rotated as guards, received all kinds of personal honours, and participated in elaborate rituals. All this symbolized their incorporation into the emperor's personal body politic. Although this may remind one of the inclusive characteristics of the erstwhile nomadic warband, the context its structure and spirit had significantly changed. Much like the usual imperial guards in other sedentary empires such as that of the Ming and the Romanovs, the unit of the *ahadis* was a relatively isolated inner army which hardly interacted with the other imperial departments and did not serve as a training ground for the *mansabdars*. The latter increasingly operated as autonomous landlords (*zamindars*) and war-jobbers in an increasingly commercialized military labour market. The divergence between pen and sword was no longer contained within the warband but covered the entire empire and operated in accordance under its own sets of rules. With the further expansion of the empire the *manzil* increasingly lost its hold on both *sipah* and *mulk*.

The later emperor Aurangzeb (r. 1658-1707), often considered the last of the 'Great Mughals', was rightly alarmed about this situation and consciously attempted to increase the crown domain at the cost of the *jagirdari* establishment. In the late seventeenth century, this triggered the so-called *jagirdari* crisis, giving rise to the failed incorporation of the Maratha gentry into the imperial system, which, for the first time, signalled that the empire had finally reached its limits. As previously mentioned, Aurangzeb's move towards the 'old ways' of Islam compares quite nicely with similar 'rescue missions' under the later Qing to counter an analogous 'bloating' of the ranks.[93] In both cases, this kind of imperial midlife crisis was typical for a (semi-)nomadic warband that had crossed over into a prosperous sedentary empire.

RUSSIA: FROM TATARS TO COSSACKS

MUSCOVY: DVOR AND BOYARS

Russian historiography was dominated for a very long time by Marxist historians for whom class was the most important analytical category. In the past two decades, however, the situation has radically changed, and there is now more room for other interpretations of Russian history. What remains, however, is the debate about whether it is part of European or Asian history. Another aspect of this debate is the question of continuity and change. Most contemporary historians of Russia stress the Byzantine legacy and see an internal continuity between Kievan and Muscovite institutions which should be studied in connection and comparison with the West (e.g. Rüss). Within this debate there is ongoing discussion about Russia's famous autocracy: did it really result in Europe's weakest aristocracy or, as is the viewpoint of recent revisionism, was it merely a façade to cover up a general consensus among the elites (e.g. Kollmann). A further, neglected part of this discussion is the work, both older and more recent, of historians that stresses a major historical break, and a consequent change, as a result of Russia's entanglements with the Mongols and Tatars (e.g. Ostrowski). Here, the debate focuses on whether or not certain Russian

---

[93] Terms taken from Elliott, *Manchu Way*, 306, 351.

institutions such as the *duma* or the *zemskii sobor* are European, Mongolian, or more indigenous phenomena.[94] This rather essentialist discussion could indeed gain from the notion of 'façade' which has already proven its usefulness when dealing with Russia's autocracy. Here it may be useful to be reminded of Korea's submission to the Mongols, which was all about outward appearance when, to gain Mongol acceptance and forestall more onerous demands for material and manpower, King Ch'ungnyŏl (r. 1298-1308) of Koryŏ donned Mongol garb, cut his hair in the Mongol fashion, and acquiesced to Mongol demands to rename the Koryŏ administrative bureaux to reflect their subordinate status within the empire.[95]

As a non-specialist and relative outsider to Russian historiography, I can but highlight some important contrasts with other regions. Firstly, in comparison with China and India, the economic conditions of Muscovy before Peter the Great remained extremely poor. Even after Muscovy had 'gathered' the other Russian lands (Novgorod, Tver and Ryazan) under its fold, funds, whether by taxing or plundering the population, were still insufficient to support a more substantial court (*dvor*), officially based in Moscow, but which for purely logistical reasons continued to lead a peripatetic existence. Due to the limited surpluses provided by Russia's poor forest land, it was not only the court that could hardly survive, but the aristocracy with hereditary estates (*votchina*) also found it hard to strike root and build strong local power centres, the more so because of the practice of divisible inheritance, which led to the multiplication of title-holders and the ongoing division, and consequent reduction in size, of estates between siblings. The lack of hereditary titles associated with posts,

[94] See D. Ostrowski, *Muscovy and the Mongols: Cross-cultural Influences on the Steppe Frontier, 1304-1589* (Cambridge: Cambridge University Press, 1998). Ostrowski builds on the earlier work of George Vernadsky and Charles Halperin (going back to the founder of the Eurasian school, Prince E. Trubetskoi). The latter, though, raised doubts about Ostrowski's major conclusion; see C.J. Halperin, 'Muscovite Political Institutions in the 14th Century' and Ostrowski's response in 'Muscovite Adaptation of Steppe Political Institutions: A Reply to Halperin's Objections', *Kritika: Explorations in Russian and Eurasian History* 1 (2000), 237-57, 267-304.

[95] Robinson, *Empire's Twilight*, 59.

or of hereditary posts of any kind, also considerably weakened the aristocracy and impeded its consolidation into an autonomous political group as was the case in Western Europe. Although in practice men became *boyars* because they were born into hereditary *boyar* families and lived to inherit that rank at court, in principle the title was not hereditary as only the family might inherit a claim to be promoted to that rank.[96] Seeking alternative sources of income, these *boyars* depended, more than any other contemporary aristocracy in Eurasia, on imperial service. But since the court itself lacked funding, it could not pay a standing cavalry army of *boyars* and so service was paid by service land or *pomest' ia*. As in the case of the Mughal *jagir, pomest' ia* was not regarded as the private property of the *pomishchik*, but it merely provided the fixed income for his maintenance and equipment, and he was not expected to concern himself with its exploitation. According to Isabel de Madariaga, 'he was not therefore a landowner in the Western sense of the word, but a land user entitled to a certain income from the land.'[97] Interestingly, as in the case of the Mughal *mansab, pomest' ia* was linked to an elaborate ranking system called *mestnichestvo*, a code of precedence which registered ranks of service. As a result, the service elites as a whole, originally of mixed social origins, were gradually sorted into those (a) who served the Grand Prince directly, as members of the *dvor*, received *pomest' ia*, and were merged into the *dvoriane*, or future servicemen, and (b) those who had served local princes and boyars and continued to carry out their service as *pomeshchiki* on a provincial basis. Despite the parallels with the Mughals, the socio-economic context of the two systems could not have been more different. Whereas in Mughal India ranking was meritocratic in principle and based on a sophisticated monetary economy, *mestnichestvo* was principally about family precedence. In other words, *pomest' ia* was no more or less than a convenient solution for the court's lack of monetary resources. As family was the main

---

[96] I. de Madariaga, *Ivan the Terrible: First Tsar of Russia* (New Haven: Yale University Press, 2005), 15. Cf. N. Shields Kollmann, *Kinship and Politics: The Making of the Muscovite Political System, 1345-1547* (Stanford: Stanford University Press, 1987), 56.

[97] Madariaga, *Ivan the Terrible*, 86.

ingredient of *mestnichestvo*, it could not be as easily manipulated by the ruler as *mansab*, which also explains why it was abolished in 1682 to give the ruler more leeway to appoint in accordance with ability and experience. [98]

In sum, from its very beginning, the Grand Duchy of Muscovy had a splintered and factionalized nobility that depended on service to the state, which was funded by *pomest'ia*. There was no room whatsoever for extensive military households – only very few served in the retinue (*druhzina*) of the ruler at the *dvor*; most of the other military elites were local landholders who were only summoned at times of danger. Hence, the *boyars* were a landed group of elites whose position lay somewhere between the rooted European nobility and the more mobile Mughal *mansabdars*. Overall, there seems to be little doubt that, in comparison to their European and Asian counterparts, the aristocracy of Muscovy was seriously abased.[99]

THE CRIMEAN WARBAND

Although its difficult ecological condition made Muscovy's court and aristocracy relatively poor, they also saved it from permanent nomadic conquest.[100] As we have seen already, Muscovy simply lacked the

---

[98] G. Alef, *The Origins of Muscovite Autocracy: The Age of Ivan III* (Berlin: Otto Harrassowitz, 1986), 222; G. Alef, 'The Crisis of the Muscovite Aristocracy: A Factor in the Growth of Monarchical Power', *Forschungen zur Osteuropäischen Geschichte* 15 (1970), 15-59; G. Alef, 'Reflections of the Boyar Duma in the Reign of Ivan III', *The Slavonic and East European Review* 45 (1967), 76-123; H. Rüss, *Herren und Diener: Die soziale und politische Mentalität des Russischen Adels. 9.-17. Jahrhundert* (Köln: Böhlau Verlag, 1994). See also the works of A.M. Kleimola: 'The Changing Face of the Muscovite Aristocracy. The 16th Century: Sources of Weakness', *Jahrbücher Geschichte Osteuropas* 25 (1977), 481-93; 'Up through Servitude: The Changing Condition of the Muscovite Elite in the Sixteenth and Seventeenth Centuries', *Russian History/Histoire Russe* 6 (1979), 210-29; 'Military Service and Elite Status in Muscovy in the Second Quarter of the Sixteenth Century', *Russian History/Histoire Russe* 7 (1980), 47-64.

[99] R. Hellie, 'Thoughts on the Absence of Elite Resistance in Muscovy', *Kritika: Explorations in Russian and Eurasian History* 1 (2000), 5-20.

[100] Which does not mean Moscow was not incidentally razed by nomadic

resources and the open landscape of northern China or northern India which facilitated the morale and logistics of (semi-)nomadic cavalry armies. As far as the landscape is concerned, though, this was not the case for the extensive Pontic Steppes beyond Russia's southern frontier, in the area now known as Ukraine, from *ukraina*, literally meaning 'borderland' or 'march' in Russian. Hence, this region was part of the early Mongol conquests which destroyed Kiev, the city which for centuries had been the main cultural and political centre of Russia. Devastated by the nomadic invasion, it actually became the new heartland of one the four Chinggisid successor empires, the so-called *ulus* of the Golden Horde, or Qipchaq Khanate. Although profiting from its commercial links with the Black and Mediterranean Seas, the area appears to have experienced a catastrophic decline as a result of both epidemics and the destruction of its economic bases at Azov, Astrakhan and Urgench under Timur's campaigns in the 1390s.[101]

From the second quarter of the fifteenth century the Golden Horde fragmented into various successor states which created a new balance of power among five political entities, each of medium economic and political might: (1) Muscovy itself; (2) the Crimean Khanate, nominally under the Ottomans after 1745; (3) the Great Horde, replaced by the Astrakhan Khanate in 1502; (4) the Kazan Khanate; and, finally, (5) the Tiumen' Khanate or Sibir. In fact, the four Tatar Khanates should not be considered fully nomadic states but instead, like so many polities in the Middle Zone that have been discussed so far, as a mixture of nomadic and agricultural communities under the leadership of so-called 'begs'. The term Tatar itself did not necessarily refer to nomads but denoted almost all non-Christian inhabitants, mostly with a Turkish background and of Muslim faith. In any case, at the interstices of these enclaves of political centralization and mixed economies, there were truly pastoral groups such as the Nogais, Bashkirs, Kalmyks and Kazakhs. Muscovy's military successes

---

armies: in 1237 by the Mongols, in 1382 by the Golden Horde, in 1571 by the Tatars of the Crimea.

[101] E.L. Keenan Jr., 'Muscovy and Kazan: Some Introductory Remarks on the Pattern of Steppe Diplomacy', *Slavic Review* 26 (1967), 553.

against Kazan and Astrakhan was in no small degree based on its alliance with these groups, and in particular the Nogais, who delivered crucial transport services and, above all, a large number of excellent warhorses. Coming from Kazakhstan, the Tibetan Buddhist Kalmyks arrived in the Volga plain only in the 1620s. Their migration instigated the building of the 800 km long Belgorod Line, the first of a long series of south Russian lines of fortification which started to constrict the available grazing space of these pastoralists and actually forced the Kalmyks to move back eastward again in the 1770s, where they were incorporated not in the Russian but in the Qing military system.[102]

Focussing on the political organization of the Tatar khanates it is remarkable that they, like the other parts of the Middle Zone discussed above, more or less retained the structure of the Turko-Mongolian warband. Even in the details, the system of the Crimean Khanate under Mehmed Giray (r. 1514-23) and Sahib Giray (r. 1532-51) comes very close to that of the early Timurids, including Babur. In elite society, there was a hierarchy at the top of which was the Khan himself, who was followed by his immediate family members: his eldest sons, other sons, and members of the royal family called *oghlans*. Many of them were sultans with positions of command. As well as kith and kin, there were the so-called begs, who were referred to as *qaracu* or heads of the four principal 'tribes', their sons or *mirzas*, and other so-called *nökörs*, who were also known as *emeldesh* or *icki* (insider) begs, all of whom served the Khan as intimate comrades in the court and elsewhere. Also very much along the lines of Turko-Mongol tradition, the so-called four 'tribes' were not strict ethnological categories but rather consisted of mixed groups in the manner of those discussed earlier in the case of the Manchu *mukun* and Timurid *tümen*. In principle, as much as *nökör* overrode tribe, the position of client overrode kinship, with the latter becoming merely fictitious and/or used metaphorically.

Interestingly, in 1533 Sahib Giray attempted to end his dependence on his traditional 'tribal' following by introducing the Ottoman model of the janissaries through the enlistment of a number of musketeers

---

[102] Khodarkovsky, *Russia's Steppe Frontier*, 132-3.

from outside these circles, from 'among the rabble of the people'.[103] As shall be seen, the Muscovite rulers would do exactly the same, at first in the form of the *pishchal'niki* or harquebuzzers, and later, from 1550, by the so-called *strel'tsy* who replaced them, and who were also musketeers in the model of janissaries, operating on foot but, unlike the latter, consisting of free men instead of slaves.[104] These personal guards, recruited from the lower levels of society, provided an attractive new counterbalance either against the *nökörs* in the case of the Crimean Khan or against the *boyars* in the case of the Tsar. It should be stressed here that although their personal nature made these guards look like the warband's *nökörs*, in principle they are different as they neither provided an imperial elite nor represented the empire's elite as a whole. As with other imperial guards in sedentary empires, they were a much more isolated and purely military force than the various forms of the nomadic warband that we came across in our northern Middle Zone.

THE WHITE TSAR

One of the Golden Horde's successor states was Muscovy itself. The Nogais called its ruler the *White* Tsar; in Mongol idiom the colour indicated its western location within the Mongol realm.[105] There are many reasons to link Muscovy's rise to empire with the emergence of the Golden Horde. For example, economically speaking, it was only within the framework of the economic demands and opportunities

[103] H. Inalcik, 'The Khan and the Tribal Aristocracy: The Crimean Khanate under Sahib Giray I', *Harvard Ukrainian Studies* 3/4 (1979-80), 445-66; see also B.F. Manz, 'The Clans of the Crimean Khanate, 1466-1532', *Harvard Ukrainian Studies* 2 (1978); U. Schamiloglu, 'The *Qaraçi* Beys of the Later Golden Horde: Notes on the Organization of the Mongol World Empire', *Archivum Eurasiae Medii Aevi* 4 (1984), 283-97; C. Kennedy, 'Fathers, Sons, and Brothers: Ties of Metaphorical Kinship between the Muscovite Grand Princes and the Tatar Elite', *Harvard Ukrainian Studies* 19 (1995), 292-301.

[104] For a discussion of the Ottoman background of these new musketeer units, through the agency of Ivan Semonovich Peresvetov, see Madariaga, *Ivan de Terrible*, 87-91.

[105] M. Khodarkovsky, 'The Non-Christian Peoples on the Muscovite Frontiers', in Maureen Perrie (ed.), *The Cambridge History of Russia*, vol. 1: *From Early Rus to 1689* (Cambridge: Cambridge University Press, 2006), 323.

created by the later Golden Horde that north-eastern Russia could recover from the earlier devastation created by the Mongols earlier.[106] Likewise, in political terms, the Khan of the Golden Horde confirmed the Grand Princes of Muscovy as legitimate but subordinate rulers. In addition, Ivan I (r. 1325-40) received a Mongol commission of gathering the tribute for the Khan from other Russian princes.[107] We also know that from the fourteenth century onwards, the Grand Princes relied on Tatar manpower to expand their territory further west, towards Tver and Novgorod. Hence, their armies consisted primarily of mounted archers with composite bows. The leaders of these Tatar forces were often members of the Chinggisid dynasty, and as Chinggisid *tsarevichi* they not only gave power but also authority to the Great Prince of Muscovy. Two sons of Ulug-Mehmed, the Khan of the Golden Horde, became vassals of Grand Prince Vasily II (r. 1415-62) and founded, under Russian protection, the Khanate of Kasimov, and its Chinggisid ruler helped Vasily II to regain his throne in the Russian civil wars of the mid-fifteenth century. Ivan IV, also known as *grozni*, literally the 'awesome', but better known as: the Terrible (r. 1547-1584), grew up among Chinggisid princes and even abdicated for a year in favour of one of them, who became known as the Grand Prince of Russia Symeon Bekbulatovich.[108]

Indeed, it seems that Ivan IV had a crucial decision to make: should he opt for the Polish-Lithuanian or the Tatar model, perhaps now in

---

[106] M. Perrie, 'North-eastern Russia and the Golden Horde (1246-1359)', and 'The Emergence of Moscow (1359-1462)', in *The Cambridge History of Russia*, vol. 1, 127-87.

[107] N.V. Riasanovsky, *A History of Russia* (Oxford: Oxford University Press, 1993), 97-101.

[108] Madariaga, *Ivan the Terrible*, 298-311; 437. One possible explanation for this act lies in the Indian story of Barlaam and Josaphat. Following that story, the physician and magus Bomelius had produced a horoscope for Ivan, showing that in the years 1575-6 a Grand Prince of Russia was to die. All this was confirmed by a comet and a planet conjunction. A similar temporary abdication happened in 1594 when the Safavid emperor Shah Abbas I abdicated for three days after his chief astrologer Jalal al-Din Munajjim Yazdi foretold the death of an Iranian ruler (S.P. Blake, *Time in Early Modern Islam: Calendar, Ceremony and Chronology in the Safavid, Mughal, and Ottoman Empires* [Cambridge: Cambridge University Press, 2003], 153).

its latest, Ottoman version? At the beginning of his reign, he had made important strides against the Tatars by the incorporation of Kazan (1552) and Astrakhan (1556); Sibir would soon follow in 1582. Even earlier, Muscovy had started to emancipate itself from the Tatar embrace after beating them at the Field of Kulikovo in 1380 under Dmitri Donskoi. A century later, after another confrontation on the river Ugra under Ivan III, Muscovy ceased to regard the Tatars as its overlord. But this did not mean their institutions immediately lost their attraction, particularly since the alternative models of the West were not yet considered good enough. Ivan IV once complained that he often felt like the Polish king: excluded from power and ordered about by his servants, a ruler in name but not in fact. In other words, the first option involved close cooperation with his increasingly powerful service *boyars*, which could itself only lead to the loss of his own position.

Whatever he decided, in late 1564 he suddenly abandoned Moscow for the small town of Aleksandrov. He abdicated, denouncing both the *boyars* and the clergy. Both begged him to return, which he did, but only on his own conditions: the creation of a special subdivision in the state, known as *oprichnina*, consisting of about one third of the realm, to be managed entirely at his own discretion. At the same time he expected an endorsement of his right to punish evil-doers and traitors as he saw fit. For our present purpose it is crucial to note that the term *oprichnina* also came to designate a new corps of about 1,000-6,000 personal servants called *oprichniki* – all dressed in black and riding black horses. It is hard to understand Ivan's motivations for what seems to be a desperate move away from the court and its 200 or so *boyar* families.[109] In a way it is reminiscent of Emperor Cheng-te's move to the so-called Leopard Quarter about half a century earlier; similarly to Ivan, the Cheng-te emperor decided to establish a brand new palace outside the established court so that he could conduct business on his own terms without the interference of civil officials. Here he followed his plan to revive the Ming military without the obstruction and criticism of officialdom. He surrounded himself

---

[109] Riasanovsky, *History of Russia*, 150-5. See also G. Opeide, 'Making Sense of Opričnina', *Poljarnyj Vestnik: Norwegian Journal of Slavic Studies* 3 (2000), 64-99.

with a court of his own choosing, comprised in large measure of foreigners and military men who, under his direction, began to retrain the imperial armies.[110] Although very similar in motivation, in China the emperor's sudden exodus did not have serious consequences for the administration as a whole. While the emperor's behaviour may have been interpreted as unconventional or even outlandish, business generally went on as usual. This was not the case in Russia, where the *oprichnina* unleashed a veritable reign of terror which actually increased the Tsar's hold on the princes and the service elites who had gradually turned into an obstreperous hereditary aristocracy.

So should the *oprichnina* be regarded as the Tatar option? Isabel de Madariaga makes a strong argument they may have been inspired by Ivan's Tatar wife Maria Temriukovna, a niece of the later, 'mock' Prince Symeon Bekbulatovic. Indeed, her brother, Prince Michael Cherkassky – another member of what seems to be an incredibly influential Qipchaq 'tribe' – became an important leader of the *oprichniki*.[111] Whatever its inspiration, the *oprichnina* represents a fascinating attempt to start an empire from scratch with a brand new personal warband. In Russia, though, it could never expand in the way it had done under Chinggis Khan, Babur or Nurhaci.

After Ivan's death, the struggle between the old and new elites continued and opened the door to the regency of Boris Godunov, who even became Tsar in 1598. After defeating both the high-born princelings and the low-born *oprichnina* guard, he consolidated the *boyar* elite, not in their landed holding but by offering positions in the expanded chancellery secretariat. His policies opened the way for the emergence of the Romanov family, but only after they had survived the last attempt to impose a Tatar-like warband on Russia during the Time of Troubles.

A Retro Warband: The Cossacks

In many ways the Time of Troubles (1603-13) was part of the more global seventeenth-century crisis. Muscovy experienced the almost

---

[110] J. Geiss, 'The Leopard Quarter during the Cheng-te Reign', *Ming Studies* 24 (1987), 1-38.

[111] Madariaga, *Ivan the Terrible*, 186-8.

natural after-effects of a relatively stable sixteenth century, which had seen economic stability, a doubling of the population and staggering territorial expansion. This was followed, though, by increasing demographic pressures that caused an economic crisis that had its beginnings in the long Livonian War (1558-82) in the West and a series of natural disasters at the turn of the century. Very much like the economy, political troubles at the early seventeenth century naturally followed an earlier period of political expansion, in particular into the southern steppes, which brought millions of new inhabitants into the empire, many of whom were not yet settled as peasants and were still moving around in search of a living.

The incorporation of the southern steppes had been made possible by two policies. One was the building of fortified lines (*zaseki*) that warded off nomadic raids and also sealed off their nomadic trails. The other was the support of highly mobile Cossack warbands which could fight the Tatars on their own terms. It is difficult to make a clear difference between these Cossacks and Tatars. Both were often organized in open, conscriptive groups of wandering traders-cum-warriors. The origin and background of the Cossacks was, however, different, as it incorporated mainly Slavic immigrants from the north, many of whom had a peasant or serf background, and most of whom were Christians. As such, these were frontier guardsmen who combined raiding with trade, but increasingly also agriculture. Tatars, on the other hand, often had a pastoralist Turkic background and were primarily Muslims. Michael Khodarkovsky sees the Cossacks as the 'mirror image' of their nomadic adversaries who, like them, chose to live off booty and pillage.[112] On the southern frontier, both groups merged into the hybrid Cossack category, standing somewhere midway between such real pastoralists like the Nogai and the more 'urban' enclaves of the Tatar Khanates.[113] In terms of spirit and structure, the Cossacks had all the elements of the nomadic warband that have been noted for the Jurchens and Turks *before* their Chinese and Indian

[112] M. Khodarkovsky, *Russia's Steppe Frontier: The Making of a Colonial Empire 1500-1800* (Bloomington and Indianapolis: Indian University Press, 2002), 224.

[113] G. Stökl, *Die Entstehung des Kosakentums* (Munich: Isar Verlag, 1953).

conquests. In Russia, however, it always remained a pre-conquest, frontier phenomenon but, as such, consciously stimulated, manipulated and controlled by the court. Hence, as in the case of the Banners, we have a sedentary empire recreating and exploiting a subsidiary warband to implement imperial control. The Russian case is different, though, because the Cossack warbands were not used as an internal pool of loyal administrators but were projected outward to the frontier to fight other warbands and to expand agriculture. Russian rulers knew very well that in the end it was the plough, not the sword, which would win the steppe, as much as 'the chicken would prevail over the horse'.[114] Obviously, this would also be the natural end of the nomadic warband and, with that, the Cossack way of life.[115] We should not forget, though, that during the Time of Troubles this outcome was far from foreseeable and it could all have been quite different if only the Cossacks had won the day.[116]

The opening of the southern frontier had caused an increasing influx of Tatar warriors into the empire. Some high-ranking Tatars became close satellites of the court, receiving various *pomest'ia* or even towns for their *kormlenie* (feeding).[117] More important than

[114] E.L. Keenan Jr., 'Muscovy and Kazan: Some Introductory Remarks on the Pattern of Steppe Diplomacy', *Slavic Review* 26 (1967), 557.

[115] B.J. Boeck, *Imperial Boundaries: Cossack Communities and Empire-Building in the Age of Peter the Great* (Cambridge: Cambridge University Press, 2009).

[116] M. Perrie, 'The Time of Troubles (1603-1613)', in *The Cambridge History of Russia*, vol. 1, 409-31; M. Perrie, *Pretenders and Popular Monarchism in Early Modern Russia: The False Tsars of the Time of Troubles* (Cambridge: Cambridge University Press, 1995); R.G. Skrynnikov, *Times of Troubles: Russia in Crisis* (Gulf Breeze FL: Academic International Press, 1988); C.S.L. Dunning, *Russia's First Civil War: The Times of Troubles and the Founding of the Romanov Dynasty* (Pennsylvania: Pennsylvania State University Press, 2001).

[117] D. Ostrowski, 'The Growth of Muscovy (1462-1533)', *The Cambridge History of Russia*, vol. 1, 213-39; D. Ostrowski, 'Troop Mobilization by the Muscovite Grand Princes (1313-1533)', in E. Lohr and M. Poe (eds), *The Military and Society in Russia, 1450-1917* (Leiden: E.J. Brill, 2002), 19-40; J. Martin, 'Tatars in the Muscovite Army during the Livonian War', in E. Lohr and M. Poe (eds), *The Military and Society in Russia, 1450-1917* (Leiden: E.J. Brill, 2002), 365-87; B.R. Rakhimzyanov, 'The Muslim Tatars of Muscovy and Lithuania: Some Introductory Remarks', in B.J. Boeck, R.E. Martin and

these Tatars, though, was the increasing Cossack element in the Russian army, mainly from the southern frontier. When economic conditions went from bad to worse at the end of the sixteenth century these Tatar and Cossack warbands could no longer be paid. During this time their numbers even increased as peasants and serfs moved away from the forest heartlands to find refuge in the south, where they were incorporated by some very charismatic Cossack warband leaders or *atamans*; as Chester Dunning so aptly expresses, more and more Russians 'went cossack'.[118]

One of the Cossack leaders was a certain Ivan Isaevich Bolotnikov. He was captured by the Turks and served as a galley-slave before escaping and returning to Russia through Poland. Another colourful figure was one Ivan Martynovich Zarutskii who, having been taken prisoner by the Crimean Tatars, escaped to the Don Cossacks, where he gained a reputation for exceptional bravery, and subsequently joined Bolotnikov's forces. Both of them could have been a Babur or a Nurhaci, carving out empires of their own, had they not lived in Russia. Instead of claiming the throne for themselves, they pushed the claims of an endless series of Tsarevich pretenders. And although they seriously threatened Moscow, they never managed to take it and in the end were overhauled by the so-called 'national militias' who, in 1613, picked Michael Romanov as their new Tsar. After the repression of the Cossacks, the latter increasingly served as mercenaries under *boyar* command or were pushed further along the basins of the rivers Don, Volga, Terek and Iaik towards the southern frontier. Here they began to set up their own organization which made them more answerable to the central authorities in Moscow until, in the eighteenth century, their leaders started to owe their authority not to their constituency but to the Tsar.[119] At the same time, however, they continued to stir social unrest and revolts. One of these revolts, under Bogdan Khel'nitskiy (1648-57), was directed against Poland-Lithuania

D. Rowland (eds), *Dubitando: Studies in History and Culture in Honor of Donald Ostrowski* (Bloomington: Slavica Publishers, 2012), 117-28.

[118] C.S.L. Dunning, 'Cossacks and the Southern Frontier in the Times of Troubles', *Russian History* 19 (1992), 59.

[119] Boeck, *Imperial Boundaries*, 190.

and enabled Muscovy to incorporate many of the Ukrainian Cossacks into its own army and large parts of the Ukraine into its territory. Much more of a threat to Muscovy itself were the revolts of Sten'ka Razin (1670-1) and Emelian Pugachev (1773-5), but now all the unrest was happening far from the capital and it was clear to all that the sedentary way of life had definitively come to stay in Russia, not only in the forest belt but also in the steppes.[120]

Why did Bolotnikov and Zarutskii fail where Babur and Nurhaci had succeeded? Why could their warbands not serve as the basis of conquering and sustaining an empire? The answer brings us back to Russia's ecological and economic conditions. Both Tatars and Cossacks simply lacked the resources to sustain the necessary initial expansion of the warband that was noted in the case of Babur and Nurhaci, who could take as much as they liked from the almost endless resources of the Chinese and Indian economies. Furthermore, their paths had been paved by their Jurchen and Afghan predecessors. The Cossack warband was never able to become large enough to beat the better organized and paid Russian army under *boyar* command; however, in the early seventeenth century the Russian victory was far from decisive and the outcome could still have been different.

What really contained the Cossacks in the following decades and centuries was, however, the gradual move towards a new, European-style infantry army supported by a growing administrative apparatus of various chancelleries or *prikazy*, the most important of which was the Military Service Chancellery (*Razriad*) established under Alexis (r. 1645-76). Indeed, the increasing number of *prikazy* gave new opportunities to *boyars* whose main task remained primarily the provisioning and funding of the army. All this had become necessary not because of threats from the Tatars and Cossacks, but in order to fight similar armies that served the Poles and Swedes on the western front. It was Peter the Great (r. 1682-1721) who tried to use this opportunity to make himself more autonomous from his *boyars* – a

---

[120] P. Longworth, *The Cossacks* (London: Constable, 1969). For a Ukrainian perspective, see L.Gordon, *Cossack Rebellions: Social Turmoil in the Sixteenth-Century Ukraine* (Albany: State University of New York Press, 1983).

desire that had already been seen under Ivan IV when he started the *oprichnina*.

Many of the reforms were implemented by German and other foreign officers. At the same time, it is important to stress that the Russian army never became a mercenary army as was the case in other European countries. It actually became increasingly Russian, increasingly based on conscription. Even non-Russian recruits, like the Ukrainian Cossacks, became increasingly Russified. Although *pomest'ia* was turned into *votchina*, both became subjected to a military service requirement as regulated by new censuses. The huge increase in the size of both the army and *prikazy* was paid for by the levying of a poll-tax on the increasingly fixed and defined population. The latter grew out of the famous 1649 *Ulozhenie*, which had codified serfdom as a compromise with the provincial middle service class cavalry who had become more and more concerned about the viability of their estates. Meanwhile, in the tradition of the *strel'tsy* but starting from scratch, Peter had created his own personal guard. In his adolescence Peter had gathered around himself the sons of *boyars* and organized them into two regiments of boys who fought mock battles under his command. These so-called toy regiments (*poteshnye polki*) were named Peobrazhenskii Polk and Semenovskii Polk after the suburbs where they originated, and became the nucleus of his new army.[121] This practice was later repeated by Peter III's Holsteiners and Paul's Gatchina regiment. At the same time, following the example of the palace schools of the Ottomans as well as the so-called *Ritterakademien* in other European polities, Peter established elite military academies which were meant to train his officers in the latest gunpowder technology and to lead them to be even more closely attached to the Tsar. Access to these guards and academies increasingly defined one's status.[122]

---

[121] A.J. Rieber, *The Struggle for the Eurasian Borderlands: From the Rise of Early Modern Empires to the End of the First World War* (Cambridge: Cambridge University Press, 2014), 201-2.

[122] For the developments of the seventeenth and eighteenth-century nobility and army, see R.F. Hellie, *Enserfment and Military Change in Muscovy* (Chicago: University of Chicago Press, 1971); R.O. Crummey, *Aristocrats and Servitors: The Boyar Elite in Russia 1613-1689* (Princeton: Princeton University Press,

Indeed, after Peter, and in particular after Catherine II (r. 1762-92), officers of the Guard Regiments increasingly started to build loyalties with their fellow officers and created a new class of military intelligentsia. Even ordinary soldiers became more tied to their regiments, as their children were forced to enter garrison schools that were meant to instil some basic education, and, above all, discipline and exercise. This horizontal kind of corporatism makes one historian even use the term Praetorianism which, in the nineteenth century, even started to threaten the Tsar himself, as was the case during the 1825 Decembrist Revolt.[123]

The new corporatism, based on the Imperial Guards and Academies, once again raises questions over the warband and *asabiya*. Indeed, Peter's creation of his own personal guard, selected from the main *boyar* families and which instilled a new regimental identity, brings to mind Chinggis Khan's *keshig*. At the very moment that officer corporatism gains weight and spreads across the empire and different levels of society, we may superficially recognize something that comes quite close, if not to the Chinggisid prototype itself, to its post-nomadic derivatives under the Manchus and Timurids. Although the Cossacks clearly elaborated on the Turko-Mongolian model, Peter's new standing army seems to be something different and much closer to early modern European examples, which may have had similar functional objectives as the Chingisid warband, but apparently had a completely different, sedentary rather than a nomadic origin. In the case of Peter's guards there seems to be nothing that compares to the flexible, all-inclusive nature of the Turko-Mongolian *keshig* which encompassed both military and administrative functions and represented the elites of all the subsidiary warbands that exploited the sedentary populations through collective appanages. Indeed, the

1983); J.L.H. Keep, *Soldiers of the Tsar: Army and Society in Russia 1462-1874* (Oxford: Clarendon Press, 1985); M. Poe, 'The Military Revolution, Administrative Development, and Cultural Change in Early Modern Russia', *Journal of Early Modern History* 2 (1998), 247-73; D. Lieven, 'The Elites', in D. Lieven (ed.), *The Cambridge History of Russia,* vol. 2: *Russia 1689-1917* (Cambridge: Cambridge University Press, 2006), 225-44; W.C. Fuller Jr., 'Chapter 25: The Imperial Army', in D. Lieven (ed.), *The Cambridge History of Russia,* vol. 2, 530-53.

[123] Keep, *Soldiers of the Tsar*, 201-50.

Petrine army seems to have grown naturally from the smaller and much more isolated and restricted imperial guard of the *strel'tsy*. It was only later, during the eighteenth century, when the economic basis of the empire increased and was managed ever more effectively by an enlarged bureaucracy that the disciplining powers of this new standing army grew at an unprecedented level and really started to make a difference against the nomadic warbands to its south and east.

THE OTTOMAN MIDDLE EAST: FROM GHAZIS TO JANISSARIES

Osman's Men: Beyliq and Sanjaq

South of Russia our northern Middle Zone continues into Anatolia and to the west faces the heartland of the Byzantine Empire. At the beginning of our period, the Asian part of that empire was undergoing a process of increasing nomadization. Although starting in the eleventh century, its peak occurred two centuries later when, due to the Mongol campaigns into the Pontic Steppe and Iran, more and more Turkic pastoralists moved into the semi-dry Anatolian plateau, penetrating deep into its still sedentary western valleys under the control of the Byzantines. Meanwhile, the Seljuq Empire, a product of the first nomadic movement, was more or less crushed by the Mongol 'hammer' battering against the Byzantine 'anvil'. After a short interlude under the Golden Horde, Anatolia was under the control of the Ilkhans from their remote summer and winter quarters, in Azerbaijan and the Mughan Plain respectively. Even more so than under the Seljuqs, the Ilkhanid period bound eastern Anatolia closely to the political fortunes of north-west Iran.[124]

Along the shifting frontier between the expanding Ilkhans and the retreating Byzantines emerged various Turkic marcher lords who made a living from raiding the countryside and lending their military services to the highest bidder, whether Christian or Muslim. Some of them

[124] C. Melville, 'Anatolia under the Mongols', in K. Fleet (ed.), *The Cambridge History of Turkey*, vol. 1: *Byzantium to Turkey 1071-1453* (Cambridge: Cambridge University Press, 2009), 101.

carved out their own little polities or *beyliqs*, a few even becoming truly cultural hubs for refugee Persian intellectuals searching for new jobs after the Mongol invasion. One of these *beyliqs* was created by Osman and his band of *nökörs* – hence called the Ottomans – in the north-west of Anatolia. One of the hidden secrets behind the Ottoman success story was that, even before the fall of Constantinople in 1453, they were able to build a bridgehead across the Dardanelles which made it possible to channel the surplus of Turkish manpower from Anatolia to the Balkans and, in addition, to recruit new manpower from the latter. As a consequence, it was due to their hold on the Balkans that the Ottomans were able retain their home base in Anatolia and even to withstand the devastating campaigns of Timur at the start of the fifteenth century.[125] As was plain to see, even for later generations who romanticized the events, the prime political instrument with which the Ottomans had achieved all this had been the Turko-Mongolian warband.

Whatever one may think of the notorious ghazi-thesis as presented by the overly romantic Austrian historian Paul Wittek, there is no doubt whatsoever that the Ottomans were one of many warbands that were active in the Anatolian frontier zone.[126] What is also clear, though, is that all our sources are of a much later date than the life and times of the founder of the warband, Osman Ghazi (c. 1290-1324). Hence, it is not at all surprising that Ottoman historians, defending their patrons against the much better genealogical claims of the Timurids, have interpreted the early events in an excessively Islamic light, as a heroic fight against the infidel. Removing the anachronistic idea of a civilizational clash between Islam and Christendom, there is still much to appreciate in the concept of the ghazi warlord who in seasonal *razzias* – from *ghaziya* or raiding – creates his own band of mounted warriors and comrades (in Turkish

---

[125] H. Inalcik, *An Economic and Social History of the Ottoman Empire*, vol. 1: *1300-1600* (Cambridge: Cambridge University Press, 1994), 11-3.

[126] For a recent discussion of Wittek's ideas, see P. Wittek, *The Rise of the Ottoman Empire: Studies in the History of Turkey, Thirteenth-Fifteenth Centuries* (London: Routledge, 2012), as edited and introduced by Colin Heywood. See also C. Kafadar, *Between Two Worlds: The Construction of the Ottoman State* (Berkeley: University of California Press, 1995).

yoldash but in the sources also called *nökör*) and brings them under his banner or flag (*sanjaq*). According to the doyen of Ottoman studies, Halil Inalcik, this process of creating the Ottoman warband entailed a dissolution of kinship ties with the exception of those of the leader's family. The Holy War ideology, as much as the success of the actual raids, reinforced ties within the band to produce a cohesive social group that centred around the leader.[127] With these two sentences we are back at the models of the Chinggisids and Ibn Khaldun with which we started this essay. Interestingly, Inalcik adds that it was not the sharia-minded *ulama* but the mystical Sufi dervishes who embodied the *ghaziya* spirit and brought to the leader's authority the spiritual sanction of Islam. Here we come very close to the story of the Sufi Safavid warlords who emerged only slightly later in the same Turkic climate so typical of eastern Anatolia, Azerbaijan and northern Iran. Nevertheless, we should be on our guard by now as we simply do not know how much of the religious fervour at the very beginning of the Ottoman and Safavid movements were projected into the events ex-post-facto by both Ottoman and Ottomanist historians, eager to give them some higher purpose.

Whatever their ideological drive, looking at the various Turkish marcher lords in Anatolia in this period, we can recognize something of the Chinggisid warband. Apart from the idea of *nökörs*, it seems that at the beginning of the fourteenth century entire provinces were given to military commanders (*beys*) as their banner or *sanjaq*, which comes close to the Timurid appanage. At this early stage, *sanjaq*, very much like the Timurid *tümen*, designated a military unit without the territorial connotation that it would receive later. *Sanjaq beys* had both civil and military functions. Similarly, the word *timar*, which later became known as the individual prebend of a cavalry trooper, still indicated large grants of land to subsidiary warband leaders at full liberty to arrange matters as they saw fit. Both *sanjaq* and *timar* were meant to incorporate the regional elites into a subsidiary warband or curb them by creating new provincial commanders coming from

---

[127] H. Inalcik, 'The Question of the Emergence of the Ottoman State', *International Journal of Turkish Studies* 2 (1981-2), 71-9.

the ruler's own warband.[128] Only later in the fifteenth century did the *sanjaq bey* become a provincial official in command of a specific group of *timar*-holders and their lands, still not neatly covering the borders of a province but, at least in theory, under the surveillance of the court.

Despite such similarities, Anatolia was not Mongolia. As cogently expressed by Rudi Paul Lindner, the *beyliqs* built their power on the nomads but without having enough of them.[129] It was only by developing and exploiting the sedentary resources in Rumelia that one of them managed to become an empire. So the Ottomans were nomadic chiefs who made their fortune in areas of relative agrarian wealth. Like China in the case of the Mongols and Jurchens, the Rumelian and Anatolian resources were initially substantial enough for the Ottomans to support the persistence of the nomadic warband. As they were soon to find out, though, operating in Europe was an entirely different ballgame. European circumstances did not allow for nomadic movement on a grand scale as was still possible in the eastern Anatolian extensions of the Arid Zone. What became increasingly important in this area were the drilled operations of footmen as explored by the first mercenary powers hired by the South and Central European powers. As they were increasingly equipped with the latest gunpowder weaponry, from the fifteenth century onward infantry troops started to make real headway against cavalry, especially in the European logistical context where the size of the latter always remained limited. Although it would at least take another two centuries and another round of new gunpowder innovations before infantry really won the day, for the Ottomans, having to operate in the European arena, it became necessary to recruit soldiers among the peasantry in

---

[128] P. Fodor, 'Ottoman Warfare, 1300-1453', in *The Cambridge History of Turkey*, vol. 1, 198; L. Darling, 'The Development of Ottoman Governmental Institutions in the Fourteenth Century: A Reconstruction', in Vera Costantini and Markus Koller (eds), *Living in the Ottoman Ecumenical Community: Essays in Honour of Suraiya Faroqhi* (Leiden: E.J. Brill, 2008), 24. For later developments, see C. Imber, *The Ottoman Empire* (Basingstoke: Palgrave Macmillan, 2009), 164-203.

[129] R.P. Lindner, 'Anatolia, 1300-1451', in *The Cambridge History of Turkey*, vol. 1, 107, 121.

the Balkans. All this gave rise to the introduction of *devşirme* or 'collection' whereby the sultans levied slaves from among their own Christian subjects, employing them, not as horsemen as in the case of the Turkic Mamluks in Egypt, but as highly disciplined new infantry units called janissaries.

So although there are many reminiscences of some earlier developments in our Middle Zone, the Ottoman version of the Turko-Mongolian warband had a relatively short lifespan. Until the fifteenth century it worked well to mobilize and incorporate nomadic manpower and channel it towards the western frontiers in Anatolia and the Balkans. Soon the Ottomans found out, though, that in its original guise the warband could not operate in the sedentary surroundings of western Anatolia and south-eastern Europe. Neither was it big enough to really make an impression on their eastern Timurid, Mamluk or Safavid neighbours. Under these circumstances, investing in a combination of infantry and gunpowder technology made perfect sense.

The idea of the warband held out longest along the shifting frontier (*uj*) of the Balkans. In the guise of the so-called *akinjis*, or frontier raiders, we recognize something of the Cossack phenomenon that we have analysed already for the Russian frontier and which represented a conscious policy of the court to stimulate both settlement and territorial raiding against the enemy. In the early stages, this frontier zone not only attracted the Ottomans and other Turkic ghazis but also European crusaders. Organized in the mercenary Company of the Catalans, the latter carved out their own fourteenth-century polity in Athens.[130] Although lacking a nomadic background and thus not accompanied by their families, the Christian Catalan Company was not at all that different from the Turkish *beyliqs*. Although very effective in their own ecological niche, both lacked the nomadic *and* sedentary resources to use the warband for conquering the extensive empire as achieved by the great world conquerors Chinggis Khan and Timur,

---

[130] A.E. Laiou, *Constantinople and the Latins: The Foreign Policy of Andronicus II: 1282-1328* (Cambridge Mass.: Harvard University Press, 1972), 127-200.

or indeed by their Mughal and Manchu colleagues Babur and Nurhaci. The Ottomans had to develop an alternative strategy.

BEYOND THE WARBAND: DEVŞIRME

The Ottomans who conquered large parts of northern Africa and south-eastern Europe in the sixteenth century were not leading a nomadic warband anymore but a salaried slave army assisted by *timar*-based cavalry troops or *sipahis*. As mentioned already, using military slaves was nothing new, certainly not in the Islamic world. In the Middle East it dates back to the ninth-century Abbasids but the Seljuqs were the first to possess a substantial army of military slaves. In the mid-thirteenth century, one of its successor states, the Ayyubids of Egypt and Syria, began to recruit large number of Turkish slaves from the Eurasian steppes. After successfully beating off both the Crusaders and the Mongols these slaves created their own Mamluk dynasty and continued to recruit slaves, primarily from the Qipchaq steppes controlled by the Golden Horde.[131] In the early sixteenth century, it was this slave army, dominated by Turkish cavalry, which was beaten by the Ottoman slave army dominated by 'European' infantry and artillery.

The one element in the sixteenth-century Ottoman army that may still vaguely remind one of a Turko-Mongolian warband was the standing army of six cavalry regiments of the household (*kapikulu sipahi*), recruited from graduates of the so-called palace schools which were specifically created to train freshly purchased slaves in order that they accompany the sultan on campaign, as well as during ceremonial occasions. The most prestigious among them were the *sipahi oglanları* (cavalry youths) and the *silahtar* (armbearers), followed by the *ulufeci* (salaried men) of the right and the left wings, and the *garib yigitleri* (foreigners) of the right and left wing. From their inception sometime in the fifteenth century, their numbers increased from about 2,000 to almost 20,000 by the beginning of the seventeenth century. Yet more important than these mounted troops was the fourteenth-century

---

[131] Peter B. Golden, *An Introduction to the History of the Turkic Peoples* (Wiesbaden: Otto Hassassowitz, 1992), 348-9.

introduction of an infantry corps. Interestingly, this part of the slave corps did not consist of primarily Turkish recruits but of European Christians who formed a new-style infantry unit of the janissaries, from *yeniçeri*, literally meaning 'new army'. Initially these were taken from amongst prisoners of war, most prominently through the various marcher lords, but soon they were primarily collected, whence *devşirme*, literally 'collection', as slaves from the Christian subjects in the Balkans. Once selected, the janissary novices received an education and military training, partly at court and partly in the provinces. Before becoming a major building bloc of the imperial army, the janissaries, like the *kapikulu sipahi* corps, served as the ruler's bodyguard and sometimes, as individuals, served as the ruler's representative in the provinces. In the sixteenth century, it was primarily this impressive standing army of about 10,000-20,000 well-trained infantry soldiers, equipped with the latest gunpowder technology, that delivered the Ottomans their empire, defeating the Safavids at Chaldiran in 1514, the Mamluks at Ridaniyya in 1517, and the Hungarians at Mohács in 1526.

Until the fifteenth century, the Ottoman slave army served as a rather small and isolated imperial bodyguard in a similar manner to those we have seen before under fully sedentary conditions. But when the janissary ranks swelled and started to serve as the infantry core of the Ottoman army at large, it seems to have been part of a general, European development to create salaried standing armies, as was the case of the so-called *bandes d'ordonnance* established in France and Burgundy around 1450. Of course, it is also reminiscent of the Russian *strel'tsy*. Indeed, looking closer at military developments, the Ottoman and Russian developments are strikingly similar. For example, the *sipahi* cavalry troopers, numbering around 80,000 in the middle of the sixteenth century, were increasingly paid by individual *timars* granted by the state. As we have seen, this was also the case with their Russian colleagues, who were paid by individual *pomest'ia*. Both *timar* and *pomest'ia* were, like the Byzantine *pronoia*, given to pay for the individual trooper and as such were different from the larger *iqta* or *jagir* which were bestowed on commanders in Mughal India. In all these cases, it was crucial that prebend-holders were constantly transferred from one place to the other so they could not strike root and carve out their own little kingdoms. From the sixteenth century onwards, more and more slaves penetrated the ranks of *timar*-holders.[132] In fact, if they did not take service at court, the *sipahis* became an impoverished lot. This process was aggravated by the Ottomans policy to replace *timars* with revenue farms or *çifliks*, which gave rise to the emergence of a new urban class of entrepreneurial elites called *ayan*, often to the detriment of the *sipahis*.

Yet this situation was not so strange, as the Ottoman-Russian parallel derives mainly from the ecological similarities between the two regions. Both the Ottomans and the Romanovs ruled empires that faced a limited nomadic power that failed to strike at their sedentary cores. It was only at the very fringe of their emerging empires that something of the warband's spirit, in the subsidiary form of Akinjis and Cossacks, could continue, albeit on a small scale and under more settled conditions. In fact, the Romanovs were more successful in this regard as they could profit from an ever retreating nomadic frontier. In contrast, due to strong Habsburg, Romanov and Safavid resistance, the Ottoman frontier was 'closed' in the seventeenth century. As a consequence, the Ottomans could not, like the Romanovs, continue to recruit huge numbers of frontiersmen for the army. In order to keep pace with the ever increasing numbers and firepower of their European adversaries, the Ottomans had no choice but to expand recruitment to the Anatolian peasantry. The result was the creation of Cossack-like warbands of a sedentary nature, not at the fringe but at the heart of the imperial realm. These bands of *sekbans*, *sarica* or *levends*, the latter quite significantly meaning 'bandit' or 'vagrant', consisted of peasants who had previously been carefully kept out of the military but were now lured by provincial officials to join their retinues or were directly enlisted into the sultan's army. The availability of firearms in the countryside facilitated this move away

[132] Fodor, 'Ottoman Warfare', 199-209; Imber, *The Ottoman Empire*, 116-30, 262-94. For an interesting Ottoman-Russian comparison, see G. Ágostan, 'Military Transformation in the Ottoman Empire and Russia, 1500-1800', *Kritika: Explorations in Russian and Eurasian History* 12 (2011), 281-319. Like the Russian *boyars*, the *timariots* could never develop into that rooted and notoriously obstreperous Polish or Hungarian aristocracy.

from land and towards military pursuits. As a result, the state became a significant client for mercenaries organized around units that were ready for hire.[133]

So, whereas the Russian Time of Troubles resulted in the projection of the Cossack warband to the frontier, the Ottoman Time of Troubles struck internally and by creating an armed peasantry – of a kind that had existed in India for many centuries – became an increasing problem. Hence, the so-called mercenary (Celali) rebellion in 1608 proved to be no brief episode but actually became rather endemic, while also undermining the already precarious agrarian balance in Anatolia. A lack of security in the countryside made peasants less willing to stay in the fertile Anatolian plains, which were increasingly left to *çiflik*-holders who concentrated on animal breeding. Flocks with a few herdsmen promised higher profits than grain or any other crop requiring more labour. It once again showed that Anatolia was part of an Arid Zone where productive sedentary agriculture could not be taken for granted.[134] By the eighteenth century, recruitment was primarily in the hands of the *çifliks*-holding *ayan* who had replaced the dysfunctional old-style *timariot sipahis*; indeed, at a time when Russia turned towards general conscription, the irregular recruitment of these provincial militias even surpassed the janissaries and became the standard procedure of Ottoman campaigning.[135]

Looking at eighteenth-century developments, one gets the feeling that the historiographical evaluation of the janissary phenomenon suffers from an exclusively military perspective. Although in a purely

---

[133] K. Barkey, 'In Different Times: Scheduling and Social Control in the Ottoman Empire, 1550 to 1650', *Comparative Studies in Society and History* 38 (1996), 478; see also her *Bandits and Bureaucrats: The Ottoman Route to State Centralization* (Itaca, NY: The Wilder House Series of Politics, History, and Culture, 1994).

[134] Wolf-Dieter Hütteroth, 'Ecology of the Ottoman Lands', in Suraiya N. Faroqhi (ed.), *The Cambridge History of Turkey*, vol. 3: *The Later Ottoman Empire 1603-1839* (Cambridge: Cambridge University Press, 2006), 30.

[135] V.H. Aksan, 'War and Peace', in *The Cambridge History of Turkey*, vol. 3, 81-117; idem, 'Whatever Happened to the Janissaries? Mobilization for the 1768-1774 Russo-Ottoman War', *War in History* 5 (1998), 23-36. The Ottomans introduced conscription only in the nineteenth century.

military sense the janissaries lost their military efficiency, we should keep in mind that the huge increase of their numbers, from about 20,000 at the mid-sixteenth century to about 400,000 (!) at the end of the eighteenth century, as well as their increasing geographical spread, must have played a crucial role in enhancing the social cohesion of the empire as a whole. The corps (*ocak*) as such had evolved from a more or less isolated imperial bodyguard into the core of the imperial army and then into an imperial network that pervaded society as a whole. All this had already begun in the sixteenth century with the allowance of janissaries to marry – only bachelors continued to live in relatively isolated barracks. Later, their ranks became more open to outsiders: (adopted) relatives, friends and clients of janissaries, many of whom were Turks and Muslims. Although the cohesion of the corps as a whole declined, what remained was a mighty pressure group within the empire that continued to share a culture of imperial service, combined with a specific branch (Bektashi) of Sufi devotion. In addition, after 1740, the demilitarized corps actually increased its financial hold on society and opened up its ranks even further as janissary salary claims started to be marketed as certificates (*esame*). Whatever the complaints about rising corruption and imperial decline, I would argue that the increasing integration of the janissaries into society actually explains a great deal of the empire's astonishing endurance.[136] Because they were everywhere, it seems that the integrated 'degenerated' janissaries, still very much the emperor's men, were even more effective than the segregated 'pure' Manchu banners in keeping the empire together.

COMING FULL CIRCLE: MUSTAFA ALI AND IBN KHALDUN

During the sixteenth century, at the height of their imperial conquests, there was a broad, almost Eurasia-wide consensus, stretching from Niccolo Machiavelli in Italy to Iskandar Muda in Aceh, that the

---

[136] G. Veinstein, 'On the Ottoman Janissaries (Fourteenth-Nineteenth Centuries)', in E. Zürcher (ed.), *Fighting for a Living: A Comparative History of Military Labour 1500-2000* (Amsterdam: Amsterdam University Press, 2013), 115-34; and V.H. Aksan, 'Mobilization of Warrior Populations in the Ottoman Context', in *Fighting for a Living*, 331-53.

Ottoman standing army of janissaries provided the model par excellence for building and sustaining empire. Apart from the sheer scale and the sophisticated logistics of this army, what made the 'Rumi' model particularly attractive was that emperors-to-be could achieve this without making unfavourable deals, either with the hereditary landed nobility or with stingy mercenary entrepreneurs. Indeed, even in the twenty-first century the military slave system makes an astonishingly modern impression, or as the late Ernest Gellner exclaimed, 'we are all mamluks now'. To go on borrowing his words, individuals were recruited into state service in an atomized manner and were torn out of their kin background by being technically slaves. Sustained religious and military training becomes a means of inducing an *esprit de corps*. Returning to our own topic, although the janissary corps was certainly not a Chinggisid type of warband, it can be perceived as an artificial, educationally-produced alternative to the warband.[137] In Ibn Khaldun's terms, it is the principle of military slavery that overrules kinship and religion as the most important ingredient of Ottoman *asabiya*.

Looking for other models further east, the sixteenth century Ottoman historian Mustafa Ali knew perfectly well that of the dynasties of the four great Islamic empires, the Ottomans alone had neither a genealogical mandate for sovereignty like that of the Uzbeks and Mughals, nor a religious one like that of the Safavids. For Ali, though, it was not military slavery, Ibn Khaldun's third ingredient, but universal justice that the Ottomans had as their unique selling point. This was based on his analysis that the old regime of a religiously-based universal caliphate broadly identified with the sharia and Arab hegemony had been replaced by a new dispensation of universal nomad dominion under the legitimate sovereignty of Chinggis Khan and his successors, who had received divine sanction to conquer the world and to distribute it to family and followers. To avoid the decline

[137] E. Gellner, 'Tribalism and the State in the Middle East', in P.S. Khoury and J. Kostiner (eds), *Tribes and State Formation in the Middle East* (London: I.B. Tauris, 1991), 115, 121. In my view Gellner makes too much of 'tribal' cohesion, which may be right for the North-African context but much less so for the Eurasian one where not the ascriptive tribe but the conscriptive warband was the dominant model before military slavery.

of empire, due to excessive wealth and degeneration, the 'modern' ruler should take care to remain true to his mandate: the dynastic commitment to justice and order made manifest in *kanun* which, more than the sharia, embodies the grace of God. For Ali, history had demonstrated that the Chinggisids were more worthy and effectual sovereigns than the irresponsible, ignorant, and morally corrupt scions of the Abbasid house. Therefore he stressed the legitimating importance of the principle of universal and impersonal justice which the Mongols had implemented in the form of the Chinggisid dynastic law or *yasa*. For Ali, the Chinggisid *yasa* was the same as the Ottoman *kanun*. It was this dynastic commitment to universal justice in two forms – Islamic sharia and dynastic in *kanun* – coupled with a strong central authority, that constituted the primary legitimating principle of the Ottoman Empire.[138] It is quite significant that a sixteenth-century Ottoman historian unconsciously paraphrases Ibn Khaldun by referring to the legacy of Chinggis Khan. Despite the introduction of various new sedentary forms of imperial rule, the latter was still considered very much alive and kicking. Having come full circle, it seems the right moment to draw some conclusions.

CONCLUSION:
THE WARBAND AS A CATEGORY IN
GLOBAL TIME AND PLACE

In an attempt to make sense of periodization in world history, the late Jerry Bentley pointed out that the first half of the second millennium should be called the age of transregional nomadic empires. Nomadic peoples established empires incorporating vast stretches of Eurasian land mass, and sponsored direct interactions between distant peoples. The migrations, conquests, and empire-building efforts of nomadic peoples guaranteed that cross-cultural interactions would take place in more intensive and systematic fashion than in earlier eras.[139] I have attempted to demonstrate that the key to this nomadic

[138] C. Fleischer, 'Royal Authority, Dynastic Cyclism, and "Ibn Khaldûnism" in Sixteenth-Century Ottoman Letters', *Journal of Asian and African Studies* 18 (1983), 198-220.

[139] J.H. Bentley, 'Cross-Cultural Interaction and Periodization in World

success story was the Chinggisid model of the Turko-Mongolian warband which reverberated far beyond the steppes of Central Asia. Actually, it was at its most effective not in the Central Zone, but in the transitional areas between the northern Middle and Outer Zones. Reviewing its main features, it was a highly disciplined, meritocratic group of military elites that was open to talented outsiders; thus, far from being an ascriptive group, it was consciously engineered to form a new warband of mixed ethnic origin that had collective rights (appanages) on the leader's revenue. Its members had no fixed or hereditary landed interests and tended to be generalists who combined military and administrative functions, aspects which overlapped with each other. Some of them represented the leader in consciously created subsidiary warbands which could take a new 'ethnic' label, but which in principle remained conscriptive. Furthermore, all of this ethnic engineering took place on a massive scale and involved all parts of the society. What we have found is that this kind of warband has a surprising coherence within the political organization of Jurchen, Mongolian, Turkic, Timurid and Tatar empires. Contemporaries would recognize these empires under the label of *ulus*, each of which consisted of a central warband or *keshig* that supported various subsidiary warbands under various denominations, such as *mukun* under the Jurchen, *tümen* under the early Timurids, or *sanjaq* under the early Ottomans. In principle, the empire as a whole was allotted as appanages to the various subsidiary tribes. As we have seen, under the sedentary conditions that followed a conquest, these institutions tended to lose their fluidity and either evaporated entirely or solidified under new ascriptive 'banners' that recall the three main ingredients of Ibn Khaldun's *asabiya*: kinship, devotion, and slavery. Hence, neither being an exclusively Central Eurasian nor a universal category, the warband described in this chapter is very much a specific phenomenon of the Arid Zone's nomadic frontier.

History', *The American Historical Review* 101, 3 (1996), 766-7. See also his more elaborate *Old World Encounters: Cross-Cultural Contacts and Exchanges in Pre-Modern Times* (Oxford: Oxford University Press, 1993), 111-65.

Index

Abbasids 323, 329
Abd al-Razzaq 78
Abul Fazl 35, 246
Abul Hasan, Sultan 130, 155
Abyssinians 169
Aceh 327
Afghan, Afghans 15-17, 20, 23, 25-6,
    32, 36-7, 41, 43-4, 46, 48-50,
    61-2, 65, 69, 72, 76, 92, 111-13,
    157-79, 180-1, 189, 201, 204-38,
    240, 244-5, 247-8, 253, 290, 315;
    see also Pathan
Afghanistan 17, 29-31, 37, 39, 43, 47,
    50, 99, 166-7, 179, 182, 212, 221,
    229, 235, 245, 247, 255, 293-4,
    296
Africa 100, 102, 106, 183
Afridis 214, 226, 233
Agra 119, 157, 176, 230, 235
Ahadis 299, 301
Ahmad Ali Khan Rampuri 49
Ahmad Khan Bangash 35, 216-39
Ahmad Shah Durrani 44, 157, 164,
    166, 168-9, 171, 173, 177, 179-80,
    234, 238
Ain Khan Sarwani 209
Aisin Gioro 285
Aiyanar, Lord 63
Akbar 35, 76, 119, 121, 186, 246,
    291, 295, 299-300
Akhlaq 266
Akinjis 325
Akuta 281

Alam, Muzaffar 15
Alavi, Seema 226
Aleksandrov 310
Alexis, Tsar 315
Ali Muhammad Khan Rohilla 235
Aligarh 15-16
Allahabad 215, 217-18, 232, 234, 236
Allsen, Thomas 104, 273, 278
Almas Ali Khan 241, 243-4
America 51, 183
Amethi 218, 232, 237
Amir al-Din, son of Qais 213
Amrit Mahal 68
Amsterdam 82
Amu Darya 26
Anatolia 253-4, 258, 295, 318-22,
    326, 330
Andhra 67, 146
Andkhui 26
Aq Quyunlu 260, 265, 294
Aquinas, Thomas 90, 137, 278
Arab, Arabia, Arabic 18-19, 31, 37,
    47, 56, 64, 68, 109, 152, 217, 258,
    260, 261, 264, 270, 328
Aravalli Range 57
Archer, archery 63, 83, 99, 103-4, 139,
    154, 163-4, 172, 191-4, 202-3,
    206-7, 250, 260, 299, 309; see also
    bows
Arcot 27, 32, 34
Arid Zone 19-21, 55-70, 73-4, 77,
    80-4, 88, 89, 91, 97, 105-7, 146,
    260, 266, 270, 275, 321, 326

Aristotle 101
Armenian 25
Armour 83, 104, 192-3, 202-3
Artillery 23, 102, 153, 157, 161-4,
    169, 174, 177-80, 183-4, 189, 196,
    198, 200, 202-8, 299, 323; *see also*
    cannon
Asabiya 153, 251, 265, 271, 276-7,
    279, 297, 317, 328, 330
Asaf al-Daula 44-5, 240
Ascetics 61-2, 70, 85, 94-5, 97-8,
    148-50
Asoka 16
Astrakhan 306-7, 310
Athens 322
Augustine 90, 137
Aurangzeb 111, 118-19, 124, 186,
    190, 199, 209, 289, 301-2
Awadh 27, 35, 44, 163, 215, 217, 225,
    229, 232-5, 238-48
Ayalon, David 184
Ayyubids 323
Azerbaijan 318, 320
Azov 306

Baizis 212
Baba Ratan 73
Babis 50
Babur 190, 211, 293, 296-9, 307, 311,
    314-15, 323
Bactrian 64, 175
Baghdad 247
Bahawalpur 27
Bahmani 109
Baihaqi 276
Baksariyas 203
Balkans 319, 322, 324
Balkh 26
Balotra 27, 28
Balrampur 30
Baltic Sea 139
Baluchi 72
Baluchistan 37, 50, 65, 167, 255
Balwant Singh 44, 234

Bamtela 228
Bangash 24, 209-21, 226, 228-39,
    242, 245, 248
Banias 25
Banjara 42-3, 62, 66, 72
Banners 286-8; *see also mukun*
Bannu, 211
Barendse René 19
Bari district 121
Bartlett, Robert 91
Bashkirs 306
Basra 33
Bayazid Ansari 209
Bayly, Christopher 15, 42
Bayqara 290, 293
Beas River 27, 71
Beckwith, Christopher 252
Beijing 82, 112-14, 128, 257
Bekbulatovich, Symeon 309, 311
Belgorod Line 307
Benares 27, 30, 35, 44-5, 57, 215,
    229, 234
Bengal 34-6, 38-40, 46-7, 109, 157,
    159, 176, 245
Bentley, Jeremy 329
Berads 203
Bernier, François 202
Bhadoi 215
Bharatpur 224
Bhatinda 27, 49, 73, 75
Bhatnir 72-3
Bhats 42-3
Bhattis 42-4, 73-7
Bhau 171
Bhima River 27, 37, 58-9, 63
Bhopal 29, 32, 50, 247
Bhure Khan Chela 218
Bibi Sahiba Bangash 215-18, 223, 228,
    232
Bihar 27, 30, 46-7, 157, 245
Bijapur 67, 146
Bikaner 27, 44, 74
Bilge 263
Bilotra 31

Bilsa 29, 31
Black Sea 306
Blake, Stephan 97
Bodyguard 123, 125, 169, 173, 222, 324, 327
Bogdanoff, Alexander 256
Bohras 25
Bolan Pass 27
Bolotnikov, Ivan Isaevich 314-5
Bombay 32, 34, 41, 50
Bo'orchu 269
Boundary 51, 76, 79, 86, 91, 107, 138, 140-1, 154; see also frontier
Bourquin, Louis 76
Bow 83, 192, 202, 206-7, 261, 287, 309; see also archer, archery
Brahmins 61-2, 95, 145, 148-50, 154-5, 221
Braudel, Fernand 79, 81
British 24, 32-4, 40-1, 45-7, 50, 66, 74, 76-7, 99, 157, 159, 161, 166, 176, 179, 181-2, 205, 225, 240-5, 248
Bu Said 247
Bukhara 26, 48
Bulliet, Richard 64
Bundelas 203, 220
Bundelkhand 49, 162, 210, 220, 227-8, 232
Burgundy 324
Butwal 30
Buyids 126
Byzantium 302, 318, 324

Calcutta 33, 82
Camel 23, 59, 64, 72, 74, 163, 174-80, 182, 194, 201; see also dromedary
Cannon 22, 163, 174, 196, 198, 200; see also artillery
Carbine 172
Carnatic 91, 190, 201; see also Karnataka
Carpathians 107

Caspian Sea 26
Catalans 322
Catherine II of Russia 317
Caucasus 258
Celtic 86, 94
Central Asia, Central Asian, Central Eurasia, Central Eurasian 17, 19, 24, 26-7, 31-2, 39, 45-6, 48, 52-5, 57, 59-60, 63-5, 70, 80, 84, 88-9, 100, 104-8, 111, 114-18, 120-2, 125, 127-9, 139, 148, 163, 172, 174-5, 190-3, 203, 206-7, 235, 251-3, 255-6, 258-60, 270, 272, 274, 276, 278, 281, 286, 294, 296, 330
Central Europe 104, 114, 321
Chaghatai 293
Chait Singh 44-5
Chaldiran, Battle of 324
Chambal 58
Chamberlain, Michael 153
Chand-Kheri 29, 31
Charhat Singh 35
Chaudhuri, K.N. 79
Chela 210-11, 214, 216-26, 228-32, 236, 239, 242, 246-8, 300
Cheng-te 310
Cherkassky, Michael 311
Chicago 184
China, Chinese 19, 21, 24, 52, 54, 80, 84-6, 99-100, 102-9, 123-9, 138, 195, 251, 253-60, 263, 266, 268-9, 274, 276, 279-80, 283, 286-9, 291, 299, 303, 306, 311-12, 315, 321
Chinggis Khan 24, 115, 250, 254, 259, 261, 263, 269, 271-4, 278, 280-2, 286, 290, 311, 317, 322, 328-9
Chinggisid 250, 252, 258, 264, 268, 272, 274-5, 284, 290-2, 295, 297-8, 306, 309, 317, 320, 328-30
Chishtis 96
Chitor 121
Chittung River 72, 76

Cholas, 54
Ch'ungnyŏl, ruler of Korea 303
Cistercians 94
Clive, Robert 157
Cochin 32
Comitatus 122, 252, 270
Confucianism 251, 267, 283, 285
Constantinople 319
Cooper, Randolph 23
Coromandel 58
Cossacks 256, 271, 302, 311-17, 322, 325-6
Crimea 305-8, 314
Cuddapah 32, 50, 57, 69
Cumans 268

Dahomey 101
Daim Khan Chela 219
Dalir Khan Chela 220, 248
Dalir Himmat Khan 216, 239
Damascus 177
Danchang 99
Dardanelles 319
Daud Khan Rohilla 48, 50
Daulat Khan Shamilzai 213
Deccan 22, 27, 29, 32, 34-5, 37, 41, 45-7, 49-50, 57, 60, 63, 67, 69, 71, 74, 91, 103, 163, 176, 190, 193, 198-9, 203, 209-10, 227-8, 232, 235, 290
Delhi 67, 70-2, 74-7, 82, 111-12, 119, 146, 157, 176, 189, 197, 211, 224, 229-30, 232, 235, 237, 245-6, 257, 260, 298, 300
Denmark 86
Derajat 27
Devagiri 67, 146
Devşirme 322, 324
Di Cosmo, Nicola 105
Dipalpur 75
Doab 235, 238, 240, 244-5
Don River 314
Donskoi, Dmitri 310

Drill 23, 83, 157-9, 165-6, 177, 180, 183, 201-3, 205-6, 321
Dromedary 22, 62, 64-5, 67-8, 70, 74, 77, 88, 103, 175, 194, 201; see also camel
Druhzina 305
Duddri 30
Dunde Khan Rohilla 35
Dunning, Chester 314
Durrani, Durranis 17, 50, 159, 162, 164, 166-70, 172-4, 177-80, 201, 235, 238-9, 245, 247
Dutch 25, 130-1, 155-6
Dvarasamudram 67, 146

East India Company (Dutch) 38, 130
East India Company (English) 32, 34, 38, 40-1, 45-6, 48, 50, 76, 157, 159, 205, 238, 241, 243, 245
East India Company (French) 38
Eastern Europe 19, 26, 107, 253, 294
Eastern Ghats 58
Eaton, Richard 22
Egypt 184, 260, 294, 322-3
Elburz Mountains 255
Elephant 78, 153, 161, 163-4, 171, 176, 189, 194, 201, 210, 218-19, 222, 227-8, 232-3, 237, 243, 299
Elias, Norbert 131-5, 141, 143, 148, 155
Elliott, Mark 286
English 25, 38, 45, 47, 157

Faizabad 176
Fakhr al-Daula 225, 239, 248
Farid al-Din 73
Farrukhabad 15, 24, 44, 209-11, 214-15, 217-18, 220-1, 223-6, 228-33, 235, 238, 240-5, 247-8
Farrukhsiyar 227-8, 235
Fars 31
Fatehgarh 234, 241
Fathabad 76

Feud, feuding 86, 89-91, 135-8, 140, 154
Feudalism 78, 101, 117, 128
Firdausi 264
Firearm 101, 164-5, 170, 173, 180, 185, 198, 201-2, 325; *see also* carbine, musket, pistol
Firuz Shah Tughluq 75
Fitna 93, 153, 188-90, 199, 207-8; *see also* feud, feuding
Fletcher, Joseph 115, 123
Foucault, Michel 134
France, French 25, 38, 79, 159, 202, 324
Franciscans 94
Freud, Siegmund 132-3, 141
Frontier 20-1, 24, 51-7, 68-71, 75, 77, 80-7, 89, 91-5, 97, 100, 102-3, 106-8, 112-15, 117, 124, 127, 129, 135, 138-40, 146-8, 150-5, 157, 186, 196, 207-8, 253-9, 261-2, 293, 306, 312-14, 318-19, 322, 325-6, 330; *see also* boundary

Galdan 118
Ganges River 29, 42, 54, 57-8, 71, 74, 228, 233, 235
Ganjam 34, 46
Garhwal hills 30
Gat, Azar 128
Gellner, Ernest 52, 97, 328
Georgia 268
German 316
Ghaggar River 71-2
Ghazan Ilkhanid 294
Ghazipur 46
Ghazi 51, 92, 270, 277-8, 319, 322
Ghaznavids 63, 84, 91, 103, 146, 167, 180, 245, 294
Ghilzai-Afghans 176
Ghulam 168-71, 173, 178-9 217, 246, 294, 300
Ghurids 63, 91, 103, 146, 167, 180, 245, 294

Gilan 264
Giocangga 285
Gobi Desert 255
Godaveri River 54, 58
Godunov, Boris 311
Gogha 32, 50
Golden, Peter 104, 271, 273
Golden Horde 306, 308-9, 318, 323
Golkonda 67, 130, 146, 155
Goody, Jack 101-2
Gosains 61-2, 65, 96
Grass(es) 18-19, 40, 42-4, 59, 63, 72
Gratian 90, 137
Great Wall 81, 85, 107, 113, 122, 127, 255, 260
Grenade 195
Grupper, S.M. 275
Gujars 43, 44
Gujarat 27, 40, 65
Gulnabad 176
Gunpowder 21-3, 101-2, 104, 114-15, 165, 183-5, 195-8, 200-1, 203-4, 207-8, 250, 316, 321-2, 324; *see also* artillery, cannon, carbine, firearm, grenade, musket, musketry, pistol, *zamburak*
Guptas 54

Habib, Irfan 16, 35-6
Habsburg Empire 173, 325
Hafiz Rahmat Khan 17, 35, 41, 234, 237
Haider Ali 34
Haji Bahadur; see Daulat Khan Shamilzai
Hajipur 29-30
Han 128, 255, 259, 287-8
Hanjun 287
Hansi 72, 75-6
Haridwar 29-30, 32, 48, 60
Harya Khail 213
Haryana 71, 73, 75-7
Hastings, Warren 244
Hathras 224

Heesterman, J.C. 15, 20, 276
Henan 99
Hengshan 99
Herat 26, 174, 293, 297
Himalayan Mountains 29-30, 48, 59, 71, 164, 222, 234
Himmat Khan Bangash, son of Malik Ain Khan 209
Himmat Khan Bangash, son of Mahmud Khan 239
Hindu Kush 26, 255
Hisar 72, 74, 76
Hodgson, Marshall 21, 102, 184
Holsteiners 316
Hong Taiji 115, 120, 286-7
Hongwu 284
Horse 17-20, 24-50, 56, 59, 62-5, 67-70, 72-4, 76-7, 83-4, 87-8, 92, 97, 99-103, 105-16, 121, 124-7, 129-30, 158-61, 169, 173, 176, 180, 182, 185, 193-6, 200, 202-3, 205-8, 227-9, 232-3, 235, 237, 244, 250, 253-4, 259-61, 268-9, 271, 279-80, 299, 307, 310, 313; see also warhorse
Hospitallers 94
Hoysalas 67, 84, 146, 190
Hua Yue 99
Huizinga, Johan 133
Humayun 298-9
Hungary, Hungarians 86, 139, 260, 268, 324
Husain, Iqbal 16
Hyderabad 32, 34, 50, 163

Iaik River 314
Ibn Hassul 264
Ibn Khaldun 52, 88, 115, 125, 152, 249-53, 258, 263, 265-7, 275-7, 279, 290, 292, 295, 297, 320, 327-30
Ilkhans 294, 318
Illig Khagan 263
Imam Khan Bangash 217

Imber, Colin 278
Inalcik, Halil 185, 320
Indus River 27, 54, 58, 70, 72, 211
Iqta 126-7, 300, 324
Iran, Iranian, Iranians 17, 26, 32, 36-8, 60, 63-4, 68, 70, 73, 84, 92, 106-9, 111, 114, 118-19, 126-7, 158, 167, 174-6, 180, 182, 184, 186, 217, 232, 236, 245, 253, 255, 258, 260, 262, 265, 275, 284, 293-5, 318, 320; see also Persian
Iraq 31, 47
Ireland 94
Irtush River 112
Irvine, William 215-16, 218
Iskandar Muda 327
Islam Khan Chela 224
Italy 100
Itimad al-Daula 235
Ivan I of Russia 309
Ivan III of Russia 310
Ivan IV of Russia 309-11, 316

Jagir 124, 127, 209, 215, 223, 228, 230-1, 304, 324
Jahan Khan Lodi 216
Jahandar Shah 227
Jahangir 119
Jaipur 27
Jaisalmer 73
Janissaries 170, 184
Japan 253, 260
Jats 35, 73, 163, 170, 234-5
Jerusalem 94, 97
Jews 25
Jin 99, 128, 260, 262, 266-7, 280-6, 289, 292; see also Jurchen
Jodhpur 28
Jullundur Doab 27, 31
Junagadh 50
Jurchen 19, 84, 103, 125, 253, 260, 262, 266-7, 279, 280-1, 283-6, 290, 297, 312, 315, 321, 330

Kabul 26, 49-50, 72, 111, 169, 209, 211, 229, 296-8
Kaghzai 212-14
Kakar, Sudhir 132, 141-3
Kakatiyas 67, 84, 146, 190
Kalat 50
Kallars 60
Kalmuks 169
Kalmyks 26, 306-7
Kamgar Khan Baluch 233
Kampil 231
Kandahar 26, 50, 173-4, 179
Kangxi 118-22, 128
Kanpur 44
Karim Khan Zand 247
Karlanis 212-14
Karnataka 57, 67, 69; *see also* Carnatic
Kasganj 224
Kashi Raj 179
Kashifi 266
Kashmir 25, 167
Kasim Khan Bangash 215
Kasimbazar 159
Kasimov 309
Kasur 50
Kathi 40-1
Kathiawar 32, 37, 39-42, 45, 47, 50, 63, 68, 175
Kautilya 141, 152
Kaveri River 54
Kazakhs 26, 306-7
Kazan 306-7, 310
Keegan, John 154
Kerala 57
Kesh 291
Keshig 269-75, 278-9, 281-6, 289-92, 297, 317, 330
Khaljis 63, 75
Khandoba 63
Khanpur-Kota 29
Khataks 212
Khirilchis 211
Khitan 84, 103, 118, 260, 262, 280
Khodarkovsky, Michael 312

Khoqand 247
Khorasan 56, 64, 167, 245, 254-5, 257, 290, 293-4, 330
Khudabanda Khan Bangash 239-40
Khudaganj 233
Khugianis 211
Khwandamir 299
Khweshgi 76
Khyber Pass 27, 211
Kiev 302, 306
Kipling, Rudyard 17
Kohat 211-12
Kolff, Dirk 20, 36, 159, 201, 219
Kollmann, Nancy 302
Konkan 57
Korea 115, 255, 280, 303
Kota 28
Krishna 57-8
Kulikovo, Battle of 310
Kumaun 30
Kurds 169
Kurnool 32, 50, 57, 69
Kurram 211
Kushana 259
Kutch 33, 37, 40, 47, 175

Lahore 72, 229-30
Lakhi Jungle 27, 31, 44, 63, 73, 76
Landars 211
Lattimore, Owen 255-6
Law, Robin 101-2
Law de Lauriston, Jean 159, 161, 169, 179
Leiden 15-16, 19, 20
Lewis, Bernard 79, 100
Liao 103, 260, 266-7, 280, 287
Lindner, Rudi Paul 278, 321
Lingayats 61, 96
Lodis 75, 111, 239
Lohanis 49
London 82
Lucknow 44, 176, 240-1

Machiavelli, Niccolò 327

Madariaga, Isabel de 304, 311
Madras 32-4, 46
Magadha 57
Magyars 139
Mahanadi 54
Maharashtra 67, 146, 201
Mahmud al-Kashghari 264
Mahmud Gawan 22
Mahmud Khan Bangash 237, 239
Mahmud of Ghazni 73
Maier, Charles 250
Majalgaon 27
Makran 167
Malik Ain Khan Bangash 209
Malik Kafur 196
Malik-Miris 212
Malwa 28-9, 31, 57-8, 71, 228, 232, 236
Malwar 35
Mamluk, mamluks 123, 189, 193, 217, 220-1, 226, 245-7, 300, 328
Mamluks 107, 109, 184, 247, 260, 294, 322-4
Manchu 100, 102, 106, 108, 110, 113-25, 128-9, 266, 271, 280-90, 307, 317, 323, 327; *see also* Qing
Manchuria 115, 254-6, 280, 290-6, 330
Mangalore 32
Manipur 171
Manohar Thana 29
Mansabdar, mansabdari, mansabdars 118, 123-5, 153, 188, 228, 231, 235-7, 299-301, 304-5
Manu 152
Manucci, Niccolao 130-1
Maratha, Marathas 27, 29, 32, 34-5, 37, 40, 46, 49, 61, 63, 69, 74, 76, 92, 162, 171, 174, 178, 189, 199, 204, 228, 232, 234-5, 240, 302
Maravars 60, 69
Masson Smith, J. 109, 110
Mathura 44, 157
Mau 209-10, 217, 220, 223-4, 227-8, 240

Maudan 220
Maula Sardar Rohilla 35
Mau-Rashidabad 209
Mauryas 16, 54
Mawarannahr, 296; *see also* Transoxania
McCord, Edward 288
McNeill, William 184
Mehmed Giray 307
Mehrabad 42
Melville, Charles 275
Mestnichestvo 304-5
Mewat 71
Meymaneh 26
Ming 109-10, 113, 127, 247, 251, 284-9, 301, 310
Mir Atai Khan 171
Mirza Khan Ansari 209
Modave, Comte de 38, 175
Mohács 324
Möngke 126, 273
Mongols, Mongolia, Mongolian 19, 83, 85, 104, 107, 109-11, 113, 115-16, 118, 122, 124-6, 172, 195, 250, 253, 255, 257, 259-63, 264, 266-9, 272-4, 280-4, 286-7, 290-2, 294, 296, 298, 302-3, 306, 308-9, 317-19, 321, 323, 329-30
Moorcroft, William 41, 47, 48
Moreland, William 299
Moscow 112, 257, 303, 310, 314
Mughal, Mughals 15-17, 20-1, 35-7, 48, 54, 76, 100, 102, 106, 108, 110-25, 127-9, 131, 153, 159-60, 162-3, 173, 184, 187-90, 193, 198-200, 202-3, 209-11, 219-20, 226-30, 236-9, 246, 248, 260, 278, 289-90, 295, 298-300, 302, 304-5, 323-4, 328
Mughan Plain 318
Muhamdabad 228
Muhammad Amin Khan Itimad al-Daula 235
Muhammad Jafar Shamlu 179

Muhammad Khan Bangash 209-10, 212, 215, 218, 220-37, 248
Muhammad Shah 231
Muhammad Wali al-Lah 213
Muhammadpur 75
Muharram 240
Mukun 279, 281, 285, 297, 307, 330; *see also* Banners
Multan 27, 74
Mundwa 28-9
Munkh-Erdene, Lhamsuren 273, 278
Muscat 247
Muscovy 112, 261, 267-8, 302-3, 305-6, 308-11, 315
Musket, musketry 157, 160, 164-5, 170-2, 204-6
Mustafa Ali 277-8, 327-9
Muzaffar Jang 225-6, 239-40, 248
Mysore 32, 34, 36, 50, 57, 65, 67-8, 146, 247

Nadir Shah 44, 76, 167-9, 173, 236
Nagapatnam 130
Nagaur 28
Najib al-Daula 35, 238
Nanking 113
Napoleon 205
Narmada 58
Nasab 210, 213-14, 216, 221, 238, 298
Nasir al-Din Tusi 266
Nation-state 117, 135, 154
Naukar, nökör 20, 117, 122-5, 128-9, 209-11, 226-7, 231-2, 238, 244, 246-8, 269, 272, 275, 282-4, 289, 298, 307-8, 319-20
Nawal Rai 233
Nayakas 61, 63, 92, 199
Needham, Joseph 199
Nepal 31
Niknam Khan Chela 223
Nizam al-Mulk 232, 235, 276, 294
Nogais 112-13, 306-8, 312
Nökör; see *naukar*

Nomadism, nomads 24, 26-7, 52-5, 59, 64, 67, 69-70, 77, 81-2, 84-5, 97-8, 104, 108-9, 115-16, 123-4, 126, 139-40, 147-8, 167, 173, 193, 254-9, 262-3, 270, 306, 321
Nomadic 19-21, 24, 43, 52-5, 66, 72, 80-9, 91-2, 97, 100, 105-8, 111-17, 120, 122-3, 125-9, 135, 139-40, 146-8, 152-5, 173, 191, 193, 206-7, 250-66, 268, 270-1, 274-6, 279-80, 284, 289, 293-5, 301-2, 305-6, 308, 312-13, 317-18, 321-3, 325, 329-30; *see also ahadis*, bodyguards, *comitatus*, *druhzina, keshig, mukun, nökörs, oprichniki*
Normans 86, 139
Novgorod 303, 309
Nurhaci 120, 285-7, 296, 311, 314-15, 323

Ogödei 126
Oprichnina, oprichniki 310-11, 316
Orakzais 212
Orissa 107
Orkhon 263
Orme, Robert 34
Osman Ghazi 319
Ostrowski, Donald 302
Ottoman, Ottomans 102, 106, 120, 127-8, 162, 170, 173, 184-5, 260, 270, 276-8, 294, 299, 300, 306-7, 310, 316, 318-26, 328-30
Ötüken 263
Ox 58-9, 62, 65-8, 70, 74, 76, 88, 103, 163, 280
Oxus River 169

Pacific Ocean 25
Pahari 175
Pampas 87
Panipat 23, 75, 162, 171, 174, 177, 178, 203, 238
Parker 23, 183, 199-200

Parsis 25
Parthian 259
Pathan, Pathans 169, 214-15, 217-18, 220-1, 224, 226, 233, 236-7, 239-43, 248
Patiala 72
Paul I of Russia 316
Peobrazhenskii Polk 316
Perlin, Frank 15
Persia, Persian, Persians 31, 54, 56, 99, 165, 168, 170, 173, 176, 180, 217, 235-7, 247-8, 261, 264, 292-3, 297-8, 319; see also Iran
Persian Gulf 31
Peshawar 169, 211
Peter III of Russia 316
Peter the Great 303, 315-18
Pistol 172
Plassey 157-8
Poland 86
Poland-Lithuania 261, 309-15
Polo, Marco 283
Polybius 251
Pomest'ia 304-5, 313, 316, 324
Pontic Steppe 256, 306, 318
Portuguese 25, 112
Post-nomadic 20, 100, 106, 108, 115-17, 124, 129, 256, 260, 271, 275, 284, 289, 317
Powindah 27
Prikazy 315-16
Prebends 320, 324; see also iqta, jagir, pomest'ia, soyurghal, timar, prikazy
Pugachev Revolt 315
Punjab 27, 30-1, 43, 46-7, 49, 63, 71-3, 99, 146, 167, 234, 255
Pusa 34, 46-8
Pushkar 27-8, 31, 49

Qaim Khan Bangash 215-16, 218, 232-3, 235, 237-8
Qaimganj 224, 228
Qais 213
Qandahar 26, 211

Qara Khitai 260
Qara Qorum 273
Qara Quyunlu 260, 294
Qara Usman 265
Qasur 76
Qazaqliq 254, 256, 270-1, 274, 296; see also Cossacks
Qi Jiguang 285
Qianlong Emperor 119-22, 124, 289
Qing 114, 260, 267, 279, 280, 285-6, 288-9, 299, 302, 307
Qipchak 260, 268, 306, 311, 323
Qizilbashes 168, 295
Quraishis 213

Radabad 75
Radhu Singh Rajput 35
Raja Jaswant Singh 215
Rajasthan 27-8, 31, 37, 46-7, 57, 65, 71, 146, 175, 201, 255
Rajputs 20, 35, 37, 61, 73, 92, 171-2, 189, 199, 214, 219, 220-1, 231, 248
Rama 145
Rampur 44, 47, 49
Rana Lakhi Bhatti 73
Ranjit Singh 32
Ranking 290, 304, 313; see also mansab(dars), mestnichestvo
Rashid al-Din 261
Rayalaseema 57, 69, 146
Razin, Sten'ka 315
Reinhardt, Walter 76
Ridaniyya, Battle of 324
Roberts, Michael 183
Roh 17, 209, 212
Rohilkhand 15, 17, 27, 30-1, 39, 41-5, 48, 50, 162, 210, 221, 229, 232, 234-6, 244-5, 247
Rohilla, Rohillas 16-17, 35-6, 41-4, 46, 48-50, 69, 76, 176, 212, 214-15, 232-4, 237-8, 240, 244-5
Roman 81, 86, 90, 137, 140, 154, 259, 274

Romanovs 301, 311, 314, 325
Roshaniyya 209, 211
Rubruck, William of 273
Rumelia 321
Russia, Russian, Russians 26, 108, 114, 127, 254-76, 302-18, 322-6
Rustam Khan Afridi 233-4
Ryazan 303

Saadalla Khan Rohilla 234
Safavids 17, 102, 119-20, 127-8, 169, 170, 173, 184, 260, 278, 284, 294-5, 299-300, 320, 322, 324-5, 328
Safdar Jang 232-4, 236
Safed Koh 211
Sahib Giray 307
Saiyid 213, 227, 235
Saiyid Abdulla Khan 227
Saiyids 213, 235
Salmond, J. 99
Samana 75
Samanids 294
Samarqand 293, 296, 298
Sanjaq 330
Sarasvati River 72
Sarbani Afghans 213
Sarin 46
Sarwani Kaghzai 213
Savanur 69
Sayyids 75
Seljuqs 260, 294, 318, 323
Shah Abbas I 119, 170
Shah Alam Khan 48
Shah Jahan 119, 216
Shahabad 46
Shahpur-Akbarpur 219
Shahrukh 293
Shaka 259
Shamilzai 213
Shamshabad 209, 219, 223-4, 231
Shensi 113
Shiva 64
Shivaji 118, 186

Shuja al-Daula 237-8, 240
Siberia 25, 112, 183, 255
Sibir 306, 310
Sikhs 29, 32, 35, 44, 49, 72, 76, 96, 176
Silk Road 279
Sind 27, 57, 146, 167, 175, 255
Sinkiang 113
Siraj al-Daula 157
Sirhind 75, 167
Sirsa 72
Siwalik Hills 71
Skinner 76, 206
Slave, slavery 23-4, 101, 152, 168-9, 171, 173-4, 180-9, 210, 214, 216-18, 221-2, 226, 237, 239, 245-8, 258, 275, 277, 279, 282, 291, 294-5, 299-300, 308, 314, 322-5, 328, 330; see also chelas, devşirme, ghulam, Janissaries, mamluk, Mamluks
Slavs 86, 139, 312
Sneath, David 278
Sölde 296
Sonepur-Hajipur 27
Song 110, 113, 262
South America 87
Southeast Asia 100, 104, 108-9, 183, 253
Soyurghal 293
Spanish 86, 109, 139
Sri Lanka 25
Steel 165
Stein, Burton 70, 185
Strel'tsy 308, 316, 318
Streusand, Douglas 162, 199
Subrahmanyam, Sanjay 23
Subsidairy 241, 271-4, 281, 313, 317, 320, 325, 330; see also ahadis, bodyguards, comitatus, druhzina, keshig, mukun, nökörs, oprichniki
Sufi, sufism 49, 61, 73, 155, 189, 229, 246, 295, 299, 320, 327

Sulaiman Mountains 27, 31, 48-9, 70, 211
Sultan Balban 73
Sunderji, horse-dealer 49
Sung 267
Sutlej River 27, 59, 71-3
Swedes 315
Swiss 203
Syria 323
Szechwan 113

Tabarhind 73
Tabriz 82
Tacitus 78, 251
Tajik 169, 264, 268
Tang 263, 267
Tangut Xi Xia 103
Tapti River 58
Tarai 234
Tarain, Battle of 75
Tatars 257, 261, 270, 302, 306-7, 309-14, 330
Telugu 60, 145
Templars 94
Temriukovna, Maria 311
Terek River 314
Thar Desert 57, 71, 175
Thomas, George 76
Tibet 113
Timar 127, 320-1, 323-5
Timur 75, 275, 290-3, 296, 298-9, 306, 319, 322
Timurids 123, 275, 290, 293, 295, 297, 307, 317, 319-20, 322, 330
Tipu Sultan 68
Tirupati 27, 29, 32-3, 60, 65, 69
Tiumen' Khanate 306
Tonk 50
Transoxania 56, 296; *see also* Mawarannahr
Tughluqabad 197
Tughluqs 75-6
Tümen 123, 274, 290-2, 298, 300, 307, 320, 330

Tungabhadra River 57
Tungunsic 260
Turan, Turani 17, 37, 167, 232, 235-6, 264
Turis 211, 212
Turkish, Turks 19, 56, 63, 83, 92, 103, 109, 116, 125, 158, 172, 189, 191-3, 203, 205-8, 217, 220, 245, 253, 258, 260-5, 268, 270, 290, 292, 294-5, 306, 312, 314, 318-20, 322-4, 327, 330
Turkistan 26, 29, 37-8, 43, 47-8, 115, 253, 255, 257, 290, 293
Turkmen 265, 294
Turkoman 26, 169
Turko-Persian 56, 70, 174, 220-1, 245-6, 294
Turner, Frederick Jackson 51
Tver 303, 309

Udaipur 35
Ugra 310
Ukraine 86, 254, 256, 306, 315-16, 330
Ulug-Mehmed 309
Ummedganj 28-9
Urgench 306
Ustarzai 210, 214
Uzbek 169, 247, 328

Vasily II, Grand Prince 309
Vedic era 53
Vienna 81-2
Vijayanagara 54, 67-9, 109, 127, 146, 185
Vikings 139
Volga River 112, 307, 314

Wagoner, Phillip B. 22
Walker, Colonel 40
Wallerstein, Immanuel 81
Wang Yangming 285
Wanli Emperor 285

Warangal 67, 146
Warband 23-4, 121, 251-5, 258-9, 268-70, 273-84, 286, 288, 292, 295-99, 301-2, 307-8, 311-15, 317, 319-23, 325-6, 328, 330; *see also ahadis*, bodyguards, *comitatus, druhzina, keshig, mukun, nökörs, oprichniki*
Warhorse 15, 18, 20-1, 99-105, 108-14, 116-17, 125, 127, 129, 252, 259, 269
Weber, Max 135
Wendel, François-Xavier 163, 230
Western Ghats 57, 59
White, Lynn 101-2
Wilayati 26
Willes, John 241-3
William, Walloon Globetrotter 273
Wink, André 15-16, 19, 93, 152-3
Wiswas Rao 171
Wittek, Paul 319
Wodyars 68
Woods, John 275

Wyatt, E. 40-1

Xiéli, see lllig Khagan
Xiongnu 259

Yadavas 60, 67, 84, 146, 190
Yamuna River 58, 71, 74, 178
Yaqut Khan Chela 216, 225
Yasa 264, 283, 329
Yasin Khan Ustarzai Bangash 210
Yongzheng Emperor 119
Yuan 104, 128, 266-7, 280, 282-4, 286, 289, 292
Yurt 117, 127-9

Zafarabad 75
Zamburak 176-7, 201
Zands 170
Zarutskii, Ivan Martynovich 314-15
Zhu Yuanzhang 284
Ziegler, Norman 219
Zulfiqar Khan 219
Zunghars 113

آزادالدوله نجف خان بهادر ذوالفقار جنگ

1. Muhammad Khan Bangash, c. 1730. *Source:* Bibliotheque Nationale de France.
2. Nawab Amir al-Umara Zabita Khan by Son of Ganga Ram, Mihr Chand, Faizabad, c. 1770. *Source:* Arthur M. Sackler Gallery, Smithsonian Institution, Washington, D.C.

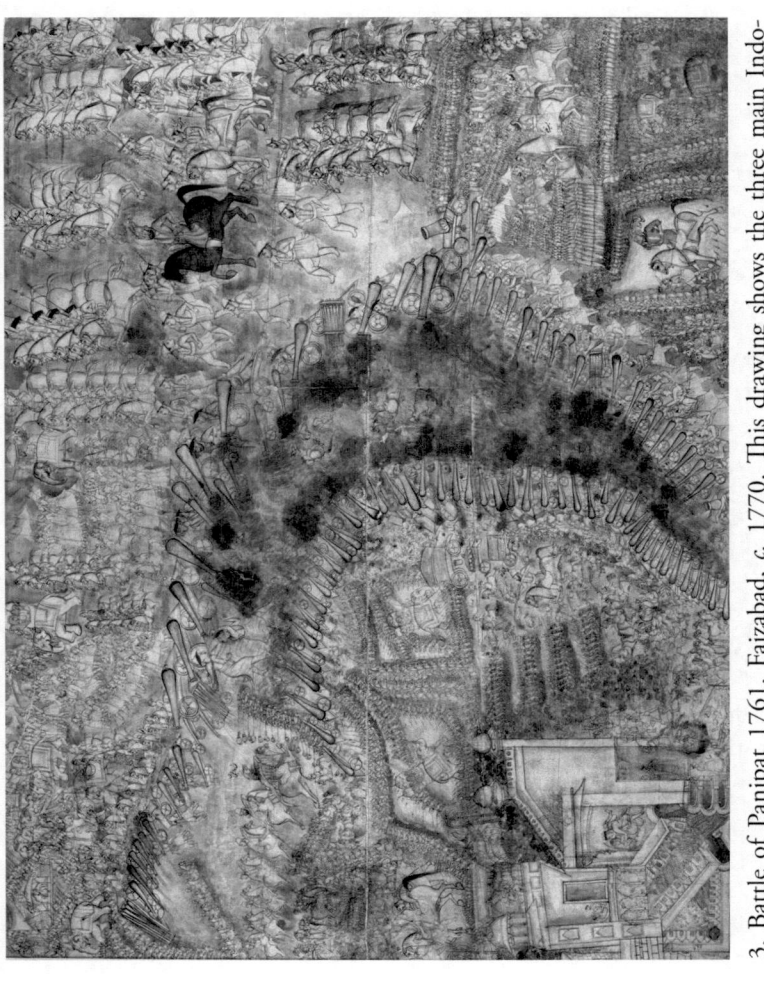

3. Battle of Panipat 1761. Faizabad, c. 1770. This drawing shows the three main Indo-Afghan allies of Ahmad Shah Durrani (on the brown stallion at the centre): Najib Khan (on the horse to his left), Hafiz Rahmat Khan (on the elephant to his right) and Ahmad Khan Bangash (in the palanquin to his right). *Source:* British Library London.

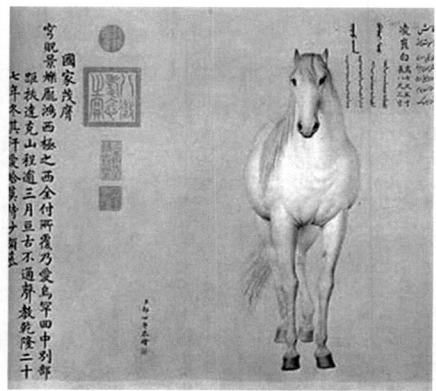

4. Four Afghan Steeds: The Afghan envoy to beijing presented the Qing emperor with four splendid horses. Milanese Jesuit missionary artist Giuseppe Castiglione (Lang Shinin in pinyin) painted the horses for the Qing emperor by the in the 18th century. *Source:* National Palace Museum in Taiwan.

5. Tomb of Ali Muhammad Khan Rohilla in Aonla (UP). *Source:* Photograph by author.

6. Tomb of Muhammad Khan Bangash in Farrukhabad (UP). *Source:* Photograph at www.khyber.org

7. Tomb of Najib al-Daula in Najibabad (UP). *Source:* Photograph by author.

(Plates 5-7: The tombs of the three main founding fathers of Indo-Afghan power show some remarkable resemblances in structure and style.)

8. Two Yusufzai infantrymen.
9. Two Rohilla men.
10. Kabuli horseman in chain mail holding a lance.
11. Yusufzai horseman.

Source: British Library London.

12. Rohilla horseman, watercolour by Sita Ram. *Source:* British Library London.

13. Watercolour of Patthargarh Fort in Najibabad by Sita Ram (1814-15).
14. Watercolour of the tomb of Hafiz Rahmat Khan in Bareilly by Sita Ram (1814-15).
Source: British Library London.